JOURNEY TO DIVERSE MICROBIAL WORLDS

Cellular Origin and Life in Extreme Habitats

Volume 2

Journey to Diverse Microbial Worlds

Adaptation to Exotic Environments

Edited by

Joseph Seckbach

KLUWER ACADEMIC PUBLISHERS

DORDRECHT / BOSTON / LONDON

A C.I.P. Catalogue record for this book is available from the Library of Congress.

ISBN 0-7923-6020-6

Published by Kluwer Academic Publishers,
P.O. Box 17, 3300 AA Dordrecht, The Netherlands.

Sold and distributed in North, Central and South America
by Kluwer Academic Publishers,
101 Philip Drive, Norwell, MA 02061, U.S.A.

In all other countries, sold and distributed
by Kluwer Academic Publishers,
P.O. Box 322, 3300 AH Dordrecht, The Netherlands.

Printed on acid-free paper

This book is dedicated to:
Professor Lawrence Bogorad (Harvard University), my Ph.D. advisor, colleague and dear friend, with our best wishes of health and happiness.

It is also devoted wholly to my versatile personal cell, to my partner, the family nucleus - **Fern**, to our organelles, **Mesha, Avigail, Raziel, Eliezer** and **Eliyahu** and all their symbiotic partners (**Efrat, Amos, Shulamit and Rachelie**) and their offspring (**Shem, Yarden, Maayan, Hillel** and **Shmuel**).

TABLE OF CONTENTS

Biodata of **James T. Staley** contributer of the *Foreword* to this volume.

Dr. Jim Staley is currently a Professor of Microbiology at the University of Washington and Vice Chairman of Bergey's Manual Trust. He received his Ph.D. at the University of California, Davis in 1967. Dr. Staley has been interested in bacterial diversity throughout his career. Early on he coined the term 'prostheca' to describe the group of cellular appendages produced by *Caulobacter*, *Prosthecobacter*, *Hyphomicrobium* and the polyprosthecate bacteria such as *Stella* and *Ancalomicrobium*. As a microbial autecologist, he has studied microorganisms in freshwater, marine, desert and polar environments. More recently he has been interested in the bacteriology of polar sea ice environments and, based on that work, proposed a set of biogeography and co-evolution postulates to aid in the determination of whether bacteria are cosmopolitan or endemic. Recent work in his lab has shown that strains of the genus *Simonsiella*, an oral commensal bacterium of mammals, have coevolved with their hosts including humans, sheep, dogs and cats. Dr. Staley has published more than 100 research papers and co-authored a textbook on general microbiology and microbial diversity.
E-mail: **jtstaley@u.washington.edu**

FOREWORD
It's the Little Things in Life that Count the Most

Most microbiologists appreciate the remarkable diversity of microbial life. After all, microorganisms were the exclusive inhabitants of Earth for a period of about three billion years (Ga). But even to those familiar with microorganisms, their vast diversity as revealed by the Tree of Life, came as a surprise (Figure 1). In contrast, plants and animals evolved only in the past 600 million years. Thus, the "microbial era" persisted for a long interval of time during which speciation and evolution led to much of the diversification of life on Earth.

Perhaps even more important than the duration of the microbial era, however, is the fact that microbial life was here first (Woese, 1998). Many of the resources available to sustain life and allow for early speciation already existed in abundance at the time of Earth's formation. A variety of energy sources including sunlight and reduced inorganic and organic compounds were available before life originated. These included chemical substances such hydrogen gas, sulfide, sulfur, ammonia, reduced iron and hydrocarbons to mention a few. Biological systems evolved to generate energy from these and other reduced substances in the environments, by oxidizing them in the absence of oxygen. Sunlight was also available and evidence for photosynthesis extends back in time to at least three Ga ago. Apart from the success that plants have had in becoming the dominant photosynthetic organisms on land, microorganisms still exclusively utilize most of these early energy sources. Thus, to rephrase an old expression, "The early bugs got the works."

Unlike animals and plants, microbial life is found in every ecosystem on Earth. Where there is life, there is microbial life. Indeed, when major climatic, astronomical and geological events impact Earth, the least affected types of life are microorganisms. In contrast, entire groups of animals, as exemplified by the dinosaurs, can become extinct in a very short period of time.

One of the most fascinating aspects of microbiology, as amply exemplified in this book, is that some microorganisms evolved to live under conditions that are too harsh for animals and plants. Extremes of temperature, pH, oxidation-reduction potentials, salinity and humidity and combinations of these are found in habitats on Earth only colonized by microorganisms. Microbial life is teeming in these unusual habitats and, although macroorganisms are more "advanced" forms of life, they have not successfully replaced the microorganisms.

The Tree of Life, produced in large part through the efforts of Carl Woese (1994), provided the first truly scientific view of biological diversity (Figure 1). This is because the Tree of Life is based on sequence comparisons of a macromolecule found in all living organisms, that of the sequence of the RNA from the small subunit of the ribosome. Sequences from some other highly conserved macromolecules such as ATPases and elongation factors produce very similar trees. Each branch in the Tree of Life represents a major phylogenetic group of life. Thus, there is a branch for plants and another for animals that are considered to be Kingdoms. But look at all the remaining branches (i.e., Kingdoms)! In addition to those shown, more than 20 other major branches of microbial life have been detected in environments but as yet, isolated strains are not yet available (Hugenholtz et al., 1999), and it is likely that several more remain to be discovered.

Despite the striking evidence provided by the Tree of Life, some biologists, such as Ernst Mayr (1998) have cast doubt on the biodiversity of microbial life. Mayr states that "Archaebacteria.even where combined with the eubacteria, as prokaryotes, this group does not reach anywhere near the size and diversity of the eukaryotes." Biologists who, like Mayr, base their views of diversity on organismal morphology have difficulty comprehending the vast physiological and metabolic diversity of microorganisms.

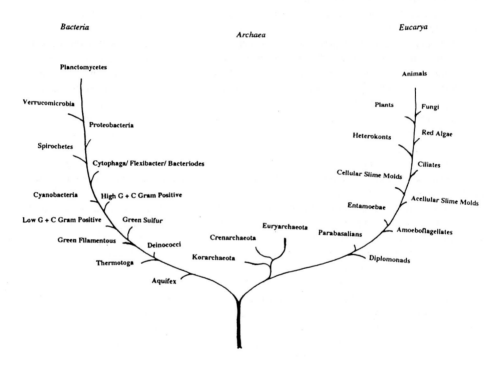

Figure 1. A diagrammatic representation of the Tree of Life based on 16S and 18S rRNA analyses showing the Bacteria, the Archaea and the Eukarya (modified after Woese et al., 1990). In addition to the branches shown, many additional microbial phyla exist most of which contain organisms that have not yet been isolated in pure culture.

Perhaps, to make the point more clear, microbiologists should ask such biologists questions like: "Where are the hyperthermophilic animals and plants that live on sulfide as an energy source?" "Why is it that the only carbon dioxide fixation pathway used by plants is the Calvin-Benson cycle which evolved from bacteria? "Do any animals have the ability to live on carbon monoxide and methane as energy sources." "Where do the plants that generate methane gas live?" "Which plants and animals produce sulfuric acid and live at pHs as low as 1?" "Are there any animals or plants that can carry out photosynthesis or chemosynthesis anaerobically?" These sorts of questions should encourage 'nay sayers' to appreciate the significance that metabolism and physiology have played as driving forces for the evolution and diversification of life.

Then, too, there is the problem of the species definition. Mayr (1998) concludes that microbial diversity is low based on the numbers of extant, described species. He states "Approximately 10,000 eubacteria have been named. The number of species of eukaryotes probably exceeds 30 million; in other words, it is greater by several orders of magnitude." This view is misleading for two very important reasons. First, most bacteria have not yet been named. The vast majority live in ecosystems still awaiting their discovery by a microbiologist. It is estimated that far less than 0.1% of the bacteria have been isolated and described. But an even greater problem with such statements is that they assume that the bacterial species definition is equivalent to the definition of plants and animals.

Two lines of evidence indicate that the bacterial species definition is much broader than that of animals and plants. Based on molecular characteristics such as DNA base composition range, DNA/DNA reassociation values and range of 16S and 18S rDNA sequence, it has been concluded that *Escherichia coli*, as a species, is defined much more broadly than that of its host mammalian species. Indeed, it is approximately equivalent to the animal Order, several taxonomic ranks above the species (Staley, 1997; Staley, 1999). Likewise, co-divergence of the bacterial genus, *Simonsiella* with its host animal species, humans, sheep and dogs, indicates that a single species of *Simonsiella* (Hedlund and Staley, submitted) exists for each animal Order (Primates, Ruminantia, and Carnivora, respectively). Thus, again the bacterial species is approximately equivalent in evolutionary divergence to the animal Order. If co-speciation can be demonstrated at the species level in the animal hosts, then *Simonsiella* would represent a genus with a single species for each mammalian species. In other words, there would be more than 4,500 species of *Simonsiella* alone!

Much of the future of biology, and astrobiology, too, lies in microbiology. Many of the great questions of science cannot be answered without understanding microorganisms. "How did life begin?" "What were the first organisms like?" "What were their energy sources and electron acceptors?" "How did the initial speciation events occur?" "How resilient is life?" "Have any bacterial lineages become extinct?" "How did the eukaryotic cell evolve?" "To what extent did genes derived from bacteria and their organelles play a role eukaryotic cell evolution?" "What are the minimal genetic and metabolic needs of a free-living organism?"

Using modest resources, microbiologists are just beginning to comprehend the vast unknown diversity of microorganisms in Earth's biosphere. They more fully appreciate the evolutionary implications of animals and plants evolving into a world dominated by microorganisms that have made the existence of all life possible. Increasingly, as microbiologists work in their labs and in the field, it is becoming apparent to all scientists, and even some lay people, that the microbial world and its importance to life on Earth and elsewhere in the Universe, can no longer be denied or ignored.

This book provides an escape from our conventional views of biology. Sit back and read about the remarkable and marvelous world of microbial biodiversity.

Acknowledgements

I appreciate the helpful comments of Brian Hedlund and Cheryl Jenkins as well as Joseph Seckbach and some anonymous editors.

References

Hedlund, B. P. and Staley, J. T. Co-speciation of the commensal bacterium *Simonsiella* with its mammalian hosts. Submitted for publication.
Hugenholtz, P., Goegel, B. M. and Pace, N. R. (1998) J. Bacteriol. **180**: 4765-4774.
Mayr, E. (1998) Proc. Nat'l. Acad. Sci. **95**: 9720-9723.
Staley, J. T. (1997) Current Opinion in Biotech. **8**: 340-345.
Staley, J. T. (1999) ASM News **65**: 681-687.
Woese, C. R. (1994) Microbiol. Rev. **58**: 1-9.
Woese, C. R. (1998) Proc. Nat'l. Acad. Sci. **95**: 11043.
Woese, C. R., O. Kandler, and Wheelis, M.C. (1990) Proc. Nat'l. Acad. Sci. USA **87**: 4576-4579.

University of Washington
Seattle, WA USA 98195
<Jtstaley@w.washington.edu>
15 February 2000

James T. Staley

Biodata of **Joseph Seckbach**, editor of this volume, and the chief editor of the book-series of **COLE** (Cellular Origin and Life in Extreme Habitats). He is the author of the chapters "*Acidophilic Microorganisms,*" "*A Vista into the Diverse Microbial World: An Introduction to Microbes at the Edge of Life*" (with co-author A. Oren), and "*Introduction to Astronomy; Origin, Evolution, Distribution and Destiny of Life in the Universe*" (with co-authors F. Westall and J. Chela-Flores) in this current volume.

Dr. Joseph Seckbach edited (and contributed two chapters to) *Enigmatic Microorganisms and Life in Extreme Environments* (Kluwer Academic Publishers, The Netherlands, 1999). See: **http://www.wkap.nl/bookcc.htm/0-7923-5492-3.** Likewise, he organized and contributed to the "*Cyanidium book*" entitled *Evolutionary Pathways and Enigmatic Algae: Cyanidium caldarium (Rhodophyta) and Related Cells* [Kluwer, 1994, see: http://www.wkap.nl/bookcc.htm/0-7923-2635-0]. He is the co-author (with author R. Ikan) of the Hebrew-language publication *Chemistry Lexicon* (1991, 1999) and co-editor of *From Symbiosis to Eukaryotism: Endocytobiology VII* (E.Wagner et al., eds.) published by the University of Freiburg and Geneva (1999).

Dr. Seckbach earned his Ph.D. from the University of Chicago (1965) and spent his postdoctoral years in the Division of Biology at Caltech (Pasadena, CA). Then he headed a team at the University of California, Los Angeles (UCLA) searching for extraterrestrial life. Dr. Seckbach has been with the Hebrew University of Jerusalem since 1970 and performed algal research and taught Biological courses. He spent sabbatical periods in Tübingen (Germany), UCLA and Harvard University. At Louisiana State University (LSU, Baton Rouge), he served (1997/1998) as the first selected occupant of the John P. Laborde endowed Chair for the Louisiana Sea Grant and Technology Transfer, and as a visiting Professor in the Department of Life Sciences.

Among his publications are books, scientific articles concerning plant ferritin (phytoferritin), cellular ultrastructure, evolution, acido-thermophilic algae, and life in extreme environments. He has also edited and translated books of popular science. Dr. Seckbach's recent interest is in the field of enigmatic microorganisms and life in extreme environments.

E-Mail: **seckbach@cc.huji.ac.il**

"You, O Lord my God, have done many things; the wonders You have devised for us; none can equal You, I would rehearse the tale of them, but they are more than can be told."
*(Psalms **40**, 6)*

PREFACE: *VOYAGE TO THE EDGES OF LIFE*

The purpose of this treatise is to introduce the teacher, researcher, and student as well as the "open minded" reader to some new aspects of uncommon and lesser-known diversity and variability among living microorganisms. Life is ubiquitous on Earth and some of these microorganisms or their products may serve not only for the textbook information but also for applied industrial products. This is the second book in the COLE (Cellular Evolution and Life in Extreme Habitats) series published by Kluwer. This new volume follows the first link entitled *Enigmatic Microorganisms and Life in Extreme Environments* (1999), see: http://www.wkap.nl/bookcc.htm/0-7923-5492-3. In the last two decades interest has grown in the field of microbial life under extreme habitats. In previous volumes—*Enigmatic Microorganism* and *Evolutionary Pathways and Enigmatic Algae* (1994) (see: http://www.wkap.nl/bookcc.htm/0-7923-2635-0)—we also dealt with aspects of cellular origins and with habitats where microbes have been detected under very exceptional living conditions.

In this volume, we take a new voyage to the edges of life. In ten sections, almost thirty researchers present current, updated reviews covering this sphere of knowledge. Our experts in these fields deal with areas from fossilized data and cruise along several lines of extremophiles up to possibilities of extraterrestrial life. This book covers a wide range of extremophiles. There are discussions on hyperthermophiles (Archaean thriving up to 113^0C) and Psychrophiles (such as Antarctic cells), microbes which live at the edges of pH ranges (acidity vs. alkalinity) and in hypersaline medium (Prokaryotes and newly discovered Fungi living in 30% salt solution in the Dead Sea, Israel). Other manuscripts deal with additional niches of severe environments; A chapter is devoted to the phenomena of symbiosis in which organisms of different species share one cellular habitat; bioluminescence of bacteria, the diversity of methanogens, rock-dwelling fungi, microbial life on petroleum, and polymorphism in bacteria. Finally, three chapters are devoted to the analogues of harsh terrestrial environments, like Antarctica and Yellowstone hot springs (Wyoming, USA), to the survival of "our" cellular candidates on Mars, Europa (the Juvian moon), and beyond. Our contributors have a wide geographical distribution, coming from the USA, Israel, Germany, Italy, Norway, UK, France, The Netherlands, and Russia.

This timely volume reflects the growing number of people who are involved in these studies and the ever increasing output of books on these subjects. It is our hope that the windows opened in this volume exposing life in various extreme habitats will lead to further discoveries of presently unknown microbial biodiversity occurring at the edge of Life.

Hebrew University of Jerusalem   **Joseph Seckbach**
Seckbach@cc.huji.ac.il
March 2000

Acknowledgements

I would like to thank our board members, Professors Aharon Oren, Ian Dundas, Russell Chapman and Raphael Ikan for their support and assistance in various steps of "making" this volume. Specific gratitude and appreciation are due to Professor **Aharon Oren** (Hebrew University) who took an active interest and involvement this volume and in the COLE (Cellular Origin and Life in Extreme Environments) book series. He also proofread several of our manuscripts and always made constructive suggestions and corrections. I thank also Professor **Woody Hastings** (Harvard University) and **Ian Dundas** (University of Bergen, Norway) for their critical reviewing of manuscripts. In addition, I express much gratefulness to our contributors and mainly to the "Early Birds" who submitted their chapters in due time and admire their understanding and patience. The publishing team of Kluwer who deal with this book is also acknowledged with appreciation. Last but not least, gratefulness and gratitude, are due to my wife, **Fern Seckbach** for assisting in proofreading, suggestions, understanding and much patience during the making of this volume.

Hebrew University   **Joseph Seckbach**
Jerusalem, Israel
seckbach@cc.huji.ac.il
March 2000

I

GENERAL DIVERSITY

Joseph Seckbach and Aharon Oren

For the Biodata and portraits of these authors,
see their other contributions. "Preface" and Chapter 8 (J.S.), and Chapter
16 (A.O.).

J. Seckbach (ed.), Journey to Diverse Microbial Worlds, 3-13.
© 2000 *Kluwer Academic Publishers. Printed in the Netherlands.*

A VISTA INTO THE DIVERSE MICROBIAL WORLD:
An Introduction to Microbes at the Edge of Life

JOSEPH SECKBACH[1] and AHARON OREN[2]
[1]P.O.Box 1132, Efrat, 90435 Israel.
[2]Division of Microbial and Molecular Ecology, The Institute of Life Sciences, and the Moshe Shilo Minerva Center for Marine Biogeochemistry, The Hebrew University of Jerusalem, 91904, Israel.

1. General Overview of Extremophiles

Our understanding of the biodiversity in the microbial world has recently grown in many aspects. One aspect on which our knowledge has greatly increased is the intriguing field of extremophilic microorganisms that live in bizarre niches on and below the global surface.

Extremophile microorganisms dwell in habitats hostile to most forms of life. Biodiversity in such biotopes is restricted due to the environmental extremes. Some microbes may live in environments in which several stress factors occur simultaneously. For example, *Sulfolobus acidocaldarius*, *Metallosphaera sedula* and *Stygiolobus azoricus* live at 80^0–90^0C and at pH 1 to 5.5. Similarly, some barophiles are also psychrophiles, growing in the ocean depth. Barophilic microorganisms may in addition have acidophilic or alkaliphilic properties. Those "extremophilic" microorganisms undoubtedly consider their own unique habitats a "Garden of Eden," while for others who enter such extreme niches the ending will be fatal.

Most microorganisms discussed in this chapter are members of the Prokaryotes, i.e., Bacteria and Archaea. Prokaryotes are unique in being small sized microorganisms that possess a simple cell morphology, most of them appearing as rods, coccoids, or spiral shapes. Their cytoplasm contains a circular naked chromosome (in some species a linear chromosome, or even multiple chromosomes were shown to occur) and ribosomes. The cell is surrounded by a single membrane and in most cases by a cell wall. These cells, being simple in ultrastructure, often have a very high rate of multiplication. The Prokaryotes display a high degree of physiological flexibility, enabling them to adapt to various extreme habitats. They mediate important processes in the cycling of carbon and other elements, thereby influencing the composition of the atmosphere, and they maintain complex relationships with other organisms, including man. The pharmaceutical and biotechnological industry extracts many products from various Prokaryotes, used in disease control and in various industrial processes (e.g. thermostable enzymes).

The bacterial domain encompasses most known prokaryote species (including certain extremophiles), while the archaeal domain includes among others acidophiles,

5

thermophiles, methanogens, and halobacteria. The domain Eukarya also contains a number of extremophiles, including psychrophilic algae, which stain ice and snow, halophiles, alkaliphiles, acidophiles, and others (see below). More information about microbial versatility can be found in the chapter by Roberts in Seckbach (1999) and in the chapters by Hoham and Ling, Valentine and Boone, Ron, Gorbushina and Krumbein, Zaritsky et al., and Heldal in the present volume.

Extremophiles may not only serve as sources for valuable natural products, they are also of great interest in the study of biodiversity on Earth and the understanding of the origin of life, as well as for the investigation of the possibility of life to occur on other stars and planets (see chapters by Seckbach and Westall, Chela-Flores, and by McKay in this volume).

2. The Extremophiles

Extreme environments (defined from our anthropocentric view) possess various factors incompatible with most forms of life. However, in spite of the apparent hostility of such environments, they may contain a much more diverse microflora than has often been presumed. In the past twenty years it has become clear that many of these environments may be inhabited by surprisingly diverse microbial communities. The number of different extremophiles known to reside or even thrive in biotopes with environmental extremes has grown rapidly in recent years. This increase in known biodiversity is based both on laboratory cultures of isolated microorganisms and on the characterization of 16S rDNA sequences recovered directly from the environments. Most of these sequences are not represented as yet by any cultured organisms, and it may thus be assumed that the number of species grown in culture grossly underestimates the true diversity of life forms, even in the most extreme of conditions. Only now are we beginning to open the "black box" of the once unknown mysterious life forms represented by the thermophilic and hyperthermophilic, halophilic, and alkaliphilic microorganisms, as well as other organisms found in habitats hostile to most forms of life. For further information see Seckbach (1997, 2000) and the chapters by Rainey and Ward-Rainey, Ollivier et al., and Madigan in this volume.

3. Temperature Effects

Temperature is one of the most important environmental factors governing growth and survival of microorganisms. Every organism has its characteristic cardinal temperatures (minimum, optimum, and maximum), defining the relationship between growth and temperature. Microorganisms may live at temperatures ranging from subzero to the boiling point of water and above. Life is regarded to exist between -5^0 and 113^0C at least, which is currently the highest temperature shown to support growth of microorganisms in culture.

Microbes may be classified according to their temperature requirement into psychrophiles (-2^0 to below 20^0C), mesophiles ($20-40^0C$) and thermophiles and hyperthermophiles ($45-113^0C$).

Thermal environments originate as the result of solar heating, geothermal activity, intense radiation, combustion processes, and human activities (power plants, industrial discharges, etc.). Hot-springs may have temperatures near the boiling point, while steam vents (fumaroles) can have temperatures as high as 500^0C in undersea hot-springs (hydrothermal vents) and active sea mounts where volcanic lava is emitted directly onto the seafloor. "Black smoker" vents emit the hottest water, forming "chimneys" built of metal sulfides that precipitate when the metal-rich hydrothermal fluid mixes with cold seawater. Between the chimney ($200-400^0C$) and the seawater (2^0C) a temperature gradient is formed, populated by a range of hyperthermophiles differing in their temperature requirements.

The psychrophiles include the snow and ice algae (belonging to the domain Eukarya), that live around 1^0C. Some thermophilic cyanobacteria (domain Bacteria) thrive up to 74-75^0C, being the upper temperature limit for photosynthesis. Heterotrophic Bacteria are also among the thermophiles, while the most thermophilic of all microorganisms are representatives of the archaeal domain. The archaeon *Pyrolobus fumarii* is currently the most thermophilic of all known Prokaryotes; it grows up to a temperature of 113^0C, and astonishingly it can survive in an autoclave (121^0C) for over 1 hour (Madigan and Oren 1999). Hyperthermophilic prokaryotes are also present deep within the earth where geothermal heating supports their lifestyle (Onstott et al. in Seckbach 1999). The upper temperature for life at such depths may be around 110^0C.

3.1. THERMOPHILES: THE HEAT LOVERS

Thermophilic microorganisms can be defined as those growing between 45 and 75^0C. Typical representatives of this category are the eukaryal thermophilic alga *Cyanidium caldarium* and its cohorts (see Seckbach 1994a, 1994b, 1995) and certain fungi that thrive in acidic media, while having an upper temperature of $55-60^0C$ (Brock 1978).

Most high temperature lovers that grow above 75^0C (so-called hyperthermophiles) are anaerobes, and they are found in hydrothermal vents, in hot springs, oil wells, volcanic lakes, compost piles, and in other hot environments. Recently archaeal cells have been observed in volcanic and geothermal ecosystems with temperatures exceeding 100^0C. Thus, microorganisms like *Pyrococcus* and *Pyrodictium*, anaerobic species isolated from marine solfataric mud, as well as *Hyperthermus butylicus,* have an upper temperature limit of $105-110^0C$ (Stetter 1998). Species of *Pyrolobus*, a genus of acidophilic hyperthermophiles have likewise a temperature range from 90^0C to 113^0C, with an optimum at 106^0C. Hydrothermal vents also provide a habitat for non-acidophilic species such as *Thermococcus* with a temperature range of $56-103^0C$. Likewise, neutrophilic *Thermococcus* species have been observed to occur at $50-107^0C$ in deep-sea sediments and in hot springs. Methanogens with a temperature range of $55-97^0C$, growing between pH 6-10, have been isolated from sludge digesters and solfataras, while *Methanopyrus*, living in hydrothermal vents, has a temperature range of $84-110^0C$. For further details, see Rachel in Seckbach (1999) and the section of "Thermophiles and Acidophiles" in this volume.

3.2. PSYCHROPHILES: THE COLD LOVERS

The oceans, which occupy half of the Earth's surface, have an average temperature of 5^0C, and water in the depth of the ocean averages 1-3^0C. In spite of these low temperatures, extensive microbial growth of a diverse microbial community may occur in pockets of liquid water in permanently frozen areas and in thaw of lake ice and sea ice during the summer period, both in the Arctic and the Antarctic.

Liquid water is the vital solvent for life, and its presence is an absolute requirement for growth. When water solidifies, growth is suspended. Although the freezing point determines the lower temperature for growth, snow can nevertheless become stained in various bright colors by the pigmentation of algae like *Chlamydomonas nivalis*, *Chloromonas*, and certain dinoflagellates [Roberts in Seckbach (1999), Hoham and Ling in this volume].

Psychrophilic microorganisms (defined as having their optimum growth below 15^0C) dominate cold environments. Such microbes are often killed by exposure to higher temperatures. Among these psychrophiles are *Polaromonas vacuolata*, growing at a temperature range of 0^0C (min.), 4^0C (opt.) and 12^0C (max.). Some heterotrophic anaerobes are found in refrigerated beef having temperatures between 1-15^0C. Other psychrophilic microorganisms may be found in saline ice lakes. It has been proposed that some organisms produce antifreeze proteins that interfere with the crystallization process of water, thus preventing the formation of ice crystals in their cells.

Recently, microorganisms have been isolated from ice cores of up to 3,600 meter depth from Vostok, the Russian Station in Antarctica. This frozen terrestrial area consists of a thousand meters depth of glacial icy layer, which covers a liquid water lake underneath (Priscu et al. 1999, Karl et al. 1999). These cores have been collected from deep drills of about 120 meters over this hidden lake. These ice fragments contain microbes, including bacteria, cyanobacteria, and algae such as diatoms, fungi, as well as protozoa, spores of different organisms, and pollen grains. Some of these organisms have been revived after a long period of dormancy, and these isolates are still being investigated and characterized. The DNA isolated from these organisms is similar to that of known recent organisms. It is assumed that this interesting lake beneath the thick ice layer may also contain microorganisms, most of which are probably anaerobes.

4. pH Effects

Environments on Earth range from acidic to alkaline. Microbial activity may occur both in extremely acidic (down to pH 0.7) and in very alkaline environments (as high as pH 12). Algal photosynthesis may take place in waters ranging from pH 3.5 to 10. The intracellular pH is in all cases maintained close to neutral, i.e. very different from the pH of the environment (see the chapter by Seckbach in this volume). Steep pH gradients across the cell membrane are thus to be generated and maintained by cells living in extremes of extracellular pH.

4.1. ACIDOPHILES: SOUR MICROBES

Acidophilic microbes grow at pH values between 0.7 and 6. Acidic environments of such low pH often originate as a result of the biogeochemical activities of bacteria, such as oxidation of SO_2 and H_2S in hydrothermal vents and sulfur springs. Acidity may also result from the production of organic acids during fermentation or from the oxidation of ferrous iron. For example, *Thiobacillus ferrooxidans* generates acid by oxidation of Fe^{2+} to Fe^{3+}. The latter precipitates out as $Fe(OH)_3$, releasing protons in the process. Similarly, *Thiobacillus thiooxidans* generates acid from the oxidation of sulfide or elemental sulfur to sulfuric acid ($HS^- + 2 O_2 \rightarrow SO_4^{2-} + H^+$).

Extreme acidophiles are microorganisms that have evolved to grow best near the lower end of the pH scale, with a minimum of near pH 0 to an optimum below 3. Among the eukaryotic microorganisms, three genera of fungi (*Acontium*, *Cephalosporium*, and *Trichosporon*) are known to grow near pH 0. Acidothermophilic algae belonging to the Cyanidiaceae (*Cyanidium caldarium*, *Cyanidioschyzon merolae*, *Galdieria sulphuraria*) thrive at high temperatures ($45-57^0C$) in very acidic solutions (pH 2-4). *Cyanidium caldarium* even tolerates a solution of 1 N H_2SO_4 (Seckbach 1994a, 1995). The halophilic green alga *Dunaliella acidophila* grows at pH values as low as 0.5-3 with an optimum of pH 1.0 (see the chapter by Pick in Seckbach 1999).

Most known anaerobic acidophiles are heterotrophic hyperthemophiles, belonging to the archaeal domain. Among the Prokaryotes, the most acidophilic species known is *Picrophilus oshimae* - a thermophilic archaeon whose pH optimum is 0.7. This organism is stable in hot (opt. 60^0C) and extremely acidic media (below pH 1). The acidophilic anaerobe *Stygioglobus azoicus* grows optimally at pH 2.5-3.

The internal pH of acidophilic microorganisms is more near-neutral than the external one. Those cells maintain their intracellular pH at the desired level by using strong proton pumping activity. In addition, the selective permeability properties of the membrane prevent the uncontrolled entrance of protons into the cells. For further discussion on acidophiles see Seckbach 1994a, the article by Pick in Seckbach 1999, and the chapters by Seckbach and by Weiss Bizzoco in this volume.

4.2. ALKALIPHILES

Alkaliphiles grow optimally at pH values above 8. Extreme alkaliphiles live in very alkaline soils and in soda lakes, where the pH can rise to values as high as 12. Anaerobic alkaliphiles have been isolated from several environments, including salt lake sediments, river and marine sediments, sewage sludge, bio-heated compost, hot springs and subterrestrial environments.

Among the alkaliphilic heterotrophs we can name *Spirochaeta* species that produce acetate, ethanol, lactate and hydrogen from sugars. *Natroniella* strains (pH range of about 7.9-10.7) ferment lactate and ethanol to acetate. There are also some alkalitolerant species such as *Anaerobranca horikoshii* that live in alkaline habitats associated with petroleum deposits. For further data, see the chapters by Roberts and by Kamekura in Seckbach 1999, and the chapters by Jones and Grant, Boussiba et al., and by Zavarzin and Zhilina in this volume.

5. Halophiles - Microbes in Saline Habitats

Many microorganisms require NaCl for growth. Such organisms may be found in marine habitats, in saline lakes and other hypersaline environments of neutral or alkaline pH. Halophiles grow in salt solutions ranging from 2-20% NaCl, while extreme halophiles live in hypersaline brines of 20-30% salt and higher, even up to NaCl saturation. Such organisms occur in inland salt lakes, soda lakes, and hypersaline microbial mats. They may also be present in highly salted foods, in saline soils, and even in underground saline deposits. Extreme halophiles can easily be isolated from salt lakes such as the Dead Sea (on the border between Israel and Jordan) and the Great Salt Lake (Utah, USA). These hypersaline environments are often colored red by dense populations of pigmented halophiles. Anaerobic halophiles may be present in subterrestrial subsurface brines, including oil field brines. Some underground salt deposits may harbor "ancient" microorganisms, which have been dormant for millions of years. Such "living fossils" have been recently revived, and they are presently under study (see the chapter by Vreeland and Rosenzweig in Seckbach 1999).

Among the microorganisms living in hypersaline environments are aerobic and anaerobic heterotrophs, sulfate reducers, methanogens, and phototrophs. The halophiles have developed a variety of strategies to cope with the high osmotic strength of their external medium. Further information can be found in the chapters by Oren, Ventosa and Arahal, Nakamura, Vreeland and Rosenzweig, and Shand and Perez in Seckbach 1999, and the contributions by Buchalo et al. and Oren in this volume.

6. Barophiles and Piezophiles (Microbial Life at Great Depths)

The barophiles (weight lovers) or piezophiles (pressure lovers) live under high hydrostatic pressure. For every 10 m of depth in water column there is an increase of 1 atmosphere in hydrostatic pressure. Microbial life has been found in the marine system up to the greatest depths of 10,898 m (equivalent to 110 MPa [1 MPa = 9.87 atmosphere]).

Many of the deep dwelling microbes grow at low temperatures ($\sim 4^0$C), and they are termed barophilic psychrophiles. Barotolerant microbes grow well at ambient pressure of 1 atmosphere, but are able to grow also at high hydrostatic pressure. The obligate barophiles, in contrast, require high pressure for optimal growth, and may not be able to grow at all at atmospheric pressure (see the chapter by Walsh and Seckbach in Seckbach 1999). Psychrophilic barophilic bacteria have been isolated from the deep sea near the Philippines from a depth exceeding 10,000 meters. These bacteria grow at 4^0C only at pressures above 500 atmospheres, with optimal growth being achieved at 700 atmospheres. The upper limit of microbial growth is probably about 130 MPa or 1,283 atmospheres (see the chapter by Bartlett and Bidle in Seckbach 1999). Additional barophilic psychrophiles may be detected in the deep ice near the Vostok Station in Antarctica (see above) and perhaps they are present also in the under icy lake covered with thousands meters of thick ice. For further studies on Antarctic icy life in the hidden lake see Priscu et al. (1999) and Karl et al. (1999).

Certain microbes thriving at very high temperature near the submarine vents are hyperthermophilic barophiles that live in temperatures of over 100^0C. The archaeon *Thermococcus barophilus* is a hyperthermophile that grows at 95-100^0C under elevated pressure and behaves as an obligate barophile. The mesophilic-barophilic sulfate reducing bacterium *Desulfovibrio profundus*, isolated from deep marine sediments of the Japanese sea, grows at a pressure of 100-150 atmospheres. These isolates are highly barotolerant rather than obligately barophilic, so they survive atmospheric pressure as well.

In contrast to most of the deep ocean high-pressure environments, which are cold, the subterranean environment is characterized by a temperature increase of 20-30^0C for every km of depth. Onstott et al. (in Seckbach 1999) discussed the possibility of the existence of a diverse barophilic or barotolerant microbial community thousands of meters deep below the Earth's surface. The investigators obtained viable microbes from samples recovered from depths as great as 3,400 meters below land surface underneath a gold mining operation in South Africa. Similarly, hyperthermophilic Archaea have been observed in water drawn from an oil reservoir located at a depth of about 3,000 meters. Subterrestrial microbial communities are probably quite diverse; both Bacteria and Archaea have been detected at depths as great as 4 km below the surface. Still deeper strata are too hot for any known form of life. Life below the sea floor may be expected to occur down to few km below the sea bottom. Additional information may be found in the chapter by Bartlett and Bidle in Seckbach 1999 and in the contributions by Deming and Huston, Haygood and Allen, and by Yayanos in this volume.

7. Further Microbial Frontiers

7.1. XEROPHILES - DESICCATION AND DRYNESS

Even extremely dry regions support both prokaryotic and eukaryotic life. Examples are the lichens, which grow on stones in the deserts worldwide. Of all eukaryotes, fungi are the best adapted to life out of liquid water. In addition, bacterial spores may survive in a desiccated state. Such spores have been revived after having been dormant for millions of years. To give a few examples: dormant bacterial spores have been revived from the stomachs of frozen mammoths and from fossilized insects' guts embedded in amber after up to 40 millions of years (Seckbach 2000). Cryptoendolithic microbial communities and Siberian permafrost microbes are also known to be desiccation-tolerant. Likewise, cells of a *Streptococcus mitis* strain which have been left inside a TV camera aboard Surveyor 3 on the surface of the moon for two and a half years, exposed there to vacuum and low temperature and without nutrients, could easily be revived upon retrieval of the camera and its transportation back to Earth by Apollo 12 (Mitchell and Ellis 1971).

Davis (1972) presented survival records of microorganisms including bacteria, algae, fungi and protozoa following desiccation. Some cyanobacteria (e.g. *Nostoc commune*) survived desiccation for no less than 107 years, while the diatom *Nitzschia palea* and two chlorophytes (*Pleurococcus* sp. and *Cystococcus* sp.) were shown to be viable after 98 years of dry storage in a herbarium. These finding are important not only for our understanding of the biological limits of survival of microorganisms, but have also important implications for space research missions.

7.2. ANAEROBIC MICROORGANISMS ("LA VIE SANS AIR")

Life originated and evolved in anaerobic environments. It may be assumed that prior to the "oxygen revolution" all microbes lived under anoxic conditions. Today most of the biosphere is aerobic, but many anaerobic niches still exist. The habitat of many of the present-day anaerobes may resemble the anoxic environment of the Earth until approximately two billion years ago. When oxygen started to accumulate in the biosphere, the pioneer microorganisms living in the presence of oxygen had to develop mechanisms to avoid oxidative damage by this potentially toxic agent.

Today, oxic and anoxic environments coexist in the biosphere, and the interrelationship between aerobic and anaerobic processes allows the effective cycling of carbon, nitrogen, sulfur, and other elements. The anaerobic decomposers involved in the food web are responsible to a large extent for the recycling of these elements within the biogeochemical cycles. The two worlds thus rely and depend on each other.

The wealth of anaerobic microorganisms found in Nature contributes much to the overall biodiversity. Anaerobic representatives are found in all three domains of life (Archaea, Bacteria and Eukarya). Among the environments with a rich diversity of anaerobes we may mention the digestive tracts of animals, including the rumen of ruminants and anaerobic sludge digesters in sewage treatment plants. Several types of eukaryal cells may grow under aerobic as well as under anaerobic conditions. Others are strict anaerobes, such as the Archaezoa, which are amitochondrian cells. Some claim that these have never possessed the ability of aerobic growth; however, they do carry genes in their nucleus characteristic of the mitochondrial genome (see the chapter by Roberts in Seckbach 1999). There is now good evidence that all types of Eukarya once possessed mitochondria, which in some cases were lost, leading to the development of secondary adaptation to life in oxygen-free or microaerophilic habitats (D. Roberts, personal communication). Another case of a eukaryal organism that does not require oxygen, at least temporarily, is presented by the phototrophic thermoacidophilic members of the Cyanidiaceae, which thrive on pure CO_2 without the need for air supply (Seckbach et al. 1970; Seckbach 1994a).

Many environments in which anaerobes live are characterized by extremes of environmental conditions such as temperature, salt concentration, pH, etc. Examples are such thermophilic Archaea dwelling around hydrothermal volcanic vents on the submarine ocean floor (Stetter 1998, and the chapters by Bartlett and Bidle in Seckbach 1999 and by Yayanos in this volume). There are also terrestrial barophiles living at high temperature in the depth of the Earth crust, some of them being obligate anaerobes (see the chapter by Onstott et al. in Seckbach 1999). Likewise, some methanogens live at extremely high temperatures in very low redox environments, deriving energy from the reduction of CO_2 with hydrogen.

For further discussions on anaerobic microorganisms, see Stetter (1998), the articles by Hackstein et al. and by Roberts in Seckbach 1999, and the chapter by Zavarzin and Zhilina in this volume.

8. Conclusions

Life is ubiquitous on Earth, and microorganisms can adapt to virtually all extreme conditions encountered on our planet. As apparent from the chapters in this volume, much information already exists on the unique adaptations of microorganisms to life in extreme environments. Further outlook of the versatility of the microbial world is also presented in our previous book entitled: *Enigmatic Microorganisms and Life in Extreme Environments* [http://www.wkap.nl/bookcc.htm/0-7923-5492-3] (J. Seckbach [ed.], Kluwer Academic Publishers, The Netherlands, 1999). However, it is to be expected that Nature has still many surprises in store for us, and we have no doubt that future studies will disclose yet unknown type of microorganisms existing at the edge of life.

9. Acknowledgement

We thank several colleagues who expressed their views on the status of anaerobic microorganisms.

10. References

Brock, T.D. (1978) Thermophilic Microorganisms and Life at high Temperatures. Springer-Verlag. New York.
Davis, J.S. (1972) The Biologist **54**: 52-93.
Karl, D.M., Bird, D.F., Björkman, K., Houlihan, T., Shackleford, R. and Tupas, L. (1999) Science **286**: 2144-2147.
Madigan, M.T. and Oren, A. (1999) Curr. Opinion Microbiol. **2**: 265-269.
Mitchell, F.J. and Ellis, W.L. (1971) in: A.A. Levinson (ed.) Proceedings of the Second Lunar Science Conference, vol. 3. The MIT press, Cambridge, MA. pp. 2721-2733.
Priscu, J.C., Adams, E.E., Lyons, W.B., Voytek, M.A., Mogk, D.W., Brown, R.L., McKay, Ch.P., Takacs, C.D., Welch, K.A., Wolf, C.F., Kirshtein, J.D. and Avci, R. (1999) Science **286**: 241-2144.
Seckbach, J. (1994a) (ed.) Evolutionary Pathways and Enigmatic Algae: *Cyanidium caldarium* (Rhodophyta) and Related Cells. Kluwer Academic Publishers, Dordrecht, the Netherlands.
Seckbach, J. (1994b) J. Biol. Phys. **20**: 335-345.
Seckbach, J. (1995) in: C. Ponnamperuma and J. Chela-Flores (eds.) Chemical Evolution: Structure and Model of the First Cells, Kluwer Academic Publishers, Dordrecht, The Netherlands. pp. 335-345.
Seckbach, J. (1997) in: C.B. Cosmovici, S. Bowyer and D. Wertheimer (eds.) Astronomical and Biochemical Origins and the Search for Life in the Universe. Editrice Compositori, Bologna, Italy, pp. 511-523.
Seckbach, J. (ed.) (1999) Enigmatic Microorganisms and Life in Extreme Environments. Kluwer Academic Publishers, Dordrecht, the Netherlands.
Seckbach, J. (2000) in: Astronomical Society of the Pacific Conference Series (in Press).
Seckbach, J., Baker, F.A. and Shugarman, P.M. (1970) Nature **227**: 744-745.
Stetter, K.O. (1998) in: K. Horikoshi and W.D. Grant (eds.) Extremophiles: Microbial Life in Extreme Environments. Wiley-Liss, Inc. N.Y. pp. 1-24.

Biodata of Dr **F. Westall** author (with co-author **Maud M. Walsh**) of the chapter *"The Diversity of Fossil Microorganisms in Archaean-Age Rocks"* and a co-author of **"Introduction to Astrobiology: Origin, Evolution, Distribution and Destiny of Life in the Universe"** (with J. Seckbach and J. Chela-Flores).

Dr Frances Westall received her Ph.D. in Marine Geology in 1984 from the University of Cape Town, South Africa. At the Alfred Wagner Institute for Polar and Maine Research in Bremerhaven, Germany (1984-1989) she studies the influence of climatic changes over the last 4 million years on sediments deposited in the Antarctic South Atlantic Ocean. Her subject of study changed to fossil bacteria during research at the University of Nantes, France (1989-1991) and then to fossilization of bacteria and the search for the oldest fossil bacteria on Earth and in Martian meteorite ALH85001 at the University of Bologna, Italy (1991-1997). Dr. Westall continued her research on the oldest forms of Life on Earth as possible analogues for extraterrestrial life as a NRC fellow at the Johnson Space Center, Houston, Texas (from 1998 to 2000). She recently joined the Lunar and Planetary Institute at Houston, TX.

E-mail: **westall@lpi.usra.edu**

J. Seckbach (ed.), Journey to Diverse Microbial Worlds, 15-27.
© 2000 *Kluwer Academic Publishers. Printed in the Netherlands.*

Biodata of Dr. **Maud M. Walsh** co-author (with author Dr. **F. Westall**) of *"The Diversity of Fossil Microorganisms in Archaean-Age Rocks."*

Dr. Maud M. Walsh is an Assistant Professor, Research, in the Institute for Environmental Studies at Louisiana State University (Baton Rouge). She earned a B. A. in English from Bryn Mawr College in 1975, a M.L.S. in Library Science from Louisiana State University in 1976, and a Ph.D. in Geology and Geophysics from Louisiana State University in 1989. Her research interests span across the field of geomicrobiology, from bioremediation of hazardous waste using microorganisms to the study of early life on Earth. She has studied some of the oldest organisms on Earth as preserved in the rock record of the Swaziland Supergroup, South Africa. Dr. Walsh is currently president of the Louisiana State University chapter of *Sigma Xi*, the Scientific Research Society, and is publicity officer of the Baton Rouge chapter of the Association for Women in Science.
E-mail: **evwals@unix1.sncc.lsu.edu**

THE DIVERSITY OF FOSSIL MICROORGANISMS IN ARCHAEAN-AGE ROCKS

F. WESTALL[1] and M.M. WALSH[2]
[1]SN2-NASA-Johnson Space Center
Houston TX 77058 USA
and
[2]Institute for Environmental Studies
Louisiana State University, 42 Atkinson Hall
Baton Rouge, LA 70803-5705 USA

1. Introduction

Archaean-age rocks (3.9-2.5 b.y.) on Earth are uncommon owing to the destructive nature of the plate tectonics which characterizes this planet. Furthermore, much of the surviving Archaean-age material has been subjected to severe metamorphism. Since this is the period when life developed, these problems pose serious limitations on the search for early life. However, the situation is not entirely gloomy and sufficient rock material has been preserved in usable form to provide a preliminary impression of Archaean life. The environmental setting for early life on Earth, however, was very different from today's environment, and even from the succeeding Proterozoic era (2.5-0.54 b.y.). The different environmental conditions had important implications for early life and its diversity.

1.1. THE ARCHAEAN ENVIRONMENT

There is still much debate as to the exact nature of the early Earth's environment owing to the lack of material preserved but certain generalizations can be made. In the first place, the length of day was much reduced because the Earth was spinning faster (it is the tidal drag due to the proximity of the moon that has caused a slowing of the Earth's rotation, Walker et al., 1983). The Archaean day was about 15 hours instead of 24 hours long. The length of day had implications for atmospheric circulation with possibly a warmer tropical zone and cooler poles (Hunt, 1979). Tides were also larger (by about 50 %, Walker et al., 1983) although it is not clear what effect this phenomenon had on life.

As far as surface temperatures are concerned, the geological evidence suggests that ambient temperatures were not much higher or lower than they are today, despite the 25-30% lower solar luminosity (Sagan and Mullen, 1972). The higher partial pressures of the CO_2 atmosphere apparently produced a desirable greenhouse effect, keeping surface temperatures above the freezing point of water but below about 60°C (Owen et al., 1979; Walker et al., 1983; Kasting, 1987). The higher CO_2 partial pressures in themselves would have had an effect on the Archaean biota, aiding carbon fixation by photosynthetic chemosynthetic autotrophs, for which there is perhaps some metabolic evidence (Walker et al., 1983).

The question of the oxidation state of the Archaean atmosphere and the levels of free oxygen in it have been taxing the minds of students of the Archaean for a number of decades. On geological grounds (the temporal distribution of banded iron formations, red beds, palaeosols and detrital uranite and pyrite), the overall Archaean environment appears to have been either neutral or slightly reducing (Walker et al., 1983). However, there is evidence for local oxidation, viz. the presence of haematite and gypsum (although the indentifications of the latter mineral as opposed to anhydrite are hotly debated) and weathered, oxidised palaeosols (Knoll, 1979; Walker et al., 1983; Knoll and Holland, 1995), and the absence of kerogen (presumably oxidised) in some littoral sandstones (Dimroth and Kimberley, 1976). If there were O_2 in the atmosphere, the levels must have been <1% present atmospheric levels (PAL) (Cloud, 1976; Canuto et al., 1983; Kasting, 1987). A slight increase to 1-2% PAL occurred in the early Proterozoic, possibly related to increased biogenic productivity stimulated by the tectonic development of stable cratons (Towe, 1990; Knoll and Holland 1995). An alternative hypothesis to the presence of O_2 as an oxidising agent was put forward by Kasting (1987) and McKay and Hartman (1991) who suggested hydrogen peroxide.

The lack of, or extremely low levels of, O_2 in the Archaean atmosphere had profound implications for the metabolism of the primitive organisms. In the first place, significant amounts of O_2 would have inhibited prebiotic chemical evolution but once life had developed, the calculated levels of O_2 would have been too low for supporting aerobic life and, furthermore, they would have been too low to provide ozone protection against gene-damaging UV radiation. Kasting (1987) notes that the level of O_2 needed to produce a sufficiently thick ozone layer is at least 10% PAL. This level was not reached until Proterozoic times (about 1700 b.y., Walker et al., 1983; Knoll and Holland, 1995). High levels of biologically-damaging UV radiation during the Archaean would have strongly affected life. Until primitive life had evolved UV blocking mechanisms, it would have been limited to protected benthic environments (Cockell, 1998) and the development of other protective strategies, such as the production of extracellular mucus and mat formation (Margulis et al., 1976; Pierson et al., 1993). The large concentration of dissolved ferrous ion in the Archaean oceans would have mitigated the situation by absorbing some of the damaging UV radiation (Olson and Pierson, 1986; Garcia-Pichel, 1998).

The tectonic setting of the Archean Earth was very different from that of the later Proterozoic and Phanerozoic Earth (Windley, 1977; Ernst, 1983; Veizer, 1983; de Wit, 1998). The planet was dominated by volcanism and, although there is some evidence for the existence of felsic crust (material of which the continents today are almost entirely made) in the Archean, it is not known how much there was and whether there were cratonal regions similar to modern continents. Estimates of the amount of continental crust in existence during the Hadean (>3.8 b.y.) and the Archaean range from between two thirds to three-quarters of the present amount of crust (Lambert, 1976; Ernst, 1983). The sediments deposited on the volcanic sequences were young and immature and directly derived from the volcanic rocks (Ernst, 1983; Lowe, 1999a,b. Whereas the

Proterozoic is characterised by abundant, differentiated sediments deposited in shallow seas on stable continental platforms, the Archaean sediments appear to have been deposited in narrow, shallow zones flanking emergent volcanic provinces (Knoll, 1979; Ernst, 1983; Lowe, 1999a).

Evidence from lunar cratering suggests that both the Earth and the Moon were subjected to heavy bolide bombardment from 4.44 to 3.8 b.y. (Maher and Stevenson, 1988; Chyba, 1990). Some of these impacts could have been large enough to volatolise the oceans and, thus, sterilise the Earth with fatal consequences for life. It is believed that either life started repeatedly and easily after such events and/or that it took refuge in environments which provided some protection, such as hydrothermal vents. In fact 16sRNA gene sequencing seems to indicate that at some stage in its early history, life did pass through some kind of hydrothermal "bottleneck" (Nisbet and Fowler, 1996).

1.2. EARLY LIFE

Given the evidence for a neutral to slightly reducing environment on early Earth, the first organisms must have had an anaerobic metabolism. Indeed, the phylogenetic tree shows anaerobes at its deepest branches (Woese, 1987). O_2 would have been a poison from which the primitive organisms needed to protect themselves. But a low level of O_2 would have been necessary to stimulate mechanisms within the organisms for coping with O_2, hence paving the way for oxygenic photosynthesis and the eventual build-up of O_2 in the atmosphere (Knoll, 1979). Thus, the earliest organisms were probably anaerobic chemotrophs. Biological diversity would have been limited by a number of factors including UV radiation (Cockell, 1998). The ability of organisms to adapt to an oxidised environment, the protection provided by increased atmospheric O_2 levels, and the tectonic changes providing stable cratonic environments (carbonate platforms) all contributed to an increase in productivity and subsequent diversity of life in the succeeding Proterozoic period (Knoll, 1979; Knoll and Holland, 1995; Knoll 1995).

The timing of the appearance of the all-important oxygenic photosynthesis is still uncertain. Indirect evidence for phototactic behaviour and photosynthesis in the Archaean comes from the existence of stromatolites (mineralised microbial mats) in the Archaean formations (in the later eras and at the present time largely constructed by oxygenic photosyntesisers, the cyanobacteria) (Walter, 1983; Knoll, 1996) and from carbon isotope studies (Schidlowski, 1988). Prokaryotic photosynthesisers, however, do not necessarily need to be oxygenic (Walter, 1983) and, although certain carbonaceous fossiliferous remains from the Early Archaean have been compared to cyanobacteria, the relatively degraded nature of the fossils makes positive identification impossible. Moreover, microbial mats are found in a wide variety of environments, including the deep marine realm (Jannash, 1984; Williams and Reimer, 1983), so the presence of mat-like laminations in Archaean rocks does not imply that photosynthesis had developed at that time. On the other hand, there is clear evidence from molecular fossils for the existence of oxygenic photosynthesisers (cyanobacteria) in Late Archaean 2.7 b.y. old sediments from the Hammersley Basin of Australia (Summons et al., 1999; Brocks, et al., 1999). More surprisingly, the molecular fossils in these sediments suggest that eukaryotic organisms had already evolved, and this was well before the period in which there was a significant increase in atmospheric oxygen, which was thought to be fundamental for the evolution of eukaryotes (Brocks et al., 1999; Knoll, 1999). The oldest morphologically identifiable eukaryote fossils occur in much younger rocks in a Mesoproterozoic formation, 2.1 b.y., in Michigan (Han and Runnegar, 1992).

1.3.THE PRESERVATION OF ARCHAEAN BACTERIA

For detailed discussions concerning the preservation and interpretation of fossil bacteria, the reader is referred to Schopf and Walter (1983), Butterfield (1990), Buick (1991), Westall et al. (1995), Westall (1997), Westall (1999a,b), and Westall et al., (1999a). Briefly, bacteria are preserved in rocks as (i) organic compressions in fine grained sediments, (ii) by permineralisation (whereby the organic tissues act as nucleation sites for silica which, on polymerisation, locks the organic template in a mineral matrix) (Figs. 1a-d), (iii) by mineral replacement (a continuation of the permineralisation process in which the mineral replaces the organic template; minerals such as silica, carbonates, phosphate, sulphides, and oxides are involved) (Figs. 1e,f), and (iv) as empty moulds in a mineral matrix.

As noted above, the bacteria may be preserved in varying states of degradation (Knoll et al., 1988): interpretations of some highly degraded individuals may be, therefore, tenuous. The apparent poor degree of preservation of permineralised fossils in Archaean stromatolites compared to much better preserved associations in similar sediments in Proterozoic rocks was attributed to the fact that the Archaean rocks have undergone lowgrade metamorphism (with consequent thermal degradation of the organic matter). However, the excellent preservation of permineralised biofilms and some minerally-replaced bacteria in Archaean rocks that have undergone lowgrade greenschist metamorphism from the Barberton greenstone belt of South Africa (Figs. 1a,b,e,f; Westall and Gerneke, 1998; Westall et al., 1999a) suggests that preservation in these low grade metamorphic rocks may reflect the original state of degradation rather than the effect of metamorphism. Many of the microbial mat communities may have flourished in the vicinity of hot springs or hydrothermal vents (of which there were an abundance during the Archaean) where organic matter is subject to thermal degradation.

The organic matter making up the organisms breaks down upon the death of the organism. Time and thermal abuse continues the breakup of the original macromolecules to result in simpler degradation products which are refractory and known as "kerogen". Some of these macromolecules can be related to specific types of microorganisms. The degradation products may be remarkably persistent and some can be identified in sediments as old as 2.7 b.y. (Summons et al., 1999; Brocks et al., 1999). Further thermal degradation of the kerogen results in the carbon mineral graphite.

Isotopic fractionation of carbon, oxygen, sulphur, nitrogen and hydrogen through biological processes allows us to draw some conclusions about the presence of life on

Figure 1. Permineralised and silica-replaced bacteria and bacterial mats from carbonaceous cherts in the 3.3-3.5 b.y.-old Onverwacht Group, Barberton greenstone belt, South Africa.
(a,b) Permineralised microbial mats consisting of very fine, tabular carbonaceous films, tens of micrimeters in thickness but laterally very extensive (decimeters to meters). (b) Scanning electron microscope (SEM) micrograph showing detail of the smooth to ropy texture of the biofilm.
(c,d) Permineralised filamentous fossil bacteria in light micrograph and SEM view of a transverse section of a filament.
(e,f) Silica-replaced short rod-shaped and coccoidal fossil bacteria. (e) The short rods are embedded in silicified biofilm. (f) Linear association of three coccoidal cells enclosed in a common, wrinkled envelope.

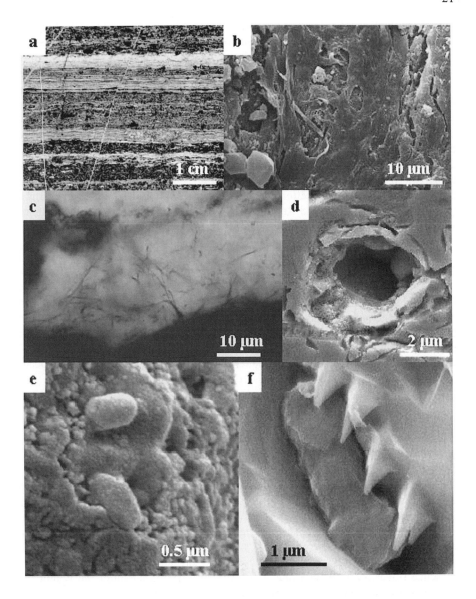

early Earth and, to some extent, the nature of the biological processes (Schidlowski et al., 1983; Schidlowski, 1988). Overall, biological processes, such as metabolism, result in the transfer of ^{12}C into the organism. Since the remains of the organism are incorporated into the sediments as organic matter, the lighter isotopic signature remains, although subsequent alteration and metamorphism of the sediments may result in a shift towrads heavier isotopic values.

Other evidence of microbial activity is preserved in the form of biologically-constructed sediments, such as biolaminites. These include stromatolites which can be columnar and hummocky edifices as well as planar laminated sediments (Fig. 1a). They are constructed from layers of microbial mats which trap and bind detritus and in which authigenic minerals are also precipitated (Walter, 1983; Krumbein, 1983). However, some of the vertical constructions called stromatolites may be purely mineral growths (Lowe, 1994; Grotzinger and Rothman, 1996), and some caution is called for in their interpretation.

2. Fossil diversity in the Archaean

Compared to the successive Proterozoic period, the Archaean is generally believed to be characterised by a poorly diverse biota (Schopf, 1994; Knoll, 1996). This is in part due to environmental controls on evolution, to the state of preservation of the fossils, and probably also to the paucity of Archaean sediments. At present Archaean rocks are being reexamined and reevaluated using a battery of modern microscopic and biogeochemical methods which are revealing more fossil evidence than heretofore suspected.

2.1. THE EARLY ARCHAEAN

The oldest sediments preserved (unfortunately in a highly metamorphosed state) occur at Isukasia in W. Greenland (3.8-3.7 b.y., Nutman et al., 1997; Moorbath et al., 1997; Whitehouse et al., 1998). They already contain convincing carbon isotope evidence for the presence of (photo)autotrophic organisms (Schidlowski et al., 1979; Schidlowski, 1988; Rosing, 1999). More recently, ion microprobe measurements of graphitic inclusions, preserved from the effects of metamorphism in apatite crystals, suggest that there may have been an archaeobacterial component (e.g. methanotrophs) to the Isua microorganisms (Mojzsis et al., 1996). Less convincing are the reports of carbonaceous microfossils (including yeast) from these highly metamorphosed sediments (Pflug and Jaeschke-Boyer, 1979; Bridgewater et al., 1981; Pflug, 1982; Schopf and Walter, 1983).

The oldest morphological evidence of fossil bacteria is to be found in the Early Archaean formations (3.2-3.5 b.y., Kröner et al., 1991; Thorpe et al., 1992; Heubeck and Lowe, 1994a,b; Byerly et al., 1996) of South Africa and Western Australia. In fact, it is believed by some that these two areas were once part of the same depositional basin (Knoll, 1996). Both these areas consist of mainly volcanic and volcaniclastic sequences with interbedded fossiliferous cherts (Viljoen and Viljoen, 1969; de Wit et al., 1982; Hickman, 1983; Lowe and Byerly, 1999) and stromatolitic structures (Walter et al., 1980; Walter, 1983; Byerly et al., 1986, Walsh, 1992, Walsh and Lowe, 1999, and Westall et al., 1999a; but see Lowe (1994) for a discussion of the interpretation of Early Archaean stromatolites).

Interpretations of carbonaceous structures resembling bacteria in the cherts have been the subject of heated debate (Walter and Schopf, 1983; Buick, 1984; Awramik et al., 1983; Buick, 1988, 1991; Schopf, 1993; Westall et al., 1999a). Part of the problem resides in the poor state of preservation of some of the permineralised carbonaceous structures. However, some of the filaments described by Brooks et al. (1973), Walsh and Lowe (1985), Schopf (1992,1993), Walsh (1992) and Walsh and Westall (in prep.) clearly resemble filamentous bacteria (Figs. 1c,d). They are a few micrometers in diameter and tens of micrometers in length, and often appear to be hollow. Such an interpretation is supported by the association of the filaments with well-preserved microbial mat biolaminae (Figs. 1a,b; Walsh, 1992; Walsh and Lowe, 1999; Walsh and Westall, in prep.). Some of the larger, seemingly segmented filaments (> 3.5 μm diameter, Schopf, 1992,1993) were compared to cyanobacteria, although their state of preservation makes such interpretations rather tenuous. An interpretation of other larger (8-21 μm diameter), coccoidal structures, also compared to cyanobacteria (Schopf and Packer, 1987), has also been called into question (Knoll, 1996), although smaller spheres (diameter about 2 μm, Knoll and Barghoorn, 1977) may be biogenic. In addition to the carbonaceous, permineralised filaments, these cherts contain well-preserved, silica-replaced microfossils. Westall and Gerneke (1998), Westall (1999a) and Westall et al. (1999a) document rounded, rod-shaped structures up to 3.8 μm embedded in silicified biofilm or microbial mat (Fig. 1e). In addition, small, silica-replaced spheres of 1 μm diameter and with wrinkled surfaces are interpreted as probable coccoid bacteria (Fig.1f).

It is difficult to interpret bacterium type from morphology alone but some inferences can be made taking into account the sedimentological environment. The association of the filamentous fossils with permineralised microbial mats representing tabular stromatolites suggests that they were mat-building organisms (and probably photosynthesising but not necessarily oxygenic). Other, vertical types of stromatolites are unfossiliferous but bear all the marks of having been constructed by phototactic (and therefore photosynthetic, but again, not necessarily oxygenic) microorganisms (Walter, 1983; Walsh and Lowe, 1999; Westall et al. 1999b). The smaller rod-shaped and coccoid fossils could represent heterotrophic bacteria. The $\delta^{13}C$ values confirm that autotrophic microorganisms were present in the early Archaean environments of South Africa and Australia (Oehler et al., 1972; Moore et al., 1974; Schidlowski et al., 1983; Awramik et al., 1983; Robert, 1988; de Ronde and Ebbesen, 1996; Westall et al., 1999a). Moreover, de Ronde and Ebbesen (1996) interpreted their data from the Fig Tree shales as indicating a photosynthetic, pelagic biota. Considering the lack of ozone protection in the atmosphere at this time one can speculate that perhaps these early planktonic organisms had already evolved some kind of gene repair apparatus.

2.2. THE LATE ARCHAEAN

Late Archaean formations in southern Africa, India and Australia also yield fossiliferous, black cherts and there are many more occurrences of biogenically-formed stromatolites in these provinces as well as in Canada than in the preceding Early Archaean period. Moreover, molecular fossils from the Late Archaean of Western Australia document the first definitive evidence of not only oxygenic photosynthesis but also eukarotic organisms (Summons et al. 1999; Brocks et al., 1999).

Late Archaean microfossils are rare. Two populations of poorly preserved carbonaceous filamentous microfossils (one 1 μm in diameter and the other 10 μm in diameter) from the 2.7 b.y.-old Fortescue Group (Tumbiana Formation, Western Australia) have been compared to cyanobacteria (Schopf and Walter, 1983), although they are also morphologically similar to sulphur oxidising and sulpher reducing bacteria (Knoll,1996). In southern Africa a poorly preserved assemblage of carbonaceous filaments (some 1 μm in diameter, others up to 18 μm in diameter and spheres (up to 30 μm in diameter), reported from the 2.6 b.y.-old Campbell Group of the Transvaal Supergroup (South Africa), has been compared to filamentous and coccoid cyanobacteria, as well as filamentous bacteria (Altermann and Schopf, 1995). Similarly, permineralised septate filaments (6-15 μm in diameter) from the >2.6 b.y.-old Donimalai Formation of the Dharwar Supergroup, India are interpreted as (oxygenic) cyanobacteria (Venkatachala et al., 1990, 1992; Raha, 1990).

There are a number of apparently biogenic stromatolites in the Late Archaean provinces of South Africa, Australia and Canada which were extensively reviewed by Walter (1983). They include the 2.9 b.y. Insuzi Group (Kaapvaal Craton, South Africa), the 2.7 b.y. Fortescue Group (Western Australia), the 2.7 b.y. Steeprock Group from the Superior Province, Canada, and the 2.6 b.y.-old Belingwe and Bulawayo greenstone belts and Ventersdorp Supergroup (southern Africa). In addition, the 2.5 b.y.-old Campbell Rand and Malmani subgroups of the Transvaal Supergroup contains a variety of stromatolites (Beukes, 1980, 1987; Sumner, 1997a, b). On the basis of textural, microfossil, mineralogical and carbon isotope evidence, both photosynthetic autotrophs (probably cyanobacteria) and heterotrophic bacteria (Walter, 1983; Buick, 1992; Sumner, 1997) have been implicated in the construction of the above-mentioned stromatolites.

This review of Late Archaean fossil bacteria concludes with the exciting biochemical evidence from 2.5-2.7 b.y.-old shales from the Hammersley Basin (Summons et al., 1999; Brocks et al., 1999). These researchers have detected not only definitive evidence of oxygenic, photosynthetic cyanobacteria but Brocks et al. (1999) document the existence at 2.7 b.y. of organisms with eukaryotic characteristics.

3. Conclusions

The Archaean was traditionally regarded as an era with poor microfossil diversity compared to the succeeding Poterozoic period, due partly to the vagaries of preservation and but mostly to the fact that it was believed that there was little biological diversity at that time. More sophisticated methods of observation and biogeochemical analysis have brought to light more evidence of life and shown that it was not only more diverse than previously believed but also that it apparently evolved more rapidly.

The Early Archaean rocks document the presence of photosynthetic and heterotrophic bacteria, as well as methanogens. The evidence for oxygenic photosynthesisers in the Early Archaean is equivocal but is clear in the Late Archaean. Early diversification of the eukaryotes during the Late Archaean is indicated by the geochemical data (Brocks et al., 1999) in a period apparently before a significant rise in the level of atmospheric oxygen. The connection between the evolution of the eukaryotes and oxygen levels in the atmospheres may need to be reexamined.

4. Acknowledgements

F.W. acknowledges the NRC, Washington and M.W. the Lousiana State University Council for Research for financial support.

5. References

Altermann, W. and Schopf, J.W. (1995) *Precambrian Research* **75**, 65-90.
Awramik, S.M., Schopf, J.W., and Walter, M.R. (1983) *Precambrian Research* **20**, 357-374.
Beukes, N.J. (1980) *Transactions Geological Society of South Africa* **83**, 343-346.
Beukes, N.J. (1987) *Sedimentary Geology* **54**, 1-46.
Bridgewater, D., Allaart, J.H., Schopf, J.W., Klein, C., Walter, M.R., Barghoorn, E.S., Strother, P., Knoll, A.H., and Gorman, B.E. (1981) *Nature* **289**, 51-53.
Brocks, J., J., Logan, G.A., Buick, R., and Summons, R.E. (1999) *Science* **285**, 1033-1036.
Brooks, J., Muir, M.D., and Shaw, G. (1973) *Nature* **244**, 215-217.
Buick, R. (1984) *Precambrian Research* **24**, 157-172.
Buick, R. (1988) *Precambrian Research* **39**, 311-317.
Buick, R. (1991) *Palaios* **5**, 441-459.
Buick, R. (1992) *Science* **253**, 74-77.
Butterfield, N.J. (1990) *Palaeobiology* **16**, 271-286.
Byerly, G.R., Lowe, D.R., and Walsh, M.M. (1986) *Nature* **319**, 424-427.
Byerly, G.R., Kröner, A., Lowe, D.R. Todt, W., and Walsh, M.M. (1996) *Precambrian Research* **78**, 125-138.
Canuto, V.M., Levine, J.S., Augustsson, T.R., Imhoff, C.L., and Giampapa, M.S. (1983) *Nature* **296**, 145-148.
Chyba, C.F. (1990) *Nature* **343**, 129-131.
Cloud, P.E. (1976) *Paleobiology* **2**, 351-387.
Cockell, C.S. (1998) *J. Theoretical Biology* **193**, 717-729.
de Ronde, C.E.J. and Ebbesen, T.W. (1996) *Geology*, **24**, 791-794.
de Wit, M.J. (1998) *Precambrian Research* **91**, 181-226.
de Wit, M.J., Hart, R., Martin, A., and Abbott, P. (1982) *Economic Geology* **77**, 1783-1801.
Dimroth, E. and Kimberley, M.M. (1976) *Canadian J. of Earth Sciences* **13**, 1161-1185.
Dungworth, G. and Schwartz, A.W. (1974) *Chemical Geology* **14**, 167-172
Ernst, W.G. (1983) in J.W. Schopf (Ed.) *Earth's Earliest Biosphere*, Princeton University Press, New Jersey, pp. 41-52.
Garcia-Pichel, F. (1998) *Origins of Life* **28**, 321-347.
Grotzinger, J. and Rothman, D.H. (1996) *Nature* **383**, 423-425.
Han, T.-M. and Runnegar, B. (1992) *Science* **257**, 232-235.
Heubeck, C. and Lowe, D.R. (1994a) *Precambrian Research* **68**, 257-290.
Heubeck, C. and Lowe, D.R. (1994b) *Tectonics* **13**, 1514-1536.
Hickman, A.H. (1983) *Western Australian Geological Survey Bulletin* **127**, 1-267.
Hunt, B,G. (1979) *Nature* **281**, 188-191.
Jannasch, H.W. (1984) in Y. Cohen. R.W. Castenholz, H.O. Halvorson (Eds.) *Microbial Mats: Stromatolites*, A.R. Liss, New york, pp. 121-131.
Kasting, J.F. (1987) *Precambrian Research* **34**, 205-229.
Knoll, A.H. (1979) *Origins of Life* **9**, 313-327.
Knoll, A.H. (1996) in J. Jansonius and D.C. McGregor (Eds.) *Palynology: principles and applications*, American Association of Stratigraphic Palynologists Foundation, **1**, 51-80.
Knoll, A.H. (1999) *Science* **285**, 1025-1026.
Knoll, A.H. and Barghoorn, E S. (1977) *Science* **198**, 396-398.
Knoll, A.H. and Holland, H.D. (1995) *Studies in Geophysics: Effects of Past Global Change on Life*, National Research Council, National Academy Press, Washington, pp. 21-33.
Knoll, A.H., Strother, P.K., and Rossi, S. (1988) *J. Paleontology* **61**, 898-926.
Kröner, A., Byerly, G.R. and Lowe, D.R. (1991) *Earth and Planetary Science Letters* **103**, 41-54.
Krumbein, W.E. (1983) *Precambrian Research* **20**, 493-531.
Lambert, R. St. J. (1976) in B.F. Windley (Ed.) *The Early History of the Earth*, Wiley, New york, pp 377-403.
Lowe, D.R. (1994) *Geology* **22**, 387-390.

26

Lowe, D.R. (1999a) In D.L. Lowe and G.R. Byerly (Eds.) *Geologic Evolution of the Barberton Greenstone Belt, South Africa*, Geological Society of America Special Paper **329**, 83-114.

Lowe, D.R. (1999b) In D.L. Lowe and G.R. Byerly (Eds.) *Geologic Evolution of the Barberton Greenstone Belt, South Africa*, Geological Society of America Special Paper **329**, 287-312.

D.L. Lowe and G.R. Byerly (1999) *Geologic Evolution of the Barberton Greenstone Belt, South Africa*, Geological Society of America Special Paper, **329**, 319 pp.

Maher, K.A. and Stevenson, D.J. (1988) *Nature* **331**, 612-614.

Margulis, L., Walker, J.C.G., and Rambler, M. (1976) *Nature* **264**, 620-624.

McKay, C.P. and Hartman, H. (1991) *Origins of Life* **21**, 157-163.

Mojzsis, S.J., Arrhenius, G., McKeegan, K.D., Harrison, T.M., Nutman, A.P., and Friend, R.L. (1996) *Nature* **384**: 55-59.

Moorbath, S., Kamber, B.S., and Whitehouse, M.J., (1997) *Chemical Geology* **135**, 213-231.

Moore, C.B., Lewis, C.F., and Kvenvolden, K.A. (1974) *Precambrian Research* **1**, 49-54.

Nisbet, E.G. and Fowler, C.M.R. (1996) *Nature*, **382**, 404-405.

Nutman, A.P. (1986) *The early Archaean to Proterozoic history of the Isukasia area, southern West Greenland*, Greenland Geological Survey, Copenhagen, Bull. **154**, pp. 80.

Nutman, A.P., Mojzsis, S.J., and Friend, C.R.L. (1997) *Geochimica Cosmochimica Acta* **61**, 2475-2484

Oehler, D.Z., Schopf, J.W., and Kvenvolden, K.A. (1972) *Science* **175**, 1246-1248.

Olson, J.M. and Pierson, B.K. (1986) *Photosynthesis Research* **9**, 251-259.

Owen, T., Cess, R.D., and Ramanathan, V. (1979) *Nature* **277**, 640-642.

Pflug. H.D. (1982) *Zublatt für Bakteriologie und Hygiene, I Abteilung Origene* **C3**, 53-64.

Pflug, H.D., and Jaeschke-Boyer, H. (1979) *Nature* **280**, 483-486.

Pierson, B.K., Mitchell, H.K., and Ruff-Roberts, A.L. (1993) *Origins of Life* **23**, 243-260.

Raha, P.K. (1990) *Geological Survey of India News* **24**, 19-28.

Robert, F. (1988) *Geochimica Cosmochimica Acta* **52**, 1473-1488.

Rosing, M.T. (1999) *Science* **283**, 674-676.

Sagan, C. and Mullen, G. (1972) *Science* **177**, 52-56.

Schidlowski, M. (1988) *Nature* **333**, 313-318.

Schidlowski, M., Appel, P.W.U., Eichmann, R., and Junge, C.E. (1979) *Geochimica Cosmochimica Acta* **43**, 189-199.

Schidlowski, M., Hayes, J.M., and Kaplan, I.R. (1983) in J.W. Schopf (Ed.) *Earth's Earliest Biosphere*, Princeton University Press, New Jersey, pp. 149-186.

Schopf, J.W. (1992) in J.W. Schopf and C. Klein, (Eds.) *The Proterozoic Biosphere: a multidisciplinary study*, Cambridge University Press, Cambridge, pp. 24-39.

Schopf, J.W. (1993) *Science* **260**, 640-646.

Schopf, J.W., Oehler, D.Z., Horodyski, R.J., and Kvenvolden, K.A. (1971) *J. Paleontology* **45**, 477-485.

Schopf, J.W. (1994) *Proceedings National Academy Sciences U.S.A.* **91**, 6735-6742.

Schopf, J.W. and Walter, M.R. (1983) in J.W. Schopf (Ed.) *Earth's Earliest Biosphere*, Princeton University Press, New Jersey, pp. 214-239.

Schopf, J.W. and Packer, B.M. (1987) *Science* **237**, 70-73.

Summons, R.E., Jahnke, L.L., Hope, J.M., and Logan, G.A. (1999) *Nature* **400**, 554-557.

Sumner, D.Y. (1997a) *Palaios* **12**, 302-318.

Sumner, D.Y. (1997b) *American J. Science* **297**, 455-487.

Thorpe, R.I., Hickman, A.H., Davis, D.W., Mortensen, J.K., and Trendall, A.F. (1992) *Precambrian Research*, **56**, 169….

Towe, K.M. (1990) *Nature* **348**, 54-56.

Veizer, J. (1983) in J.W. Schopf (Ed.) *Earth's Earliest Biosphere*, Princeton University Press, New Jersey, pp. 240-259.

Venkatachala, B.S., Shukla, M., Sharma, M., Naqvi, S.M., Srinivasan, R., and Udairaj, B. (1990) *Precambrian Research* **47**, 27-34.

Venkatachala, B.S., Shukla, M., Sharma, M., Naqvi, S.M., Srinivasan, R., and Udairaj, B. (1992) *Precambrian Research* **57**, 167-168.

Viljoen, M.J. and Viljoen, R.P. (1969) *Geological Society of South Africa Special Publication* **9**, 1-20.

Walker, J.C.G., Klein, C., Schidlowski, M., Schopf, J.W., Stevenson, D.J., and Walter, M.R. (1983) in J.W. Schopf (Ed.) *Earth's Earliest Biosphere*, Princeton University Press, New Jersey, pp. 260-290.

Walsh, M.M. (1992) *Precambrian Research* **54**, 271-293.

Walsh, M.M. and Lowe, D.R. (1985) *Nature* **314**, 530-532.

Walsh, M.M. and Lowe, D.R. (1999) In D.L. Lowe and G.R. Byerly (Eds.) *Geologic Evolution of the Barberton Greenstone Belt, South Africa*, Geological Society of America Special Paper **329**, 115-132.

Walsh, M.M. and Westall, F (in prep.) High resolution electron microscope studies of ancient microfossils.

Walter, M.R. (1976) In M.R. Walter (Ed.) *Stromatolites*, Elsevier, Amsterdam, pp. 273-310.

Walter, M.R. (1983) in J.W. Schopf (Ed.) *Earth's Earliest Biosphere*, Princeton University Press, New Jersey, pp. 187-213.

Walter, M.R., Buick, R., and Dunlop, J.S.R. (1980) *Nature* **284**, 443-445.

Westall, F. (1997) in C. Cosmovici, S. Bowyer and D. Werthimer (Eds) *Astronomical and Biochemical origins and the Search for Life in the Universe*, Editori compositrici, Bologna, pp. 491-504.

Westall, F. (1999a) *J. Geophysical Research* **104**, 16,437-16,451.

Westall, F. (1999b) in J. Seckbach (Ed.) *Enigmatic Microorganisms and Life in Extreme Environments*, Kluwer, Amsterdam, pp. 73-88.

Westall, F. and Gerneke, D. (1998) *Proceedings SPIE, Internatinal Society for Optical Engineering* **3441**, 158-169.

Westall, F., Boni, L., and Guerzoni, M.E. (1995) *Palaeontology* **38**, 495-528.

Westall, F., de Wit, M.J., Dann, J., van der Gaast, S., de Ronde, C.E.J., and Gerneke, D. (1999a) *Precambrian Research*, in press.

Westall, F., de Wit, M.J., Walsh, M.M., Folk, R.L., Chafetz, H., and Gibson, E.K. (1999b) *International Society for the Study of the Origin of Life*, San Diego, July 1999, p. 47 (Abst.)

Williams, L.A. and Reimers, C. (1983) *Geology* **11**, 267-269.

Whitehouse, M.J., Moorbath, S., and Kamber, B.S. (1998) *Mineralogical Magazine* **62A**, 1649-1650.

Windley, B.F. (1977) *The Evolving Continents*, Wiley, New York, 385 pp.

Woese, C.R. (1987) *Microbiological Reviews* **51**, 149-174.

Biodata of Dr. **Frederick Andrew Rainey** author (with co-author Dr. Naomi Ward-Rainey) of the chapter "*Prokaryotic Diversity.*"

Dr. Fredrick A. Rainey is an Assistant Professor in the Department of Biological Sciences at the Louisiana State University (Baton Rouge, LA). He obtained his Doctor of Philosophy at the University of Waikato, Hamilton, New Zealand in 1991. Dr. Rainey served as a Scientific Officer (1980-1985) at the Bacterial Research Division in Department of Agriculture in Northern Ireland. His postdoctoral period (1991-1993) was performed in the Center for Microbial Diversity and Identification at the University of Queensland, Brisbane, Australia. Then Dr. Rainey served (1993-1996) as a Scientist and Group Leader at the German Collection of Microorganisms and Cell Cultures at Braunschweig, Germany. He is a member of International committee on Systematic Bacteriology and on editorial boards and acts as a reviewer of a few Journals. His main interest is in the Prokaryotic Diversity.

E-Mail: **frainey@unix1.sncc.lsu.edu**

Biodata of **Dr. Naomi Ward-Rainey** (co-author with F.A. Rainey) of the chapter on *"Prokaryotic Diversity."*

Dr. Naomi Ward-Rainey is an Instructor at the Department of Biological Sciences, Louisiana State University, Baton Rouge, LA. USA. She obtained her Ph.D. from the University of Warwick and serves as a member of committees and a regular reviewer for two Journals.

E-Mail: **nrainey@unix1.sncc.lsu.edu**

J. Seckbach (ed.), Journey to Diverse Microbial Worlds, 29-42.
© 2000 *Kluwer Academic Publishers. Printed in the Netherlands.*

PROKARYOTIC DIVERSITY

FRED A. RAINEY and NAOMI WARD-RAINEY
Department of Biological Sciences
Louisiana State University, Baton Rouge, La. USA

1. What is a Prokaryote?

The classical definition of a prokaryote is encapsulated in the word itself (prokaryote, from the Greek for "before nucleus"), and refers to the simple cell structure that lacks a nucleus, along with compartmentalization and specialization of the cytoplasm. This fundamental feature of prokaryotic cells has been challenged by the finding that some prokaryotes exhibit cellular organization resembling that of eukaryotic cells. The most prominent example is that of *Gemmata obscuriglobus*, an aquatic microorganism that has been shown to possess membrane-bounded nuclear material (Fuerst and Webb, 1991). In relatives of *Gemmata obscuriglobus* (members of the order *Planctomycetales*), cellular compartmentalization, in which the ribosomes are separated from the rest of the cell, has been reported (Lindsay et al., 1997).

Additional characteristics that were thought to distinguish prokaryotic from eukaryotic cells have also been challenged. Several organisms have been found to possess more than the single chromosome thought to be characteristic of prokaryotes. The cells of *Rhodobacter sphaeroides*, a photosynthetic bacterium, contain two chromosomes (Suwanto and Kaplan, 1989a,b), as do those of *Brucella melitensis* (Michaux et al., 1993), and *Leptospira interrogans* (Zuerner et al., 1993). The status of both elements as chromosomes, rather than episomes, has been supported by the finding of essential genes e.g. *rrn* on both (Krawiec and Riley, 1990). The ionizing radiation-resistant prokaryote *Deinococcus radiodurans* is multigenomic, with at least four identical copies of the chromosome per stationary phase cell (Hansen, 1978).

Disparity in cell size has been one of the most obvious differences in prokaryotic vs. eukaryotic cells, but this previously clear-cut demarcation has been blurred by the discovery of bacterial giants up to 700μm in length, such as *Epulopiscium fischelsoni*, isolated from the intestinal tract of surgeonfish (Angert et al., 1993), and *Thiomargarita namibiensis*, a sulfur-oxidizing organism from marine sediment samples (Schulz et al., 1999).

1.1. THE PHYLOGENETIC DEFINITION OF A PROKARYOTE

Before the advent of molecular phylogenetic analyses, the word "prokaryote" was equivalent to "bacterium", and many scientists used the words interchangeably. However, the phylogenetic studies of bacteria and eukaryotes using the small subunit ribosomal RNA (16S rRNA) as a marker molecule indicated the existence of three domains of life, which were formally described in 1990 (Woese et al., 1990). According to this scheme, prokaryotes were classified into the domains Bacteria (formerly eubacteria) and Archaea (formerly archaebacteria), while the third domain, Eukarya, contained all the eukaryotes. The most striking aspect of this classification was that the two classically prokaryotic domains (Bacteria and Archaea) are no more related to each other than each was to the members of the domain Eukarya. In fact, some analyses suggest a closer relationship between the Archaea and the Eukarya than between the Archaea and the Bacteria (Pühler et al., 1989; Creti et al., 1991).

The implications of these phylogenetic relationships, together with the challenges to the classical morphological and cellular organization of prokaryotes described above, has altered our interpretation of the word "prokaryote". The end result is that a phylogenetic definition of the prokaryotes, encompassing members of the domains Bacteria and Archaea irrespective of their phenotypic properties, is the most practical and tenable option.

2. Diversity

An exploration of prokaryotic diversity must define the diversity that is to be examined. Biological diversity, or biodiversity, is usually defined as the total variability of life on Earth, although attempts to establish the existence of microbial life on other planets e.g. Mars may in the future lead us to revise this definition to include extraterrestrial biodiversity. A discussion of prokaryotic diversity should include our total current knowledge of prokaryotic life in all terrestrial and aquatic environments. In most fields of biology, determination of biodiversity is aided by the existence of a comprehensive taxonomy for the group of organisms under investigation. E. O. Wilson, in "The Diversity of Life" (Wilson, 1992) has broken down the 1.4 million living species currently known into major groups, showing that the insects and higher plants constitute the majority of described species (around 1 million), and that the Monera, into which the prokaryotes fall, constitute only 4,800 known species. Most of these species, represented by type strains and additional strains, are deposited in large culture collections, which constitute the main resource for cultivated prokaryotic diversity. The remainder of the cultured prokaryotic species are maintained in private laboratory culture collections, the diversity of which we are just beginning to appreciate through the application of phylogenetic analyses.

In contrast to eukaryotic macroorganisms, whose relatively larger size and more complex organization provides easily recognizable characteristics for species delineation, most prokaryotes are small in size and possess relatively simple morphology that allows few characters to be used for species description. In addition, many prokaryotes that share highly similar morphology are phylogenetically very distantly related. The small size of prokaryotes also leads to a unique spatial relationship in prokaryotic diversity studies. Consider a single hectare of forest that may contain 25 plant species, five mammals, and hundreds of insect species; each of these macroorganisms comprises numerous microbial habitats, in which hundreds, if not thousands, of prokaryotic species exist. The numerous microhabitats of the soil within this one hectare present even more niches for prokaryotes to speciate.

Unlike most macroorganisms, the majority of prokaryotes must be maintained in pure culture on artificial growth media in the laboratory in order to perform the experiments needed for species description. The difficulty of formulating culture media that reproduce exactly the organism's natural habitat means that not all Bacteria or Archaea in an environmental sample can currently be cultivated. The relatively recent development of culture-independent methods (see below) has allowed us to detect many yet to be cultivated or described prokaryotic species.

2.1. REASONS TO STUDY PROKARYOTIC DIVERSITY

Students in an introductory microbiology course will encounter during their studies only a fraction of all the prokaryotic diversity in existence; how can we persuade them that the rest is worthy of study? The foremost reason is that prokaryotes and their processes are the foundation of the biosphere and are integral to the viability of every macroorganism. Microbes mediate carbon and nutrient cycling, influence the composition of the atmosphere, and are associated with other life forms in a variety of symbiotic relationships. We are a long way from realizing the complexity of the interactions between microbes and their biotic and abiotic environment.

Understanding the diversity of prokaryotes that perform a specific function in a biotope has even greater importance if that biotope is threatened by natural or manmade change. For every species of macroorganism that is threatened with extinction, a host of prokaryotes that have adapted to commensal or parasitic relationships with that macroorganism are also at risk. Our extremely limited knowledge of existing prokaryotic diversity implies that we are not able to measure changes in that diversity that may occur as a result of fluctuations in the populations of larger organisms.

A further justification for measuring and characterizing prokaryotic diversity is one that has been recognized by the pharmaceutical industry for many years, namely the enormous potential of prokaryotic diversity to encode the synthesis of natural products useful in industrial processes and disease control.

2.2. TYPES OF PROKARYOTIC DIVERSITY

Prokaryotic diversity is exhibited in a variety of forms, the most obvious of which is probably morphological diversity. The well-known rod, coccus, and spirochete of classical medical microbiology are accompanied by their more intriguingly-shaped relatives, including prosthecate bacteria such as *Caulobacter* and *Hyphomicrobium* species, the truly star-shaped genus *Stella*, the filamentous hyphal masses of *Streptomyces* and other actinobacteria, the bleb forming thermophiles of the *Thermotogales* and the warty, fimbriated cell of the genus *Verrucomicrobium* (literally "warty microbe"). Unusual cellular morphology, perhaps the most easily detected manifestation of prokaryotic diversity, is still a powerful prompt for the description of new prokaryotic species.

Physiological diversity provides another facet of total prokaryotic diversity, and has led to the recognition of prokaryotes as the supreme inhabitants of environments that we consider to be extreme. The ability of prokaryotes to survive in these environments suggests a physiological diversity that surpasses that of the members of the domain Eukarya.

The introduction and development of techniques for physical mapping of prokaryotic chromosomes has revealed a third category of diversity, that of genome size and topology. The dogma of a single circular chromosome in all prokaryotes has been eroded by the discovery of multiple chromosomes in *Rhodobacter sphaeroides* and *Deinococcus radiodurans* (Suwanto and Kaplan, 1989a,b), and linear chromosomes in *Borrelia burgdorferi*, the Lyme disease spirochete (Baril et al., 1989; Ferdows and Barbour, 1989), *Streptomyces "lividans"* (Leblond et al., 1993), and *Rhodococcus fascians* (Crespi et al., 1992). Prokaryotic genomes have been found to range in size from 0.6 Megabases (Mb) in the intracellular parasites of the genus *Mycoplasma* (Colman et al., 1990) to 9.5 Mb in *Myxococcus xanthus*, the fruiting-body forming myxobacterial species (Kündig et al., 1993). Large genome size is thought to encode diverse gene products that confer physiological flexibility on the prokaryotic cell, allowing it to respond to fluctuations in nutrient concentrations and other environmental parameters (Roussel et al., 1994). In contrast, the small genomes of the mycoplasmas represent the "minimal genome" that suffices for intracellular parasites dependent on host cell machinery (Fraser et al., 1995). Prokaryotes also exhibit diversity of genomic organization, as exemplified by structural variation in the ribosomal RNA operon. The classical model of the rRNA operon, as found in *Escherichia coli* and the majority of prokaryotes investigated so far, comprises the rRNA genes linked in a transcriptional unit in the order 5' 16S-23S-5S 3'. Exceptions to this rule have been found in *Thermus thermophilus* (Menke et al., 1991), *Verrucomicrobium spinosum* (Menke et al., 1991), *Planctomyces limnophilus* (Menke et al., 1991; Ward-Rainey et al., 1996), certain *Deinococcus* species (unpublished data), and others. Ribosomal RNA operon structure has diverged to produce 16S rRNA genes that are physically separated from the 23S and 5S rRNA genes, and in some cases unequal numbers of 16S and 23S rRNA genes. The implications of this diversity of gene arrangement for the transcriptional control of these genes and the assembly of the ribosome are yet to be determined.

The burgeoning field of complete genome sequencing promises to further reveal the extent of diversity in prokaryotic genome organization. The publication of 20 complete and 14 partial genome sequences, representing both Bacteria and Archaea (www.tigr.org; www.ncbi.nlm.nih.gov), allows identification and comparative analysis of homologous genes between prokaryotic taxa. Many of the early microbial genome sequencing projects focused on medically significant bacteria such as *Mycobacterium tuberculosis* and *Haemophilus influenzae*. However, the more recent inclusion of genome sequencing data from extremophiles such as *Aquifex aeolicus*, *Thermotoga maritima*, *Methanococcus jannaschii* and *Deinococcus radiodurans* has made it possible to examine conserved gene sequences in organisms representative of a wider range of prokaryotic diversity.

3. Methods of Studying Prokaryotic Diversity

3.1. THE CULTURING APPROACH

The prokaryotic diversity present in an environmental sample can be measured and characterized using a variety of methods that can be divided into three broad groups (Figure 1). The first group of methods can be categorized as the culturing approach, in which different prokaryotic cells are separated from each other by isolation of colonies on solid culture media, and these colonies propagated to result in a pure culture. The organisms in the sample can be recovered by direct plating of the sample, or plating after an initial enrichment phase that selects for an organism, or group of organisms, of interest. Alternative culturing techniques such as serial dilution to extinction or roll tube methods can be applied in the isolation of strict anaerobes. The isolates obtained from the culturing approach can be identified by a number of methods, the most rapid and unambiguous of which is currently partial 16S rRNA gene sequence analysis (Rainey and Stackebrandt, 1996), due to the ubiquitous distribution of 16S rRNA genes in all prokaryotes, and the extensive 16S rRNA sequence database available for comparison.

The main advantage of the culturing approach is that strains are maintained and are available for further characterization, which may provide insight into their ecological role. Over many years, the culturing approach to assessment of prokaryotic diversity has yielded an enormous number of bacterial isolates, which, as well as providing the raw material for the description of approximately 4800 prokaryotic species, have also served as the experimental animals in investigations of the morphological, physiological, and genomic diversity described above. Our knowledge of the metabolic processes of prokaryotes and their functional roles in various ecosystems is almost entirely derived from studies of organisms in pure culture. However, the culturing approach cannot be considered to furnish a complete picture of prokaryotic diversity because of the biases inherent in the methods and materials used. Every facet of the enrichment and isolation process imposes a selective pressure on the natural prokaryotic populations present in the sample. Some of these pressures include the nutrient composition of the culture medium, the temperature and gaseous conditions of incubation, exposure to ambient light, competition between organisms, and the human bias associated with the selection of colonies for further study. The cumulative effect of these selective pressures is to reduce the spectrum of prokaryotic diversity that can be recovered from a sample by the culturing approach. It is difficult to ascertain the extent of this reduction, although the results of alternative approaches to prokaryotic diversity assessment (see below) provide some hints.

The influence of various biases on the enrichment and isolation process can be seen in Figure 2, which depicts the relative proportions of taxa within the main lineages of the 16S rRNA gene sequence-based tree of the two prokaryotic domains. The number of species within the Bacteria (4,422), as compared to the Archaea (179), reflects the fact that Archaea are generally more difficult to isolate than Bacteria, with the provision of strictly anaerobic conditions necessary for groups such as the methanogens. Within the Archaea, the distribution of taxa is influenced by the search for extremeophiles. Nearly half of all archaeal genera are extreme thermophiles, reflecting the industry-driven quest for thermostable enzymes. Within the Bacteria, the vast majority of species belong to the Proteobacteria and low G+C Gram positive bacteria. Isolation of so many members of these groups is due in part to anthropocentric causes: many disease-causing and food-related bacteria fall within the confines of these lineages, and these organisms grow quickly on nutrient-rich media. Another large slice of the species pie belongs to the Actinobacteria, furnishing evidence for the efforts of the pharmaceutical industry in isolating secondary metabolite producers such as *Streptomyces* species, which constitute nearly half of the species in this class. The remaining lineages, some containing as few as one genus and species, could reflect not a low occurrence of these organisms in the environment, but rather the investment of relatively less research time and effort in a small number of specialized laboratories.

3.2 THE CULTURE-INDEPENDENT APPROACH

Our knowledge of prokaryotic diversity has expanded enormously as a result of applying 16S rRNA-based culture-independent approaches. These strategies circumvent the cultivation step, and its associated biases, by extracting prokaryotic DNA or RNA sequences directly from an environmental sample. There are several variations on this theme, but the currently most commonly applied method (Figure 1) involves the bulk extraction of total DNA from a sample, and amplification of the 16S rRNA genes by PCR, resulting in a mixture of 16S rRNA gene sequences from various organisms present in the sample. At this stage, the PCR product can be analyzed by Amplified Ribosomal DNA Restriction Analysis (ARDRA), involving digestion of the PCR product with a frequently cutting restriction enzyme followed by electrophoresis of the resulting fragments. The banding pattern obtained provides an indication of the diversity of sequences present in the PCR product, and can be used for comparative analyses of environmental samples, e.g. variation in diversity on a spatial or temporal scale (Figure 1). The mixture of PCR products can also be separated into individual sequences by cloning into a plasmid vector, and the resulting 16S rDNA library analyzed by either sequence analysis of the individual genes and subsequent identification of the organisms from which they originated, or a higher-throughput screening technique such as ARDRA (Figure 1). ARDRA can also be applied to a set of environmental isolates obtained by the culturing approach.

The data obtained from the culture-independent approach has had an enormous impact on our view of prokaryotic diversity. This approach has provided insight into the extent of microbial diversity and exposed the inability of culturing approaches to recover the full range of diversity (Amann et al., 1995; Ward-Rainey et al., 1995). The sequences obtained using these methods has indicated the presence of organisms that have never been recovered by culturing; the ability to detect novel taxa in this way is arguably the greatest strength of the molecular approach. The phylogenetic tree of the domain Bacteria has expanded to include 36 lineages, of which approximately one-third are composed entirely of environmental sequences, with no cultured representatives (Hugenholtz et al., 1998). Other lineages containing only a few cultured taxa, such as the Acidobacteria and Verrucomicrobia, have been demonstrated to be much more widely distributed in different habitats than was previously thought (Rheims et al., 1996; Hugenholtx et al., 1998). The ubiquity of members of these cosmopolitan

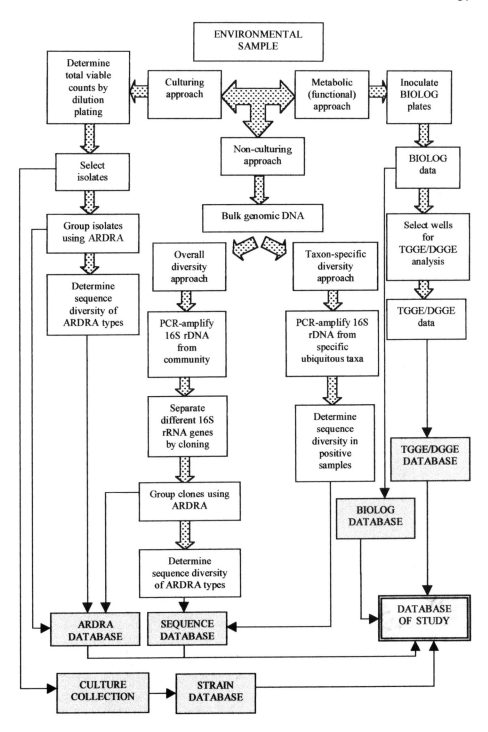

Figure 1. Approaches to the determination of prokaryotic diversity in environmental samples

divisions in different environments suggest that they possess a wide range of metabolic capabilities.

The culture-independent approach also introduces biases into the diversity assessment process. Differences in lysis efficiency due to the method used to extract DNA from the sample (e.g. enzymatic vs. physical disruption), the effect of the genome size and rRNA gene copy number of the target sequences (Farrelly et al., 1995), the composition of the primers used (Reysenbach et al., 1992; Rainey et al., 1994), and the concentration of the DNA template (Suzuki and Giovannoni, 1996; Wilson and Blitchington, 1996) are some of the selective factors at work in the culture-independent approach. These influences provide a partial explanation of why the same components of prokaryotic diversity are not recovered by the culturing and culture-independent approaches. Because of the biases intrinsic to both the culturing and culture-independent approaches, use of both strategies in parallel for determination of prokaryotic diversity in a given sample would appear to be the best solution.

3.3 THE FUNCTIONAL/METABOLIC APPROACH

The diversity of a prokaryotic population can also be measured in terms of the metabolic capabilities of the constituent organisms. One method for detecting these activities the BIOLOG system (Figure 1), which tests the ability of the population to oxidize a range of carbon sources. The system was originally designed for identification of bacterial isolates, mostly from clinical sources, but is increasingly being used to characterize prokaryotic communities. The wells of BIOLOG microtiter plates contain various carbon sources, other nutrients, and a tetrazolium dye that indicates oxidation of the substrate. Using BIOLOG, temporal and spatial differences in microbial communities from various environments have been detected (Garland and Mills, 1991; Zak et al., 1994; Bossio and Scow, 1995; Wünsche et al., 1995). The contribution of various members of the community to the BIOLOG patterns obtained can be determined by direct analysis of the BIOLOG wells, usually via Denaturing/Temperature Gradient Gel Electrophoresis (DGGE/TGGE). 16S rRNA gene PCR products amplified from the well material can be separated according to their differing base compositions, based on the differential melting behavior of the double-stranded DNA during migration through an increasing gradient of either denaturant or temperature (Muyzer et al., 1993). The number, intensity and migrated distance of the individual bands reflect both the number and relative abundance of major 16S rRNA types within the samples. This approach provides data on the phylogenetic diversity of the organisms metabolizing the substrate. Although it has been demonstrated that fast-growing microorganisms have a competitive advantage in the BIOLOG plates and that this may influence the apparent metabolic diversity of the community (Smalla et al., 1998), the method remains a valuable tool for community analysis because of its rapidity and applicability to studying community change in a large number of samples.

Functional prokaryotic diversity can also be examined using components of the culture-independent approach, with the aid of DNA probes and PCR primers that target specific functional genes such as those encoding proteins involved in nitrogen fixation (nif), and methane metabolism. These methods can be used in assays to detect the presence of the particular genes in a given environmental sample. This approach can be

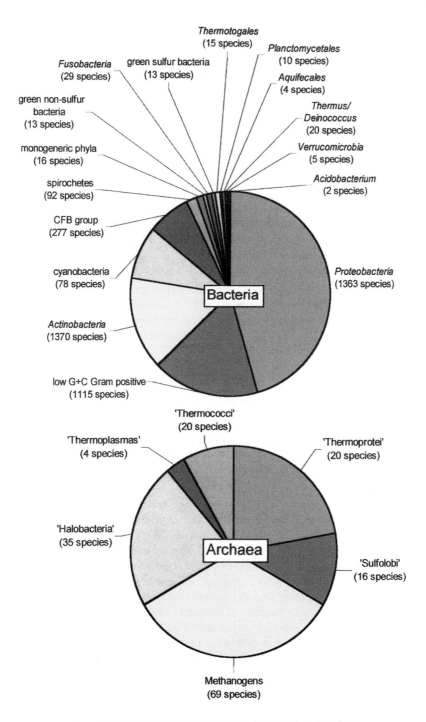

Figure 2. Distribution of genera in the domains Bacteria and Archaea

expanded to provide an indication of the diversity of the organisms responsible for this activity, either by cloning and sequencing functional gene PCR products, or subjecting them to DGGE/TGGE analyses (Ben-Porath and Zehr, 1994; McDonald et al., 1995; Okhuma et al., 1996; Rosado et al., 1998; Henckel et al., 1999).

4. New Approaches to Detecting and Characterizing Prokaryotic Diversity

Our recently expanded knowledge of prokaryotic diversity, in most part due to the results of the culture-independent approach (Hugenholtz et al., 1998), has stimulated efforts to develop culturing methods that can recover this diversity as novel living organisms. Some of these endeavours have focused on the design of new culture media that more closely simulate the natural environment of the organisms, and/or employ a novel selective agent. As an example, by devising new media with N-acetyl glucosamine as a sole carbon source and containing ampicillin as a selective agent, Schlesner (1994) succeeded in isolating 257 new strains of budding bacteria morphologically resembling members of the order *Planctomycetales*. The effectiveness of ampicillin in these isolation studies was due to the nature of the planctomycete cell wall, which contains no peptidoglycan and is almost entirely proteinaceous in composition. Subsequent phylogenetic analyses of representatives of this strain set showed it to contain as much phylogenetic diversity as all previously described members of the order (Ward-Rainey et al., 1995). Data obtained from the culture-independent methods have indicated that planctomycetes were more widespread than their classical aquatic habitats, and the novel strains obtained from the study of Schlesner (1994), isolated from a range of environments that included cattle manure and compost heaps, supported this idea.

Other attempts to obtain novel isolates have used standard culture media, but introduced a selective pressure prior to plating of the samples. For example, in a study of the prokaryotic diversity of desert soils (Rainey et al., unpublished), different soil types were exposed to various doses of gamma ionizing radiation, ranging between 0.1 and 3.0 MRad. These soils were then plated on routine culture media and the numbers and the phylogenetic diversity of the survivors determined. Survivors were obtained after exposure to all doses, and resulted in the addition of novel species to already existing genera and the discovery of new genera. To the classical ionizing radiation resistant genus *Deinococcus*, this selective enrichment technique added at least eight additional species, six of which came from a single soil sample. The single species genus *Geodermatophilus* was also expanded, at both the species and strain level. In addition, strains representing novel genera within the cytophaga and alpha-Proteobacteria lineages were isolated. The success of this isolation strategy can be attributed to the fact that ionizing radiation eliminates fast-growing and potentially inhibitory competitors within the microbial community, allowing the novel taxa to proliferate. As well as providing novel strains and species of existing genera, the study demonstrated that ionizing radiation resistance may be more prevalent across the domain Bacteria than was previously thought.

The future study of prokaryotic diversity, and our success in understanding its extent, clearly lie in the combination of culture and culture-independent approaches.

Without the cultured organisms, it is still difficult to determine their role in the ecosystem, but as molecular technologies advance we may the reach the stage of not only knowing the phylogenetic position, but also the genome composition and thus the physiological make-up, of uncultured prokaryotes. From a taxonomic stance, these data may help us design new ways to culture the as yet uncultured microorganisms and thus greatly increase the number of validly described prokaryotic taxa.

5. References Cited

Amann, R., Ludwig, W. and Schleifer, K. H. (1995) Microbiol. Rev. 59: 143-169.

Angert, E. R., Clements, K. D. and Pace, N. R. (1993) Nature 362: 239-241.

Baril, C., Richaud, C., Baranton, G. and Saint Girons, I. (1989) Res. Microbiol. 140: 507-516.

Ben-Porath, J. and Zehr, J. P. (1994) Appl. Environ. Microbiol. 60: 880-887.

Bossio, D. H. and Scow, K. M. (1995) Appl. Environ. Microbiol. 61: 4043-4050.

Colman, S. D., Hu, P. C., Litaker, W. and Bott, K. F. (1990) Mol. Microbiol. 4: 683-687.

Crespi, M., Messens, E., Caplan, A. B., Van Montagu, M. and Desomer, J. (1992) EMBO J. 11: 795-804.

Creti, R., Citarella, F., Tiboni, O., Sanagelantoni, A. M., Palm, P. and Camarano, P. (1991) J. Mol. Evol. 33: 332-342.

Farrelly, V., Rainey, F. A. and Stackebrandt, E. (1995) Appl. Environ. Microbiol. 61: 2798-2801.

Ferdows, M. S. and Barbour, A. G. (1989) Proc. Natl. Acad. Sci. USA 86: 5969-5973.

Fraser, C. M., Gocayne, J. D., White, O., Adams, M. D., Clayton, R. A., Fleischmann, R. D., Bult, C. J., Kerlavage, A. R., Sutton, G., Kelley, J. M., Fritchman, J. L., Weidman, J. F., Small, K. V., Sandusky, M., Fuhrmann, J., Nguyen, D., Utterback, T. R., Saudek, D. M., Phillips, C. A., Merrick, J. M., Tomb, J.-F., Dougherty, B. A., Bott, K. F., Hu, P.-C., Lucier, T. S., Peterson, N., Smith, H. O., Hutchinson III, C. A. and Venter, J. C. (1995) Science 270: 397-403.

Fuerst, J. A. and Webb, R. I. (1991) Proc. Natl. Acad. Sci., USA 88: 8184-8188.

Garland, J. L. and Mills, A. L. (1994) A community-level physiological approach for studying microbial communities, p. 77-83. In K. Ritz, J. Dighton, and K. E. Giller (ed.), Beyond the biomass: compositional and functional analysis of soil microbial communities. John Wiley & Sons Ltd., Chichester, United Kingdom.

Hansen, M. T. (1978) J. Bacteriol. 134: 71-75.

Henckel, T., Friedrich, M. and Conrad, R. (1999) Appl. Environ. Microbiol. 65: 1980-1990.

Hugenholtz, P., Goebel, B. M. and Pace, N. R. (1998) J. Bacteriol. 180: 4765-4774.

Krawiec, S. and Riley, M. (1990) Microbiol. Rev. 54: 502-539.

Kündig, C., Hennecke, H. and Göttfert, M. (1993) J. Bacteriol. 175: 613-622.

Leblond, P., Redenbach, M. and Cullum, J. (1993) J. Bacteriol. 175: 3422-3429.

Lindsay, M. R., Webb, R. I. and Fuerst, J. A. (1997) Microbiology 143: 739-748.

McDonald, I. R., Kenna, E. M. and Murrell, J. C. (1995) Appl. Environ. Microbiol. 61:116-121.

Menke, M. A. O. H., Liesack, W. and Stackebrandt, E. (1991) Arch. Microbiol. 155: 263-271.

Michaux, S., Paillisson, J., Carles-Nurit, M. J., Bourg, G., Allardet-Servent, A. and Ramuz, M. (1993) J. Bacteriol. 175: 701-705.

Muyzer, G., De Waal, E. C. and Uitterlinden, A. G. (1993) Appl. Environ. Microbiol. 59: 695-700.

Okhuma, M., Nosa, S., Usami, R., Horikoshi, K. and Kudo, T. (1996) Appl. Environ. Microbiol. 62: 2747-2752.

Pühler, G., Leffers, H., Gropp, F., Palm, P., Klenk, H. P., Lottspeich, F., Garrett, R. A. and Zillig, W. (1989) Proc. Natl. Acad. Sci. USA 86: 4569-4573.

Rainey, F. A. and Stackebrandt, E. 1996. Molecular methods and the identification of bacteria, p. 169-178. *In* A. Cimerman and N. Gunde-Cimerman (ed.), Biodiversity, International Biodiversity Seminar, ECCO XIV. Meeting, June 30-July 4, Gozd Martuljek, Slovenia. Slovenian National Commission for UNESCO, Ljubljana.

Rainey, F. A., Ward, N., Sly, L. I. and Stackebrandt, E. (1994) Experientia 50: 796-797.

Reysenbach, A.-L., Giver, L. J., Wickham, G. S. and Pace, N. R. (1992) Appl. Environ. Microbiol. 58: 3417-3418.

Rheims, H., Spröer, C., Rainey, F. A., and Stackebrandt, E. (1996) Microbiology 142: 2863-2870.

Rosado, A., Duarte, G. F., Seldin, L. and van Elsas, J. D. (1998) Appl. Environ. Microbiol. 64: 2770-2779.

Roussel, Y., Pebay, M., Guedon, G., Simonet, J.-M. and Decaris, B. (1994) J. Bacteriol. 176: 7413-7422.

Schlesner, H. (1994) Syst. Appl. Microbiol. 17: 135-145.

Schulz, H. N., Brinkhoff, T., Ferdelman, T. G., Hernández Mariné, M., Teske, A. and Jørgensen, B B. (1999) Science 284: 493.

Smalla, K., Wachtendorf, U., Heuer, H., Liu, W.-T. and Forney, L. (1998) Appl. Environ. Microbiol. 64: 1220-1225.

Suwanto, A. and Kaplan, S. (1989a) J. Bacteriol. 171: 5840-5849.

Suwanto, A. and Kaplan, S. (1989b) J. Bacteriol. 171: 5850-5859.

Suzuki, M. T. and Giovannoni, S. J. (1996) Appl. Environ. Microbiol. 62: 625-630.

Ward-Rainey, N., Rainey, F. A., Schlesner, H. and Stackebrandt, E. (1995) Microbiology UK 141: 3247-3250.

Ward-Rainey, N., Rainey, F. A. and Stackebrandt, E. (1996) J. Bacteriol. 178:1908-1913.

Wilson, E. O. (1992) The diversity of life. Harvard University Press.

Wilson, K. H. and Blitchington, R. B. (1996) Appl. Environ. Microbiol. 62: 2273-2278.

Woese, C. R., Kandler, O. and Wheelis, M. L. (1990) Proc. Natl. Acad. Sci. USA 87: 4576-4579.

Wünsche, L., Brüggemann, L. and Babel, W. (1995) FEMS Microbiol. Ecol. 17: 295-306.

Zak, J. C., Willig, M. R., Moorhead, D. L. and Wildman, H. G. (1994) Soil. Biol. Biochem. 26: 1101-1108.

Zuerner, R., Hermann, J. L. and Saint Girons, I. (1993) J. Bacteriol. 175: 5445-5451.

Biodata of Dr. **Vigdis Torsvik** author of *"Microbial Biology and Genetic Diversity of Microorganisms.*

Dr. Vigdis Torsvik is an Associate professor of General Microbiology at Department of Microbiology, University of Bergen, Norway. She earned her Cand. real. degree at the University of Bergen in 1973 with a thesis on growth and nitrogenase activity of *Azotobacter* species from acid soils.

Dr. Torsvik started her work on quantification and characterization of soil bacterial communities through analysis of community DNA in 1977. She has developed methods for DNA based determination of total genetic diversity in microbial communities. Currently she is the supervisor and leader of a research group in molecular microbial ecology.

Her main research areas are: Microbial diversity and characterisation of microbial communities through DNA and rRNA analysis; The effect of climate and environmental perturbation on microbial populations and communities; Gene transfer in microbial communities; Antibiotic and heavy metal resistance of bacteria in soil and sediments; Methane transformation in tundra soil from Svalbard. At present Dr. Torsvik is engaged in studies of microbes in basaltic glass from a rift valley of the Mid-Atlantic Ridge, at 76.5° N and approximately 3500m below the sea level.

E-mail: vigdis.torsvik@im.uib.no

J. Seckbach (ed.), Journey to Diverse Microbial Worlds, 43-57.

MOLECULAR BIOLOGY AND GENETIC DIVERSITY OF MICROORGANISMS

VIGDIS TORSVIK, FRIDA LISE DAAE, RUTH ANNE SANDAA and LISE ØVREÅS
Department of Microbiology, University of Bergen
Jahnebakken 5, N-5052 Bergen, Norway

1. Introduction

Microbial diversity and biodiversity are used synonymously to describe the variability among all microorganisms in an assemblage or a community. In the present review the term microorganisms encompasses prokaryotic organisms belonging to the domain *Bacteria* and *Archaea*.

Rationales for studying microbial diversity are that microorganisms play a key role in energy and matter transformation and are fundamental for sustaining ecosystems. A comprehensive understanding of the role of microbes in an ecosystem requires knowledge about the extent of their genetic, taxonomic and functional diversity. The diversity is a result of genetic variability within microbial taxa, environmental factors influencing gene expression, and the interactions between populations. It is assumed that the ability of an ecosystem to resist environmental fluctuations and maintain its functional capability during periodical stress is dependent on its biodiversity. It has been observed that collapse in community stability due to perturbation leads to reduced diversity. Thus, diversity can be used as indicator of community stability and enables monitoring and prediction of environmental changes. Microorganisms show the greatest variability with respect to living conditions and adaptations to life in extreme living habitats. Knowledge about microbial diversity is therefore important for an increased understanding of the strategies and limits of life. Despite their fundamental role in sustaining ecosystems and their great value as an untapped resource for biotechnological products and processes, our knowledge about microbial diversity is scant.

2. The Microbial Diversity Concept and Estimation of Microbial Diversity

Biodiversity can be defined at different levels of biological organization. The levels range from genetic diversity within species, via organism diversity embracing the variability between taxa, to community and ecological diversity (Bull et al., 1992, Harper & Hawksworth, 1994). Ecological diversity encompasses community parameters such as the variability in community structure, the complexity of interactions, number of trophic levels, and number of guilds (functional diversity). Diversity can also be regarded as the amount of information in a community. Commonly the diversity concept includes both a richness and an evenness component. According to that definition both the amount of information and how the information is distributed among the individuals in a community is taken into account (Atlas, 1984). This definition adopted from information technology can be applied directly to genetic diversity. Diversity is expressed in different ways: as inventories of taxonomic groups, as single numbers (diversity indices), as phylogenetic trees, or number of functional guilds.

For higher (eukaryote) organisms diversity is based on the species variability. Using the species as a unit to quantify and estimate microbial diversity causes problems as no universally accepted species definition exists for microorganisms (O'Donnell et al., 1994, Stackebrandt & Goebel, 1994). It has been agreed that microbial species can be defined at the molecular level as bacterial strains with 70% or more chromosomal DNA-DNA homology (Wayne et al., 1987). Another problem when using traditional methods to estimate diversity is their dependence on cultivation and isolation of bacteria. Most of the microorganisms from natural communities as observed in the microscope do not grow under laboratory conditions, although they are metabolically active (Fægri et al., 1977). This implies that traditional methods for diversity measurements explore only a small part of the genetic information in a community.

Applying growth independent molecular methods has eliminated the culturing problem, and changed the research strategies from analyzing isolates to analyzing the entire community. Phenotypic diversity measurements are restricted to the subset of genetic information expressed under given conditions. Genetic diversity measures the genetic potential in the microbial community independent of environmental conditions.

3. Molecular Approaches for Investigating Microbial Diversity

3.1. ANALYSING GENETIC DIVERSITY AT DIFFERENT RESOLUTION LEVELS

The information in DNA and RNA molecules of microorganisms can be used to investigate and compare diversity at all biological organization levels. Analyses of nucleic acids from environmental samples provide information about both cultured and uncultured bacteria.

High-resolution analyses like DNA fingerprinting and sequencing are suitable for identification and classification of microbial strains at the species and subspecies level. In restriction analysis and restriction fragment length polymorphism (RFLP), fragments obtained by restriction enzyme digestion are separated by agarose gel electrophoresis. In RFLP analysis the gel is blotted and hybridized with probes derived from cloned DNA fragments of related bacteria. Analysis of repetitive extragenous palindromic (REP) sequences is performed using specific oligonucleotide primers to amplify the DNA region between the sequences. High resolution fingerprinting methods have been used to monitor specific populations in microbial communities and assessing the diversity of bacterial isolates and cloned genes (de Bruijn, 1992, Massol-Deya et al., 1995, Stackebrandt & Rainey, 1995).

Base sequence differences of conserved genes like those coding for ribosomal RNA (rRNA) provide information at somewhat lower resolution. rRNA homology can not be used to define species, but is useful for discrimination between species. Sequence differences of rRNA genes (rDNA), however, are well suited to investigate diversity at the genus and higher taxon levels. Community fingerprinting techniques based on rRNA and rDNA molecules are used to analyze the structure and dynamics of microbial communities that are not too complex. The methods can resolve the composition within specific parts (organism groups or guilds) of the community. Southern blot hybridization with phylogenetic probes or sequencing has proved particularly useful to study changes in communities and to identify the numerically dominating community members.

Low-resolution methods allow analyses of broad scale differences and total genetic diversity in microbial communities. Information about the gross community composition can be gained by measuring the distribution of nucleotide bases (mole percent guanine+cytosine) in community DNA. Other low resolution methods are the application of phylogenetic probes, either in slot blot hybridization of community DNA or in fluorescence *in situ* hybridization (FISH) of intact cells (Amann et al., 1990, DeLong et al., 1989, Stahl & Amann, 1991). Such methods are used to quantify particular microbial populations and determine the overall taxon composition of microbial communities. The total genetic diversity in microbial communities can been determined by analyzing the complexity of DNA isolated from the communities (Torsvik et al., 1990a). The DNA complexity is determined by measuring the reassociation rate of denatured single stranded DNA. When applying this method, the genetic diversity is defined as the total amount of genetic information in the community and the distribution of this information among the different genetic types.

3.2. ANALYSIS OF INFORMATION IN RIBOSOMAL RNA GENES.

Ribosomal RNA genes and their transcripts, the rRNA molecules, have become the most widely used markers for phylogenetic classification of microorganisms. The rRNA based methods have revealed that the microorganisms are divided into highly diverging phylogenetic lineages, and that they comprise the largest biodiversity on Earth. Culture independent rRNA analyses have demonstrated how microorganisms from similar habitats in geographically widely separated regions can be related. On the other hand they have also demonstrated that microorganisms belonging to the same physiological group or guild may belong to different phylogenetic lineages, reflecting differences in their habitats.

3.2.1. Denaturing gradient gel electrophoresis (DGGE) community profiling.
In denaturing gradient gel electrophoresis (DGGE) or temperature gradient gel electrophoresis (TGGE), DNA fragments with the same length but different nucleotide sequences are separated. In DGGE the separation is based on difference in mobility of PCR amplified DNA molecules in polyacrylamide gels with a linear concentration gradient of denaturing chemicals (urea and formamide). DNA molecules with different sequences will differ in their melting behavior. As DNA molecules migrate in a denaturant gel, they will start melting in particular melting domains and become partly

single stranded. The partly denatured molecules will nearly stop to migrate in the gel. Thus DNA fragments stop migrating at different positions in the gel dependent on their base composition and sequence variation.

Variable regions of rRNA genes (rDNA) from microbial community DNA are PCR amplified using primers annealing to conserved regions that flank the variable regions (Muyzer et al., 1993). When the abundance of the target microorganisms is low, as is often the case for *Archaea*, nested amplification is used to increase the specific template concentration. First nearly the entire rRNA gene is amplified using an outer set of primers targeting conserved regions near the end. Then an inner primer set is used to amplify the variable region (230-500 bp) subjected to the DGGE analysis.

DGGE analyses provide community profiles with several discernible bands on the gel. These bands most probably represent the numerically predominant microbial populations in the community. Further information about the taxon composition of the community can be obtained by blotting the gel and hybridizing to phylogenetic probes, targeting the main phylogenetic subclasses of *Bacteria* and *Archaea* (Amann et al., 1995, Raskin et al., 1994). Well separated DGGE bands can be punched out from the gel and subjected to DNA sequencing. By comparing the sequences of DGGE bands with those in a database, the phylogenetic affiliation of the predominating microorganisms can be obtained (Øvreås et al., 1997). Comparison of rRNA sequences can also be used to design phylogenetic oligonucleotide probes.

Another community fingerprinting method is based on the length polymorphism of the spacer region between the 16S and 23S rRNA genes. In the ribosomal internal spacer analysis (RISA) (Acinas et al., 1999, Borneman & Triplett, 1997), PCR amplified products are separated by agarose gel electrophoresis. Community fingerprinting techniques are rapid, and allow parallel analyses of multiple samples. The structure of bacterial communities in unperturbed soil was too complex to be resolved by DGGE analysis (Øvreås et al., 1998). In a meromictic lake however, both bacterial and archaeal communities at different depth could be successfully analyzed with this method (Øvreås et al., 1997). Thus, the DGGE method is suitable for revealing the structure in communities with moderate to small complexities. In complex communities, the method can only reveal fractions or subsets of the entire diversity. This can be done by applying primers targeting specific phylogenetic or functional groups of microorganisms. Another approach is to apply specific enrichments to enhance the growth of the microorganism of interest. This strategy is particularly useful in studies of functional groups or guilds of microorganisms.

3.2.2. Analysis of cloned rRNA genes.
Cloning techniques have been widely used to analyze microbial communities. Clone libraries of PCR amplified rRNA genes in DNA from environmental samples are made in cloning vectors (Borneman et al., 1996). Bacterial and archaeal rRNA genes are amplified in separate PCR reactions using kingdom specific and universal primers(Lane et al., 1985, Torsvik et al., 1993). The cloned amplicons can be compared by fingerprinting methods like amplified ribosomal DNA-restriction analysis (ARDRA) (Vaneechoutte et al., 1992). Clones are classified by slot blot hybridization with phylogenetic probes (Amann et al., 1995). Sequencing of the cloned rRNA genes and

comparing the sequences with those obtained from databases provides information about the phylogenetic affiliation of the cloned sequences.

3.2.3. Fluorescence in situ hybridization (FISH) with phylogenetic probes.

Oligonucleotide probes homologous to conserved or variable sequences of rRNA target phylogenetic classes and subclasses of microorganisms. Phylogenetic probes labeled with fluorescent dyes are used for *in situ* detection of single cells in environmental samples (Amann, 1995, Chatzinotas et al., 1998). After fluorescence *in situ* hybridization (FISH) the microorganisms can be visualized by epifluorescence microscopy. In order to obtain a sufficiently strong hybridization signal to visualize the cell, the number of ribosomes per cell must be above a certain limit. The limit depends on the label of the probe and the fluorescence of the background. In actively growing cells the number of ribosomes per cell may be several tens of thousands. FISH has been used to identify uncultured microorganisms, studying the distribution of, and quantifying microbial populations. By counting in the microscope after hybridization to a set of different phylogenetic probes, the number of phylogenetic taxa and the distribution of individuals among taxa can be determined.

3.3. INFORMATION IN TOTAL GENOMIC DNA

3.3.1. Base composition profiling

Information about the overall composition of microbial communities can be obtained from the base composition of community DNA. The base composition is expressed as mole % guanine + cytosine (mole % G+C). The mole % G+C varies from 25 to 75% among the microorganisms. Single stranded DNA has approximately 35% higher absorbency than double stranded DNA at 260 nm. Measuring the absorbency at 260 nm with increasing temperature will give a melting profile for DNA. The melting profiles can be converted to mole % G+C profiles by calculating the first derivative of the melting curve (Ritz et al., 1997, Torsvik et al., 1995). A DNA base composition profile can also be obtained by isopycnic centrifugation of bisbenzimide-DNA complex in a CsCl gradient (Harris, 1994, Holben et al., 1993). Bisbenzimide binds preferentially to AT pairs and will decrease the buoyant density of DNA. Thus community DNA can be separated according to the base composition in a density gradient with bisbenzimide.

3.3.2. Total genetic diversity

The total genetic diversity can be estimated from the complexity of DNA isolated from microbial communities. The DNA complexity is determined by measuring the reannealing (reassociation) rate of single stranded DNA in solution under defined conditions (Britten & Kohne, 1968). For homologous DNA the reassociation in solution follows a second order reaction, and the rate decreases with increased DNA complexity, or number of different DNA types in the solution. The DNA complexity is defined as the number of base pairs in non-homologous DNA, and is equivalent to the size of a single-copy genome (Britten & Kohne, 1968, Wetmur & Davidson, 1968).

The reassociation kinetics are measured in terms of Cot values, where Co is the concentration of DNA in mole nucleotides per liter multiplied with the time in seconds (t). When the DNA concentration is held constant, the number of DNA molecules with a particular sequence decreases with increasing DNA complexity. The value $Cot_{1/2}$, when 50% of the DNA has reassociated is proportional to the DNA complexity. At defined reassociation conditions (temperature and monovalent cation concentration), $Cot_{1/2}$ is used to estimate the genome size. DNA from a bacterial community is a mixture of DNA from different bacterial types that are present in different proportions. When the number of DNA types increases, the overall reaction deviates from the ideal second order kinetics, and the curve will have a flatter slope. For such reactions the $Cot_{1/2}$ does not have any precise meaning, but nevertheless it provides information about the DNA complexity. If we regard the microbial community as super-species, the $Cot_{1/2}$ is an estimate of the "genome" size of this species. To estimate the number of bacterial species in the community, the community genome can be divided by the size of a standard bacterial genome (the *Escherichia coli* genome or the average genome size of bacteria isolated from the environment). $Cot_{1/2}$ is used as a diversity index (Torsvik et al., 1995), and includes both the amount of information in DNA and how this information is distributed (how much is found in few DNA types and how much is distributed among many types). This means that two communities with different structure can have identical $Cot_{1/2}$. The $Cot_{1/2}$ value is measured relative to DNA with known complexity like the *E. coli* genome.

To estimate prokaryotic diversity, the DNA must be substantially free from eukaryotic DNA. Therefore, the bacteria are separated from the soil by a fractionated centrifugation method (Fægri et al., 1977) before lysis. To obtain highly purified DNA, the crude bacterial lysate can be purified by hydroxyapatite chromatography (Torsvik, 1980). DNA fragments used in reassociation analysis must have a uniform size, and DNA is sheared to an average fragment size of 650 bp. Reassociation of complex DNA is slow, and to increase the rate it is measured in a solution of 6xSSC (standard saline citrate) and 30% DMSO (dimethylsulfoxide) at 25°C below the average melting temperature (Tm). Reassociation is measured as the decrease in absorbency at 260 nm with time in an UV-visible spectrophotometer (Torsvik et al., 1995).

4. Application of Molecular Methods

4.1. ESTIMATING GENETIC DIVERSITY IN MICROBIAL COMMUNITIES.

The genetic diversity of microbial communities in different environments estimated by the DNA reassociation technique is summarized in Table 1. Using this technique we have demonstrated that the microbial diversity in pristine and moderately managed, non-extreme environments is very high (Torsvik et al., 1990a, Torsvik et al., 1996). The $Cot_{1/2}$ of DNA isolated from bacteria in forest soil and pristine sediments ranged from 47,00 to approximately 9,000 moles/l.s. (in 6xSSC, 30% DMSO). This corresponds to DNA complexities of 2.7×10^{10} to 4.8×10^{10} bp. Taking the genome size of *E. coli* B as a unit (4.1×10^6 bp, 2.71×10^9 Dalton), the DNA complexity is equivalent to 6,500 – 11,500 entirely different genomes. We can only speculate on the number of species that contribute to this DNA complexity. The minimum number will be approximately 6,000

– 11,000 (with zero % homology). If 70% DNA-DNA homology level is used as a species limit (Wayne et al., 1987), the number of species would be roughly 20,000 – 37,000. The soil bacterial DNA used to estimate the genetic diversity was isolated from 100 g soil with a total bacterial number of about 1.5×10^{12}. If there were 10,000 – 20,000 species and we assume an even species distribution, each species would consist of $7.5 \times 10^7 - 1.5 \times 10^7$ individuals.

TABLE 1. Bacterial diversity in soils and marine environments as determined by reassociation kinetics ($Cot_{1/2}$; moles/l.s. at 50% reassociation), and number of «genomes». If not mentioned otherwise DNA was extracted from the bacterial fraction of environmental samples.

DNA source	$Cot_{1/2}$ [a]	No. of genomes[b]
E. coli	0.79	1
Forest soil	4,500-4,700	6,500
Forest soil, plate count community[c]	28	33
Pristine sediment	9,000	11,400
Fish farm sediment	40-50	50-70
Sewage sludge amended soil; low metal contaminated	3,700	4,700
Sewage sludge amended soil; high metal contaminated	1,200	1,500
Model experiment; control soil	5,700	8,000
Model experiment; anaerobic perturbed soil	2,500	3,200
Model experiment; methane perturbed soil	270	380

[a]In 6xSSC, 30% DMSO [b]Equivalent to E.coli genome; 4.1×10^6 bp [c]DNA from a mixture of 206 isolates

The microbial diversity in polluted and perturbed environments was often notably reduced as compared to pristine environments. The microbial diversity in sediments under a marine fish farm where organic wastes had accumulated was approximately 250 times lower than in pristine sediment with similar amount of organic matter. When the $Cot_{1/2}$ in DNA from the microbial community in pristine sediments was approximately 9,000 moles/l.s., the DNA from fish farm sediments had a $Cot_{1/2}$ of 40-50. This corresponds to DNA complexities of approximately $2.2-2.8 \times 10^8$ bp or 50-70 E. coli genomes.

The microbial genetic diversity in agricultural soils showed great differences. The $Cot_{1/2}$ of DNA from a sandy loam soil subjected to intensive agricultural management was 250 moles/l.s., whereas the $Cot_{1/2}$ of DNA from an organic soil in a pasture field was 6300 moles/l.s.

4.2. COMPARING CULTURED AND UNCULTURED MICROORGANISMS

The genetic diversity of total microbial community in soil was compared to that of a collection of 206 bacterial isolates from the same soil (Torsvik et al., 1990b). DNA reassociation was measured in 4x SSC, 30% DMSO. The $Cot_{1/2}$ of the DNA mixture from the isolates was 28 moles/l.s. This corresponds to a DNA complexity of $1.4x10^8$ bp, or 33 E. coli genomes with no homology, approximately 100 species with 70% DNA-DNA homology. The total genetic diversity of soil microbial community was $2.7x10^{10}$ bp, equivalent to 6,500 genomes with the same size as the E. coli genome. Thus the genetic diversity of the total bacterial population was approximately 200 times higher than that of the isolated bacteria. This indicates that the bacterial types that were isolated by conventional plating technique comprise only a small fraction of the soil bacterial population. Our findings are in agreement with analysis of 16S rRNA genes derived from environmental samples. Analyses of rRNA clones derived directly from microbial communities of marine bacterioplankton, hot springs and sewage sludge (Fuhrman et al., 1992, Giovannoni et al., 1990, Schuppler et al., 1995, Ward et al., 1990, Weller et al., 1992) have demonstrated that most of the cloned sequences differed from sequences of isolated microorganisms. With a few exceptions, no close relationship between rRNA sequences from isolates and cloned environmental rDNA was observed. The conclusion of these investigations was that the microbial diversity within natural microbial communities must be higher than measured by analysis of cultured bacteria. The magnitude of the diversity, however, was not investigated.

Microbial communities in metal contaminated and uncontaminated soils were compared by a polyphasic approach in which different molecular methods were applied on cultured microorganisms and total microbial communities. The control fields had received "uncontaminated" sewage sludge, whereas the contaminated fields received heavy metal amended sewage sludge at two rates of application (low and high metal contamination) (McGrath et al., 1995). Applying the reassociation technique we found that DNA from the control soil had a $Cot_{1/2}$ of 7,800 moles/l.s. The $Cot_{1/2}$ values of DNA from the low and high metal contaminated soils were 3,700 and 1,200 moles/l.s., respectively (Table 1).

The analyses of microbial community composition in the metal contaminated soils were based on isolated bacteria and cloned rDNA fragments, derived directly from community DNA by PCR amplification. Altogether, approximately 300 cloned amplicons and 300 bacteria isolated from the two soils were compared. Filter (dot blot) hybridization with a set of phylogenetic probes targeting bacterial classes and subclasses was applied. (Amann et al., 1995, Chatzinotas et al., 1998, Raskin et al., 1994). The hybridization revealed that the clones and the isolates gave a very different community composition (Figure 2). The most abundant subclass of isolates in both soils was Gram+ with low mole % G+C (LGC). With respect to the clones, this group was among the smallest. There were also profound discrepancies in the abundance of isolates and clones belonging to the α-Proteobacteria and β-Proteobacteria subclasses in the two soils. The clones revealed greater differences in the community composition of the two soils than the isolates. Clones with sequences from bacteria belonging to the α-Proteobacteria were numerically dominant in the high metal contaminated soil where they comprised nearly 40% of the clones. In the low metal contaminated soil this subclass comprised about 17% of the clones. The distribution of the *Cytophaga-Flexibacter-Bacteroides* subclasses between the soils was the opposite. It comprised 17% of the clones in low metal contaminated soil, but only 6% in the high metal contaminated soil. In the other subclasses there were only minor differences between the two soils. The differences in community composition based on isolates were relatively small. An exception was the *β-Proteobacteria,* which comprised 11% of the isolates from high and only 3% from the low metal contaminated soil. Interestingly, no sequence from *β-Proteobacteria* was detected in clones from the high metal contaminated soil.

Isolates

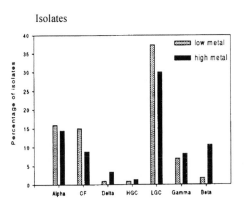

Clone

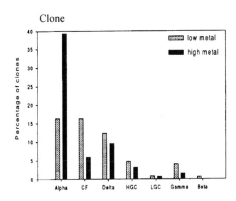

Figure 1. Changes in community structure in soil contaminated with low and high heavy metal concentrations. Group specific 16S and 23 S rRNA phylogenetic probes were hybridized against amplified 16S and 23S rRNA genes in DNA extracted directly from the soils. Phylogenetic probes targeting the following bacterial classes and subclasses were applied: Bacteria,α-Proteobacteria (Alpha), β-Proteobacteria (Beta), γ-Proteobacteria (Gamma), δ-Proteobacteria (Delta), *Cytophaga-Flexibacter-Bacteroides* (CF) subclass, Gram+ with high mole % G+C (HGC), Gram+ with low mole % G+C (LGC) (Sandaa et al., 1999).

Quantification of specific microorganisms in the soil by fluorescence *in situ* hybridization and microscopic counting (Sandaa et al., 1999) was in agreement with the results obtained with cloned amplicons. Both methods indicated a low abundance of Gram+ bacteria with low mole% G+C (LGC). The discrepancy between the number of isolates and number of clones can reflect a bias in the DNA extraction from the LGC-bacteria, which may exist as endospores in the soils. Another explanation can be that they grow well on solid media in the laboratory. Quantification by FISH confirmed the marked increase in *α-Proteobacteria* and decrease in *Cytophaga-Flexibacter-Bacteroides* subclass with increased heavy metal contamination as observed for the cloned amplicons (Sandaa et al., 1999).

A total of 78 cloned amplicons from the two metal contaminated soils that did not hybridize with phylogenetic probes were sequenced (Lane, 1991). The sequences (approximately 500 bp) were compared to sequences in the databases retrieved by the BLAST (NCBI) program. Approximately 50% of the sequences belonged to the group of Gram+ bacteria with high mole % G+C (HGC), and 26 % of the clones showed sequence similarities with *α-Proteobacteria*. This indicates either that the phylogenetic probes did not target all the members within these groups, or that there was some bias in the PCR reaction. Many of the HGC sequences showed similarities to unidentified actinomycetes from forest soil (Stackebrandt et al., 1993). Some of the α-*Proteobacteria* sequences were related to (87-96% homology) already known bacterial genera like *Sphingomonas* and *Rhodopseudomonas*. Three groups of sequences belonging to the *Bacteria* did not cluster within any of the major taxa described until now (Olsen et al., 1994).

Microbial communities in water originating from hydrothermal vents under the glacier at Sørkapp, Spitsbergen were investigated with molecular methods (unpublished). The hot water was cooled to approximately 3°C as it passed the ice. Molecular methods were applied directly on microorganisms from water samples and from enrichment cultures for methanogenic, methanotrophic and sulfide oxidizing bacteria. At the sites where the water emerged from the glacier there were substantial amount of microbes, most of them found within white slimy flocks. PCR amplified fragments (approximately 230 bp) of 16S rRNA genes from *Archaea* and *Bacteria* were separated by DGGE. Distinct DGGE bands were sequenced and compared to sequences found in databases (GenBank). Many sequences showed high homology with those from bacteria found in thermal vent communities. Dominant sequences in the water samples were similar to those from sulfide and sulfur oxidizing bacteria (*Sulfospirillum, Geospirillum, Thiothrix*). In mineral medium with methane as sole carbon-source, sequences similar to methane oxidizing bacteria of type I were detected in addition to sequences related to a mussel methanotrophic gill symbiont. In enrichment medium for sulfide and sulfur oxidizing bacteria, sequences similar to those in the water samples were detected. In anaerobic enrichment cultures on H_2+CO_2, sequences related to a hydrothermal vent eubacterium, *Geospirillum and Shewanella* were detected. Sequences derived from *Archaea* were not detected in the water samples, but a few sequences were detected in one of the enrichment cultures. These sequences were related to *Methanogenium* sp., and to an uncultured archaeon. The investigation demonstrated that the microorganisms in the water, although it was cooled and mixed with marine water, could be tracked back to a hydrothermal vent microbial community that was sustained by the energy from the depth of the Earth.

4.2. MONITORING CHANGES IN MICROBIAL COMMUNITIES

Molecular methods have been used to compare communities from pristine and perturbed habitats. It has been demonstrated that perturbations and extreme environmental conditions like increased temperature, changed soil atmosphere and pollution decrease the microbial diversity.

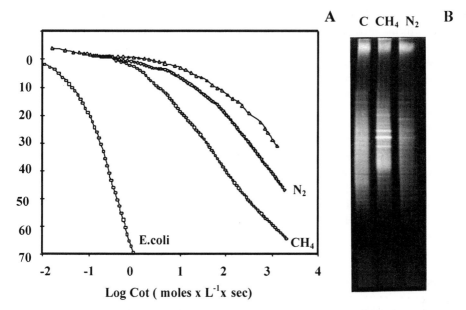

Figure. 2. A. Reassociation of DNA from bacterial fractions of control soil (Δ), anaerobic soil (◊), methane-perturbed soil (o), and *E. coli* (□). The DNA was reassociated in 6xSSC and 30% DMSO at 49°C.
B. DGGE analyses of PCR-amplified fragments of 16S rRNA genes from the same soils as in A. Lane 1 (C) shows the DGGE profile from control soil, lane 2 (CH₄) the profile from methane-perturbed soil, lane 3 (N₂) the profile from anaerobic soil (Øvreås et al. 1998)

In a model experiment the effect of environmental changes, such as a shift to anaerobic conditions and addition of a sole carbon source (methane), on a soil microbial community was investigated (Øvreås et al., 1998). Analyses of *in situ* activity of methanotrophic and methylotrophic bacteria were combined with molecular methods to study community structure and diversity. Samples from an organic agriculture soil were incubated for 3 weeks at 15°C. Two types of perturbations were applied. One was incubated under anaerobic conditions (with N_2 gas), and the other was incubated with air containing 17% methane. Striking changes in the structure and diversity of the soil communities were observed after the perturbations. DNA from the undisturbed control soil had a $Cot_{1/2}$ of 5,700 moles/l.s., in 6x SSC, 30% DMSO. Under anaerobic conditions the DNA complexity decreased to less than the half ($Cot_{1/2}$ of 2,500 moles/l.s.). In the methane amended soil the $Cot_{1/2}$ was reduced approximately 20 times, to 270 moles/l.s. Decrease in genetic diversity as revealed by reduction in $Cot_{1/2}$ values, may reflect reduced "species" richness, or reduced evenness because some bacterial types were numerically dominant.

PCR-DGGE analysis of community DNA showed that more than 100 bands, covering the entire gradient, were present in all the soil samples. The community profile of DNA from the control soil contained mainly weak bands, indicating that there were no numerically dominant populations. In the soil subjected to anaerobic condition only minor changes in the community profile was observed, but two bands became slightly more intense. In the methane-amended soil however, some strong bands were seen on top of the community profile, indicating that there were some numerically dominating bacterial populations (Figure 3). Three of the strongest bands were sequenced and aligned to sequences in a database. The alignment showed that they were derived from bacteria related to methane oxidizing bacteria of type I, belonging to the γ-*Proteobacteria*.

Our investigations have demonstrated that molecular methods are valuable tools for obtaining information about microorganisms present in the environments. A most availing strategy is to combine low-resolution, broad-scale methods like DNA reassociation and FISH, with methods having higher resolution like PCR-DGGE analysis or other community fingerprinting methods. Such methods complement each other, and together they allow assessment of the overall microbial diversity, and provide qualitative and quantitative information about the major groups of microorganisms in the environment.

5. References

Acinas, S. G., Antón, J. and Rodriguez-Valera, F. (1999) *Appl. Environ. Microbiol.,* **65**, 514-522.

Amann, R. I. (1995) In: A. D. L. Akkermans, J. D. v. Elsas, and F. J. d. Bruijn (eds.) *Molecular microbial ecology manual,* Kluwer Academic Publishers. Dordrecht. pp 3.3.6: 1-15

Amann, R. I., Krumholz, L. and Stahl, D. A. (1990) *J. Bacteriol.,* **172**, 762-770.

Amann, R. I., Ludwig, W. and Schleifer, K. H. (1995) *Microbiol Rev.,* **59**, 143-169.

Atlas, R. M. (1984) In: K. C. Marshall (ed.), *Advances in microbial ecologym.* Plenum Press. New York, Vol. 7, pp. 1-47.

Borneman, J., Skroch, P. W., O'Sullivan, K. M., Palus, J. A., Rumjanek, N. G., Jansen, J. L., Nienhuis, J. and Triplett, E. W. (1996) *Appl. Environ. Microbiol.,* **62**, 1935-1943.

Borneman, J. and Triplett, E. C. (1997) *Appl. Environ. Microbiol.,* **63**, 2647-2653.

Britten, R. J., and Kohne, D. E. (1968) *Science,* **161**, 529-540.

Bull, A. T., Goodfellow, M. and Slater, J. H. (1992) *Annu Rev Microbiol,* **46**, 219-252.

Chatzinotas, A., Sandaa, R.-A., Schönhuber, W., Amann, R., Daae, F. L., Torsvik, V., Zeyer, J. and Hahn, D. (1998) *System. Appl. Microbiol.,* **21**, 579-587.

de Bruijn, F. J. (1992) *Appl. Environ. Microbiol.,* **58**, 2180-2187.

DeLong, E. F., Wickham, G. S. and Pace, N. R. (1989) *Science,* **243**, 1360-1362.

Fuhrman, J. A., McCallum, K. and Davis, A. A. (1992) *Nature,* **356**, 148-149.

Fægri, A., Torsvik, V. L. and Goksøyr, J. (1977) *Soil Biol. Biochem.,* **9**, 105-112.

Giovannoni, S. J., Britschgi, T. B., Moyer, C. L. and Field, K. G. (1990) *Nature,* **345**, 60-62.

Harper, J. L. and Hawksworth, D. L. (1994) *Phil. Trans. R. Soc. London B.,* **345**, 5-12.

Harris, D. (1994) In: K. Ritz, J. Dighton, and K. E. Giller (eds.) *Beyond the biomass.* John Wiley Sons, Chichester, New York, Brisbane, Toronto, Singapore, pp, 111-118.

Holben, W. E., Calabrese, V. G. M., Harris, D., Ka, J. O. and Tiedje, J. M. (1993) In: R. Guerrero and C. Pedrós-Alió (eds.) *Trends in microbial ecology,* Spanish Society for Microbiology. pp, 367 -370.

Lane, D. J. (1991) In: E. Stackebrandt and M. Goodfellow (eds.) *Nucleic acid techniques in bacterial systematics.* John Wiley & Sons, New York. pp. 115-175

Lane, D. J., Pace, B., Olsen, G. J., Stahl, D. A., Sogin, M. L. and Pace, N. R. (1985) *Proc. Natl. Acad. Sci. USA,* **82**, 6955-6959.

Massol-Deya, A. A., Odelson, D. A., Hickey, R. F. and Tiedje, J. M. (1995) In: A. D. L. Akkermans, J. D. v. Elsas, and F. J. d. Bruijn (eds.), *Molecular microbial ecology manual.* Kluwer Academic Publishers, Dordrecht, pp. 3.3.2: 1-8.

McGrath, S. P., Chaudri, A. M. and Giller, K. E. (1995) *J. Indust. Microbiol.,* **14**, 94-104.

Muyzer, G., de-Waal, E. C. and Uitterlinden, A. G. (1993) *Appl. Environ. Microbiol.,* **59**, 695-700.

O'Donnell, A. G., Goodfellow, M., and Hawksworth, D. L. (1994) *Phil. Trans. R. Soc. London. B*, **345**, 65-73.

Olsen, G. J., Woese, C. R. and Overbeek, R. (1994) *J. Bacteriol.*, **176**, 1-6.

Raskin, L., Stromley, J. M., Rittmann, B. E., and Stahl, D. A. (1994) *Appl. Environ. Microbiol.*, **60**, 1232-1240.

Ritz, K., Griffiths, B. S., Torsvik, V. L., and Hendriksen, N. B. (1997) *FEMS Microbiol. Lett.*, **149**, 151-156.

Sandaa, R-A,. Torsvik, V., Enger, Ø., Daae, F.L., Castberg, T., and Hahn, D. (1999) *FEMS Microbiol. Ecol.*, **30**, 237-251.

Schuppler, M., Mertens, F., Schon, G. and Gobel, U. B. (1995) *Microbiology*, **141**, 513-521.

Stackebrandt, E. and Goebel, B. M. (1994) *Int. J. Syst. Bacteriol.*, **44**, 846-849.

Stackebrandt, E., Liesack, W. and Goebel, B. M. (1993) *Faseb. J.* **7**, 232-236.

Stackebrandt, E. and Rainey, F. A. (1995) In: A. D. L. Akkermans, J. D. van Elsas and F. J. de Bruijn (eds.) *Molecular microbial ecology manual*. Kluwer Academic Publishers, Dordrecht,. pp. 3.1.1: 1-17.

Stahl, D. A. and Amann, R. (1991) In: E. Stackebrandt and M. Goodfellow (eds.) *Nucleic acid techniques in bacterial systematics*. John Wiley & Sons. Chichester, New York, Brisbane, Toronto, Singapore. pp, 205-248

Torsvik, V., Daae, F. L. and Goksøyr, J. (1995) In: J. T. Trevors and J. D. van Elsas (eds.) *Nucleic acids in the environment: Methods and applications*. Springer Verlag, Berlin. pp. 29-48.

Torsvik, V., Goksøyr, J. and Daae, F. L. (1990a) *Appl. Environ. Microbiol.*, **56**, 782-787.

Torsvik, V., Salte, K., Sorheim, R., and Goksoyr, J. (1990b) *Appl. Environ. Microbiol.*, **56**, 776-781.

Torsvik, T., Torsvik, V., Keswani, J. and Whitemen, W.B. (1993) Abstract at the Soc. Microbiol. 93the General Meeting, Atlanta. Georgia. USA.

Torsvik, V., Sørheim, R., and Goksøyr, J. (1996) *J. Indust. Microbiol.* **17**, 170-178.

Torsvik, V. L. (1980) *Soil Biol. Biochem.*, **12**, 15-21.

Vaneechoutte, M., Rossau, R., De, V. P., Gillis, M., Janssens, D., Paepe, N., De, R. A., Fiers, T., Claeys, G. and Kersters, K. (1992) *FEMS Microbiol.Lett.* **93**, 227-233.

Ward, D. M., Weller, R. and Bateson, M. M. (1990) *Nature*, **345**, 63-65.

Wayne, L. G., Brenner, D. J., Colwell, R. R., Grimont, P. A. D., Kandler, O., Krichevsky, M. I., Moore, L. H., Murray, R. G. E., Stackebrandt, E., Starr, M. P. and Trüper, H. G. (1987) *Int. J. Syst. Bacteriol.*, **37**, 463-464.

Weller, R., Bateson, M. M., Heimbuch, B. K., Kopczynski, E. D. and Ward, D. M. (1992) *Appl. Environ. Microbiol.*, **58**, 3964-3969.

Wetmur, J. G. and Davidson, N. (1968) *J. Mol. Biol.*, **31**, 349-370.

Øvreås, L., Forney, L., Daae, F. D. and Torsvik, V. (1997) *Appl. Environ. Microbiol.*, **63**, 3367-3373.

Øvreås, L., Jensen, S., Daae, F. L. and Torsvik, V. (1998) *Appl. Environ. Microbiol.*, **64**, 2739-2742.

II

INTRODUCTION TO EXTREMOPHILES

Biodata of Dr. **Michael T. Madigan** author of *"Bacterial Habitats in Extreme Environments"*.

Dr. **Michael Madigan** is a Professor in Southern Illinois University at Carbondale IL. He received his Ph.D. at the University of Wisconsin in 1976 and obtained since then a few awards. He is involved in University activities as well as in professional services, such as reviewing manuscripts and books, and, journal editorships.

The research in his laboratory in focused on the microbiology, ecology, molecular biology, and biotechnological uses. The projects include:

Phototrophic bacteria from extreme environments; Biodiversity and biogeochemistry of Antarctic photosynthesis bacteria; Biodegradable polymers from soy oil and soy molasses; Nitrogen fixation in phototrophic bacteria; and, Production of polyhydroxyalkanoates by phototrophic bacteria.

Dr. Madigan is the co-author (with Professor. T.D. Brock) of four books on *Biology of Microorganisms*, and, he published three additional books (with co-authors), one on the *Anoxygenic photosynthetic Bacteria*, and two are the *Brock Biology of Microorganisms*. He published close to one hundred scientific Journal articles including book chapters.

E-mail: madigan@micro.siu.edu

J. Seckbach (ed.), Journey to Diverse Microbial Worlds, 61-72.
© 2000 *Kluwer Academic Publishers. Printed in the Netherlands.*

BACTERIAL HABITATS IN EXTREME ENVIRONMENTS

MICHAEL T. MADIGAN
Department of Microbiology and
Center for Systematic Biology
Southern Illinois University
Carbondale, Illinois 62901-6508   USA

1. Introduction

As our knowledge of prokaryotic diversity has increased so has our understanding of microbial habitats. Indeed, it has become clear in recent years that few natural habitats on Earth are sterile and that most habitats contain a great diversity of microorganisms, especially of prokaryotes (Bacteria and Archaea). When considering microbial habitats one should also recall the concept of the "microenvironment" (Madigan et al., 2000). Although habitats themselves can vary dramatically, for example, lake water and soil, conditions *within* any given habitat can also vary in both a spatial and temporal context. In fact, it is likely that the great physiological and phylogenetic diversity of prokaryotes we see today is the result of evolutionary pressure on these organisms to colonize every habitat and every microenvironment within each habitat that will support life.

1.1. EXTREME ENVIRONMENTS AND THEIR PROKARYOTES

In the last 30 years there has been great interest in probing the microbiology of "extreme" environments—environments in which one or more physiochemical variables are so far removed from that humans would find "normal" as to render the habitat unsuitable for human life. Organisms residing, and often thriving, in extreme environments are called *extremophiles* (Madigan and Marrs, 1997), and the list of such organisms has grown rapidly in recent years with some spectacular examples now existing in laboratory culture (see the book by Horikoshi and Grant, 1998, and Table 1).

For the purpose of this overview of the habitats of extremophiles, the latter term will be defined as prokaryotes that grow *optimally* under one or more of the following conditions: temperatures greater than 45°C or lower than 15°C; a pH of greater than 8 or lower than 6; salinities of greater than 10% NaCl; or hydrostatic pressure of greater than 200 atm. This definition is admittedly rather arbitrary but encompasses the consensus of extremophilic microorganisms being studied today. Some might argue that the atmosphere is also an extreme environment (because of drying, radiation effects, etc.), but air as a bacterial habitat will not be considered here.

How extreme can extreme environments get and still support life? A feeling for this can be obtained from the data of Table 1 that lists the current "record-holders" in terms of prokaryotes inhabiting extreme environments. For each of the categories mentioned above (temperature, pH, salinity, and pressure) the organism with the most dramatic requirement for a particular environmental extreme for optimal growth is listed. As can be seen, organisms with growth temperature optima as high as 106°C or as low as 4°C; pH optima at 10 or at 0.7; salinity optima near saturation; or pressure optima of over 700 atmospheres, have all been characterized in recent years. These are, of course, the special cases, as most extremophiles do not have physiological requirements this extreme. However, the data of Table 1 clearly reveal the amazing capabilities of prokaryotes to colonize any environment that is compatible with life. And as new extreme environments are discovered and explored it is likely that the physiochemical limits to life will be found to be even broader than those shown in Table 1.

TABLE 1. Classes and examples of extremophiles[a]

Extreme	Descriptive Term	Genus/ species	Minimum	**Optimum**	Maximum	Reference
Temperature						
High	Hyperther- mophile	*Pyrolobus fumarii*	90°C	**106°C**	113°C	Blöchl et al. (1997)
Low	Psychrophile	*Polaromonas vacuolata*	0°C	**4°C**	12°C	Irgens et al. (1996)
pH						
Low	Acidophile	*Picrophilus oshimae*	−0.06	**0.07 (60°C)**[c]	4	Schleper et al. (1995)
High	Alkaliphile	*Natronobacterium gregoryi*	8.5	**10 (20% NaCl)**[d]	12	Tindall et al. (1984)
Pressure	Barophile	MT41 (Mariana Trench)[b]	500 atm.	**700 atm. (4°C)**	>1000 atm	Yayanos et al. (1981)
Salt (NaCl)	Halophile	*Halobacterium salinarum*	15%	**25%**	32% (saturation)	Grant and Larsen (1989)

[a] In each category the organisms listed is the current "record holder" for requiring a particular extreme condition for growth
[b] Strain MT41 does not yet have a formal genus and species name
[c] *P. oshimae* is also a thermophile, growing optimally at 60°C
[d] *N. gregoryi* is also an extreme halophile, growing optimally at 20% NaCl

1.2   ADAPTATION TO EXTREME ENVIRONMENTS

It is necessary to distinguish between organisms *optimally adapted* to a particular extreme (or extremes) and those that merely *tolerate* the extreme condition. It is a tenet of extremophilic microbiology that true extremophiles are those that actually *require* the extreme for growth and survival (Madigan and Marrs, 1997). Thus, one can contrast the extremophile *Halobacterium salinarum*, which requires at least 15% NaCl for growth (however 25% NaCl is optimum, see Table 1), with *Staphylococcus aureus*, a nonextremophile that can tolerate up to 12% NaCl but which shows neither a growth requirement for, nor growth stimulation by, salt. Moreover, it is likely that most extremophiles, especially the rather spectacular examples (Table 1), are organisms that experience the physiochemical extreme on a *constant basis*. Indeed it is in large part this constancy that restricts the indigenous microflora of an extreme environment to an extremophilic lifestyle; organisms not optimally adapted to the conditions in the habitat cannot be maintained there.

Examples of the latter principle are many. Although *Bacillus stearothermophilus,* a thermophile capable of growth between 35° and 65°C, can be isolated from virtually any soil, *Pyrolobus fumarii*, a hyperthermophilic archaeon (Table 1), is restricted to volcanic habitats where heat is both abundant and constant. Psychrotolerant microorganisms, both prokaryotes and eukaryotes, can easily be isolated from various environments but true psychrophiles, organisms with growth temperature optima below 15°C (Brenchley, 1996), cannot; the latter appear limited to constantly cold environments such as deep alpine lakes and the Arctic and Antarctic (Bowman et al., 1997; Russell, 1990). In addition, significant populations of strongly acidophilic or alkaliphilic prokaryotes (see Table 1) are also, with rare exception, restricted to habitats that are extremely acidic or alkaline (and constantly so), respectively. Halophiles may pose an exception to this rule, as even organisms as halophilic as *Halobacterium* (Table 1) have been reported from marine (3–4% NaCl) environments (Oren, 1999). Presumably, such cells are present in only small numbers there and survive in microenvironments of higher salt content.

Thus, in contrast to truly extreme environments that are colonized by extremophiles the likes of those described in Table 1, non-extreme environments or environments that experience an extreme condition only on a transient basis, rarely contain a truly extremophilic microflora. This facts suggests that the evolution of extremophiles to the point of absolutely requiring one or more physiochemical extremes for their metabolism, gene expression, and reproduction, can only occur in environments in which the extreme condition is an integral and constant part of the ecosystem.

2. Temperature as an Environmental Extreme

Temperature is one of the most, if not *the* most, important environmental factor influencing the growth and survival of microorganisms (Brock, 1967; 1978). This is especially true of high temperatures because of the inherent thermal lability of the biomolecules of life, in particular, proteins. For every organism there is a set of temperatures called the *cardinal temperatures* (minimum, maximum, and optimum) characteristic of that particular organism. Table 1 lists the cardinal temperatures for two dramatic extremophiles, the hyperthermophile *Pyrolobus fumarii* and the psychrophile *Polaromonas vacuolata*. We consider the nature of their hot and cold habitats here.

2.1 THERMAL HABITATS

Thermal environments are the result of one or more of the following: solar heating, geothermal activity, intense radiation, combustion, or the activities of humans such as power plant and industrial discharges and residential and commercial hot water heaters. Temperatures in solar heated soils can reach as high as 60–70°C, and combustion processes, such as occurs in compost piles or in silage, are often in this same range due to the physiological activities of the microorganisms themselves (Brock, 1978). However, it is geothermally heated environments that yield the most spectacular examples of thermophiles (optima between 45–80°C) and hyperthermophiles (optima greater than 80°C) (Table 1 and Stetter, 1998; 1999), and we focus on these here.

Many hot springs have temperatures near boiling and steam vents (fumaroles) can have temperatures as high as 500°C (Brock, 1978). In addition to temperature, hot springs can also vary dramatically in terms of their pH and chemical composition (Brock, 1978). Thus the water chemistry of a 90°C hot spring in Yellowstone (USA) may or may not resemble that of a 90°C hot spring in Iceland or New Zealand. Although hot springs in a given location with similar underlying geology and water chemistry usually support a common microflora, the fact that thermal springs can vary dramatically in both temperature and chemistry (Brock, 1978) allows for great variability in the precise characteristics of any particular spring.

Many hot springs also develop overflow channels that route water away from the source. In these cases the outflow forms a natural thermal gradient where different populations of thermophilic prokaryotes can thrive within the temperature range that supports their growth. If one takes into account both the variable chemistry of hot springs along with the fact that many of them form steep thermal gradients, it is easy to see that the number of distinct habitats possible in thermal regions is virtually unlimited. Therefore, the large number of thermophilic and hyperthermophilic prokaryotes that have already been cultured (Stetter, 1998; 1999) or are otherwise known to exist (Barns et al., 1994; Hugenholtz et al., 1998), should not be too surprising.

In addition to terrestrial hot springs, *undersea* hot springs (hydrothermal vents and active seamounts) are habitats for prokaryotes. The laboratory of Karl Stetter in Regensburg (Germany) has lead the way in isolating new prokaryotes from these environments (Stetter, 1998; 1999). "Black smoker" vents emit the hottest water, forming large structures called "chimneys" from the metal sulfides that precipitate when the metal-rich hydrothermal fluid mixes with cold seawater. Within the rather thin chimney walls lies a huge thermal gradient (hydrothermal fluid on the inside of the chimney ranges from 200–400°C while seawater on the outside is about 2°C), and this gradient is a natural incubator for distinct populations of hyperthermophiles differing in their temperature requirements (Harmsen et al., 1997). Descriptions of the organisms isolated from hydrothermal vents (as well as hyperthermophiles from terrestrial hot springs) can be found in any of the recent reviews by Stetter (1998; 1999) (an excellent review of habitats and organisms is present in Stetter, 1999). Currently, the hyperthermophile *Pyrolobus fumarii* (Table 1) is the most thermophilic of all known prokaryotes; *P. fumarii* grows up to a temperature of 113°C and amazingly, can survive in an autoclave (121°C) for over 1 hour (Blöchl et al., 1997).

Active seamounts are locations where volcanic lava is emitted directly onto the sea floor. Despite the fact that the lava itself is too hot for life, Stetter (1999) reports the presence of large numbers of hyperthermophiles in seawater immediately surrounding active seamounts. Presumably these organisms inhabit the hot seawater that mixes with the lava. In addition to deep sea vents and seamounts, it is likely that hyperthermophilic prokaryotes are also present deep within the earth where geothermal heating supports their lifestyle (Stetter, 1999). Although gaining access to such putative organisms has proven problematic, deep drilling projects planned for the future should open this exciting microbial habitat to molecular probing and enrichment culture exploration.

Although the upper temperature for microbial life is not yet known, there is good reason to believe that it will be around 140–150°C (Stetter, 1999). Above about 150°C many critical biomolecules, for example ATP, are highly unstable, and thus life at 200°C or 300°C, at least *life as we know it*, seems highly unlikely (Stetter, 1999; Wiegel and Adams, 1998). Thus, environments above about 150°C are likely sterile and are perhaps the only naturally sterile habitats on Earth.

2.2. COLD HABITATS

Much of the earth's surface experiences low temperatures. For example, the oceans, which make up over half of the earth's surface, have an average temperature of 5°C and water in the depths of the oceans average 1–3°C (Russell, 1990). Vast land masses in the Arctic and Antarctic are permanently frozen or thaw for only a brief period in summer (Priscu, 1998). The upper few centimeters of soil in temperate regions can also freeze in winter and beneath this layer, soil can remain cool on a permanent basis. Some lakes in northern temperate regions remain frozen for half the year or more as well. All of these naturally cold environments have been shown to contain microbial populations, primarily prokaryotes and various algae (eukaryotic phototrophs) (Brenchley, 1996; Russell, 1990). Even permanently frozen liquids such as lake ice or sea ice can contain diverse microbial populations (Bowman et al., 1997), and microbial growth can occur here in pockets of liquid water. In addition to these naturally cold environments, artificially cooled environments such as household and commercial refrigerators can be teeming with microbial life depending on cleanliness practices and the nature of the stored products.

Psychrotolerant and *psychrophilic* microorganisms dominate cold environments. Psychrotolerant organisms, whose growth temperature optima are generally above

25°C, are widely distributed and include common food spoilage organisms and the typical bacterial isolates obtained from enrichment cultures using temperate soil or water as inocula and incubated at temperatures of 15°C (Russell, 1990; 1993). However, as previously mentioned, truly psychrophilic microorganisms (optima below 15°C) are more rare and seem, as far as is known, to be limited to constantly cold environments (Bowman et al., 1997).

The distribution and ecology of psychrophiles remains poorly understood, possibly because these organisms are easily killed by warming; this is especially likely for species with growth temperature optima near 0°C (Table 1). Until recently, the majority of the scientific literature on microbial growth in the cold has dealt with applied problems, in particular, food storage and spoilage (Brenchley, 1996). Naturally, this has placed the emphasis of our understanding on psychrotolerant rather than psychrophilic microorganisms. However, with a greater appreciation for the potential caveats of isolating psychrophiles (for example, by not letting samples warm up) and the greater accessibility of appropriate inocula (especially from Antarctic environments), great strides in our understanding of microbial psychrophiles will likely emerge in coming years; indeed some significant progress in this direction has already been made (Bowman et al., 1997).

3. Acidic and Alkaline Habitats

Some extremophiles have evolved to grow best near the ends of the pH scale: these are the *acidophiles* (Norris and Johnson, 1998) and the *alkaliphiles* (Horikoshi, 1998). Many environments on Earth are moderately acidic or alkaline from the physiological activities of microorganisms. For example, the production of organic acids in fermentations or the intense phototrophic CO_2 fixation by cyanobacteria and algae can drive pH values to as low as 3.5 or as high as 10, respectively. However, one rarely finds significant populations of extreme acidophiles or alkaliphiles (Table 1) in such environments. Instead, pH extremophiles are restricted to environments in which the presence of mineral acids or abundant carbonates push the pH to extremely low or extremely high values, respectively, and maintains the pH at these extreme values on a constant basis. In such environments one finds a variety of prokaryotes thriving in solutions the equivalent of battery acid or soda lime.

3.1 ACIDIC ENVIRONMENTS

Highly acidic environments result naturally from the biogeochemical activities of bacteria (Fenchel et al., 1998; Madigan et al., 2000), such as from the oxidation of SO_2 and H_2S produced in hydrothermal vents and sulfur springs, and from the oxidation of ferrous iron. For example, the extremely acidophilic iron-oxidizing bacterium *Thiobacillus ferrooxidans* generates acid by oxidizing Fe^{2+} to Fe^{3+}; the latter precipitates out as $Fe(OH)_3$, releasing protons in the process ($Fe^{3+} + 3H_2O \longrightarrow Fe(OH)_3 + 3H^+$). *T. ferrooxidans*, and the related bacterium *T. thiooxidans*, can also generate acid from the oxidation of sulfide or elemental sulfur to sulfuric acid ($HS^- + 2O_2 \longrightarrow SO_4^{2-} + H^+$). Both of these chemolithotrophs are particularly active in surface coal mining operations where exposure of pyrite (FeS_2) in the coal seam to oxygen triggers acid production from the metabolic activities just described. Runoff from surface mines can have a pH of less than 2 and is also metal rich, fueling conditions for further acidophile activity (Brock, 1978).

The most acidophilic of all known prokaryotes is the archaeon *Picrophilus oshimae*, whose pH optimum for growth is just 0.7 (Schleper et al., 1995; see also Table 1). *P. oshimae* is also a thermophile (temperature optimum, 60°C) so this organism must be stable to hot as well as acidic conditions. Cultures of *P. oshimae* were isolated from an extremely acidic (< pH 1) solfatara in Italy, and studies of the cytoplasmic membrane of cells of this organism have shown that it is capable of maintaining its structural integrity only in solutions having a pH of 4 or below (Van de Vossenberg et al., 1998); obviously, *P. oshimae* is well adapted to its strongly acidic environment.

3.2 ALKALINE ENVIRONMENTS

Extreme alkaliphiles live in soils laden with soda (natron) or in soda lakes where the pH can rise to as high as 12 (Jones et al., 1998). These environments typically contain only traces of Mg^{2+} and Ca^{2+}, cations that would otherwise complex with the carbonate to form insoluble precipitates. By contrast, the dominant cation in these environments is Na^+ and the highly soluble Na_2CO_3 drives the pH up to extreme values. *Natronobacterium gregoryi* (Table 1), for example, was isolated from Lake Magadi, a soda lake of about pH 11 located in the Rift Valley of Africa; *N. gregoryi* grows optimally at a pH of about 10 (Table 1) (Horikoshi, 1998). *N. gregoryi* is also a halophile, growing optimally at about 20% NaCl (Table 1).

Many other alkaline lakes exist in the African Rift Valley, some of which contain little or no NaCl (Jones, 1998). Because primary productivity in these ecosystems is extremely high from the photosynthetic activities of purple bacteria (Imhoff, 1992) and cyanobacteria (Jones, 1998), these lakes are likely habitats for a wide variety of prokaryotes differing in their pH and/or salinity requirements depending on the particular soda lake they inhabit. Indeed, as the microbiology of these lakes becomes better understood, this appears to be the case; cultures of several Bacteria and Archaea have now been isolated from various Rift Valley soda lakes (Jones et al., 1998), and other interesting prokaryotes have been detected by molecular probing (Grant et al., 1999). In the author's laboratory several anoxygenic phototrophic purple bacteria have been isolated from alkaline environments, including one purple nonsulfur bacterium capable of growth up to pH 12.5; this discovery indicates that other anoxyphototrophs besides *Ectothiorhodospira* (Imhoff, 1992) may play a role in the productivity of these lakes as well.

Alkaline soda lakes are also present in the Wadi Natrun region of Egypt (Imhoff, 1992), in eastern Russia, and in the western United States. In the latter connection, Mono Lake (California) and Big Soda Lake (Nevada) are large, moderately saline (~5–7% NaCl) and alkaline (pH 9–10) lakes, and studies of major biogeochemical events in these lakes (Oremland et al., 1993) strongly suggest that diverse and highly active microbial populations are present.

4. Hypersaline Habitats

Many prokaryotes have an absolute requirement for NaCl. This is true of virtually all species indigenous to marine environments and saline lakes as well as to many organisms isolated from highly alkaline environments (see Section 3.2). However, extreme halophiles like *Halobacterium* show *very high* salt requirements, even thriving in solutions saturated with NaCl (Table 1). The requirement for salt by extreme halophiles is such that these organisms often die in solutions containing less than 15% NaCl (Madigan and Oren, 1999).

Extremely halophilic microorganisms abound in hypersaline waters such as the Dead Sea, the Great Salt Lake, and solar salt evaporation ponds. Indeed, these environments are often colored red by the dense microbial communities of pigmented halophiles such as *Halobacterium* (Javor, 1989; see also photos of halophile habitats in Madigan et al., 2000) that are present. Other habitats for halophilic microorganisms include highly salted foods, saline soils, and underground salt deposits (Madigan and Oren, 1999).

A very large number of extremely halophilic Bacteria, Archaea, and Eukarya have been isolated and grown in laboratory culture (Grant et al., 1998; Kamekura, 1998). From studies of the molecular phylogeny of extreme halophiles it is clear that these organisms are highly diverse and that extreme halophily has arisen independently in several phylogenetic lineages. The biology of halophilic microorganisms has recently been reviewed (Ventosa et al., 1998) and the microbiology and biogeochemistry of halophiles is well described in an excellent recent edited volume (Oren, 1999).

5. Hydrostatic Pressure and Microbial Life

For each increasing ten meters of depth in a water column, pressure increases by about one atmosphere; thus, an organism growing at a depth of 5000 meters experiences hydrostatic pressure of some 500 atmospheres. Like for the other environmental extremes discussed thus far, organisms that merely tolerate the extreme, pressure in this case, as well as organisms that grow best or actually *require* pressure, are known. Organisms that grow best under pressure are known as *barophiles*, and extreme barophiles are the most interesting in this regard as they require extremely high pressures for growth (Table 1). Strain MT41, for example, a bacterium isolated from marine sediments in the Mariana Trench near the Philippines (a depth of greater than 10,000 meters), requires a minimum of 500 atmospheres for growth and grows optimally at a pressure of 700 atmospheres (and at a temperature of 4°C because strain MT41 is also a psychrophile, see Table 1) (Yayanos et al., 1981).

Moderate barophiles growing optimally at 200–300 atmospheres abound in the oceans in surface waters to several thousand meters depth, but most of these lack an obligate growth requirement for pressure. Extreme barophiles, at least from what is known thus far, seem restricted to great depths in the ocean (Yayanos, 1998). Curiously, however, organisms like strain MT41 survive complete decompression for relatively long periods (Yayanos and Dietz, 1983), indicating that although extreme barophiles require high pressure for growth, the absence of hydrostatic pressure does not kill cells instantly. A nice summary of the effects of pressure on microbial life in the deep sea is available in a chapter by Yayanos (1998).

6. Conclusions

The 3.8 billion years that prokaryotes have lived on Earth have given them ample time to evolve the characteristics necessary to colonize every habitat compatible with the biomolecules of life. And while enrichment culture microbiology has given us remarkable examples of "life on the fringe" (for example, see Table 1), molecular probing using phylogenetic probes (Barns et al., 1994; Hugenholtz et al., 1998) has pointed to the staggering conclusion that the organisms we now have in culture are "only the tip of the iceberg".

Indeed, prokaryotic diversity is far greater than current microbial culture collections suggest. And, as this chapter has tried to emphasize, this enormous diversity is a product of the huge variability in bacterial habitats coupled with the efforts of prokaryotes to colonize them. As we enter the new millennium, most prokaryotes, indeed probably greater than 99% of all prokaryotic species, have still evaded the enrichment culture microbiologist. Obtaining cultures of these elusive prokaryotes will be a major challenge for the next generation of microbiologists, and for the next decade or two, "culture the uncultured" should be the motto of microbiologists interested in bacterial diversity.

In my opinion, a full assessment of bacterial diversity will require at least four milestones beyond just molecular probing—although the latter is a spectacular new tool for understanding the magnitude of bacterial diversity, it unfortunately doesn't yield laboratory cultures. These milestones are: (1) a better understanding of the physiochemical conditions of bacterial habitats and how best to reproduce such in the laboratory (for example, are the "limits" shown in Table 1 really the limits to microbial life, and how important to bacterial diversity are microenvironments in microbial habitats?); (2) a better appreciation for natural inter-microbial interactions that may be a key to the culture of some (perhaps many?) prokaryotes; (3) an attitude among microbiologists that no prokaryote is inherently unculturable; and (4) more microbiologists willing to devote some or all of their efforts to the culturing of bacteria from nature.

The molecular ecologists have clearly opened the door to bacterial diversity and now it is time to probe environments for real live organisms. And, because in extreme environments the prokaryote is preeminent, it is likely that the continued study of these environments will yield many more examples of prokaryotes with interesting and perhaps previously unrecognized physiological properties and evolutionary histories.

7. Acknowledgements

The research of the author on extremophilic prokaryotes is supported by United States National Science Foundation grant OPP 9809195.

72

8. References

Barns S.M., Fundyga, R.E., Jeffries, M.W., and Pace, N.R. (1994). *Proc. Nat. Acad. Sci.* (USA). **91**: 1609–1613.

Blöchl, E., Rachel, R., Burggraf, S., Hafenbradl, D., Jannasch, H.W., and Stetter, K.O. (1997) *Extremophiles.* **1**:14–21.

Bowman, J.P., McCammon, S.A., Brown, M.V., Nichols, D.S., and McMeekin, T.A. (1997) *Appl. Environ. Microbiol.* **63**: 3068–3078.

Brenchley, J.E. (1996) *J. Indust. Microbiol.* **17**: 432–437.

Brock, T.D. (1978) *Thermophilic Microorganisms and Life at High Temperatures*, Springer-Verlag, New York.

Fenchel, T., King, G.M., and Blackburn, T.H. (1998) *Bacterial Biogeochemistry*, Academic Press, San Diego, CA.

Grant, W.D., Gemmell, R.T., and McGenity, T.J. (1998) in: K. Horikoshi and W.D. Grant (eds.) *Extremophiles—Microbial Life in Extreme Environments*, Wiley-Liss, New York, pp. 93–132.

Grant, W.D., and Larsen, H. (1989) in: J.T. Staley, M.P. Bryant, N. Pfennig, and J.G. Holt (eds.), *Bergey's Manual of Systematic Bacteriology*, Vol. 3, Williams and Wilkins, Baltimore, pp. 2216-2219.

Grant, S., Grant, W.D., Jones, B.E., Kato, C., and Lina, L. (1999) *Extremophiles* 3:139–145.

Harmsen, H.J.M., Prieur, D., and Jeanthon, C. (1997) *Appl. Environ. Microbiol.* 63:2876–2883.

Horikoshi, K. (1998) in: K. Horikoshi and W.D. Grant (eds.), *Extremophiles—Microbial Life in Extreme Environments*, Wiley-Liss, New York, pp. 155–179.

Horikoshi K. and Grant, W.D. (1998) *Extremophiles—Microbial Life in Extreme Environments*, Wiley-Liss, New York.

Hugenholtz P., Pitulle, C., Hershberger, K.L., and Pace, N.R. (1998) *J. Bacteriol.* 180:366–373.

Imhoff, J.F. (1992) in: A. Balows, H.G. Trüper, M. Dworkin, W. Harder, and K-H. Schleifer (eds.), *The Prokaryotes, 2nd Edition,* Springer-Verlag, New York, pp. 3222–3229.

Irgens, R.L., Gosink, J.J., and Staley, J.T. (1996) *Intl. J. Syst. Bacteriol.* 46:822–826.

Javor, B.J. (1989) *Hypersaline Environments–Microbiology and Biogeochemistry*, Springer-Verlag, Berlin.

Jones, B.E., Grant, W.D., Duckworth, A.W., and Owenson, G.G. (1998) *Extremophiles* 2: 191–200.

Kamekura, M. (1998) *Extremophiles* 2:289–295.

Madigan, M.T., and Marrs, B.L. (1997). *Scientific American* 276:82–87.

Madigan, M.T., Martinko, J.M., and Parker, J. (2000) *Brock Biology of Microorganisms*, 9th edition. Prentice Hall, Upper Saddle River, NJ.

Madigan, M.T., and Oren, A. (1999) *Curr. Op. Microbiol.* 2:265–269.

Norris, P.R., and Johnson, D.B. (1998) in: K. Horikoshi and W.D. Grant (eds.), *Extremophiles—Microbial Life in Extreme Environments*, Wiley-Liss, New York.

Oremland, R.S., Miller, L.G., Colbertson, C.W., Robinson, S.W., Smith, R.L., Lovely, D., Whiticar, M.J., King, G.M., Kiene, R.P., Iversen, N., and Sargent, M. (1993) in: R.S. Oremland (ed.), *Biogeochemistry of Global Change: Radiatively Important Trace Gases*, Chapman and Hall, New York, pp. 704–744.

Oren, A. (ed.) (1999) *Microbiology and Biogeochemistry of Hypersaline Environments*, CRC Press, Boca Raton, FL.

Priscu, J.C. (ed.) (1998) *Ecosystem Dynamics in a Polar Desert—The McMurdo Dry Valleys, Antarctica*, American Geophysical Union, Washington, D.C.

Russell, N.J. (1990) *Phil. Trans. R. Soc. Lond.* **B 326**: 595–611.

Russell, N.J. (1993 in: R.Guerrero and C. Pedrós-Alió (eds.), *Trends in Microbial Ecology*, Spanish Society for Microbiology, Barcelona. Schleper, C., Puhler, G., Holz, I., Gambacorta, A., Janekovic, D., Santarius, U., Klenk, H-P., and Zillig, W. (1995) *J. Bacteriol.* 177:7050–7059

Stetter, K.O. (1998) in: K. Horikoshi and W.D. Grant (eds.), *Extremophiles—Microbial Life in Extreme Environments*, Wiley-Liss, New York.

Stetter, K.O. (1999) in: J. Marti and G.J. Ernest (eds.), *Volcanoes and the Environment*, Cambridge University Press, Cambridge, England, in press.

Tindall, B.J., Ross, H.N.M., and Grant, W.D. (1984) *Syst. Appl. Microbiol.* **5**:41–57.

Van de Vossenberg, J.L.C.M, Driessen, AJ.M., Zillig, W. and Konings, W.N. (1998) *Extremophiles.* 2:67-74.

Ventosa, A., Nieto, J.J., and Oren, A. (1998) *Microbiol. Mol. Biol.Rev.* **62**:504–544.

Wiegel, J., and Adams, M.W.W. (eds.) (1998) *Thermophiles–The Keys to Molecular Evolution and the Origin of Life?* Taylor and Francis, London.

Yayanos, A.A. (1998) in: K. Horikoshi and W.D. Grant (eds.) *Extremophiles - Microbial Life in Extreme Environments*, Wiley-Liss, New York, pp. 47-92.

Yayanos, A.A. and Dietz, A.S. (1983) Sci. **220**: 497-498.

Yayanos, A.A., Dietz, A.S. and van Boxtel, R. (1981) Proc. Natl.Acad. Sci. USA **78**: 5212-5215.

Biodata of **Dr. Bernard Ollivier**, senior author of *"Anaerobes from Extreme Environments"*(with B.K.C. Patel and J.-L. Garcia).

Dr. B. Ollivier is a Research Director and Head of Program in the Laboratory of Microbiology of the Research Institute for Development - IRD in the University of Provence, Marseilles, France. He obtained his Ph.D. degree in Marseilles University, France in 1978. His areas of interest are: Microbial ecology of anaerobic digestion; Anaerobic extremophiles (halophiles and thermophiles); Microbiology of oil reservoirs. His specific research topic include: Methonogenesis and sulfate-reduction and Taxonomy of strictly anaerobic bacteria.
E-Mail: **ollivier@esil.univ-mrs.fr**

B. Ollivier

B.K.C. Patel

Biodata of **Dr. Bharhat K. C. Patel** co-author of *"Anaerobes from Extreme Environments"* (with B. Ollivier and J. -L. Garcia).

Dr. B. K.C. Patel in an Associate Professor at the School of Bimolecular and Biomedical Sciences, in Nathan Campus, Griffith University, Brisbane, Queensland, Australia. He received his Ph.D. from Otago University, NZ in 1980 and from Waikato University, NZ in 1984. His areas of interest and research are: Microbial ecology and biochemistry, Molecular microbiology, immunology and cell culture. The specific topics are: Molecular taxonomy of aerobic and strictly anaerobic bacteria, Microbiology of subterrestrial ecosystems (oil field reservoirs and the Great Australian Basin). Dr. Patel published 100 papers in International Journals.

E-mail: **B.Patel@sct.gu.education.au**

J. Seckbach (ed.), Journey to Diverse Microbial Worlds, 73-90.
© 2000 *Kluwer Academic Publishers. Printed in the Netherlands.*

Biodata of **Dr. Jean-Louis Garcia**, co-author of *"Anaerobes from Extreme Environments."* (with B. Ollivier and B.K.C. Patel).

Dr. Jean-Louis Garcia is the Research Director of the Research Institute for Development, and serves as the Head of Laboratory of Microbiology at the University of Provence, Marseilles, France. He received his DSc degree from Marseille University in 1978. Dr. Garcia is interested in microbial ecology of anaerobic digestion, denitrification in soil. His specific research topics include: Denitrification in rice paddy soils, metanogenic fermentation, taxonomy of strictly anaerobic bacteria, fermenters, sulfate reducers, methanogens and extremophiles. Dr. Garcia has published 100 papers in International Journals, and has published one book

E-Mail: **garcia@esil.univ-mrs.fr**

ANAEROBES FROM EXTREME ENVIRONMENTS

B. OLLIVIER,[1] B. K. C. PATEL[2] AND J.-L. GARCIA[1]

[1]*IRD, Laboratoire de Microbiologie, Université de Provence, CESB-ESIL case 925, 163 avenue de Luminy, 13288 Marseille cedex 9, France*
[2] *School of Biomolecular and Biomedical Sciences, Faculty of Science, Griffith University, Nathan 4111, Brisbane, Queensland, Australia*

1. Introduction

Extremophilic microorganisms belong to various physiological groups comprising thermophiles, hyperthermophiles, alkaliphiles, acidophiles, halophiles, psychrophiles, and barophiles. Research on extreme environments and their microflora was greatly stimulated by the dicovdry of the Archaea as a third form of life (Woese et al. 1990), as many extremophiles belong to the archaeal domain. Moreover, such environments, especially hydrothermal vents exhibit the physicochemical conditions considered as compatible with the origin of life. Research has focused on thermophilic and hyperthermophilic groups because they possess enzymes with enhanced thermostability and hence may have potential industrial applications (Adams 1993; Lowe et al. 1993). This paper describes the microbial diversity of anaerobes isolated from extreme environments. Fig. 1 and Fig. 2 show the phylogenetic distribution of extremophilic anaerobes amongst the *Bacteria* and *Archaea* domains, respectively.

2. Extreme Temperatures

Most prokaryotes are mesophilic with an optimum growth temperature between 25 and 40°C, but some can grow at higher or lower temperatures. Those that grow at temperatures between 45°C to 113°C (the upper known limit of life) are known as thermophiles, whereas those that grow at temperatures between 0 to 30°C are known as psychrophiles (Table 1). Thermophilic anaerobic prokaryotes are widespread in a variety of high temperature natural habitats that include (i) volcanic and geothermal ecosystems with temperature often higher than 100°C (Stetter et al. 1990), (ii) self heated environments such as compost and farm yard manures in which temperatures can reach values as high as 80-90°C (Canganella and Wiegel 1993) and (iii) mesophilic environments which includes soils and digestors in which their activities are probably restricted (Canganella and Wiegel 1993).

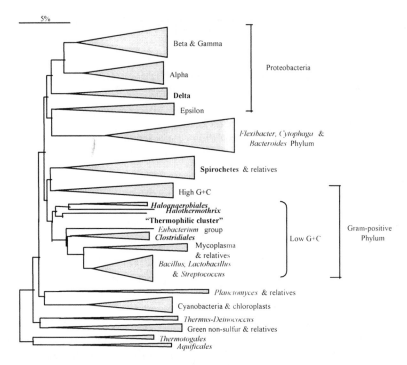

FIG. 1. Phylogenetic tree of Domain *Bacteria*. Extremophilic anaerobes are distributed in *Thermotogales*, *Clostridiales*, *Haloanaerobiales*, thermophilic cluster of low GC Gram positive phylum, Spirochetes and Delta Proteobacteria

Thermophiles can be divided into thermophiles *stricto sensu*, extreme thermophiles and hyperthermophiles (Table 1).

TABLE 1 . Definitions of extremophiles

Thermophiles		$T_{opt} > 50°C$	$T_{max} > 60°C$
Extreme thermophiles	$T_{min} > 35°C$	$T_{opt} \geq 65°C$	$T_{max} > 70°C$
Hyperthermophiles	$T_{min} > 60°C$	$T_{opt} \geq 80°C$	$T_{max} > 85°C$
Acidophiles	$pH_{min} > 0$	$2.5 \leq pH_{opt} \leq 3.0$	
Slightly acidophiles		$4.2 \leq pH_{opt} \leq 6.5$	
alkalitolerant organisms		$pH_{opt} < 8.5$	$pH_{max} \geq 9.0$
Alkaliphiles		$pH_{opt} \geq 8.5$	$pH_{max} \geq 10.0$
Halophiles			
slightly halophiles		$2\% \leq NaCl_{opt} \leq 5\%$	
moderate halophiles		$5\% \leq NaCl_{opt} \leq 20\%$	
extreme halophiles		$20\% \leq NaCl_{opt} \leq 30\%$	
Halotolerant organisms	$NaCl_{min} = 0\%$	$0\% \leq NaCl_{opt} \leq 5\%$	
Psychrophiles	$T_{min} = 0°C$	$T_{opt} < 25-30°C$	
Barophiles	$P_{min} > 0.1 MPa$	$P_{opt} = 10-50 MPa$	$P_{max} \leq 100 Mpa$

From Larsen (1962) and Wiegel (1998)

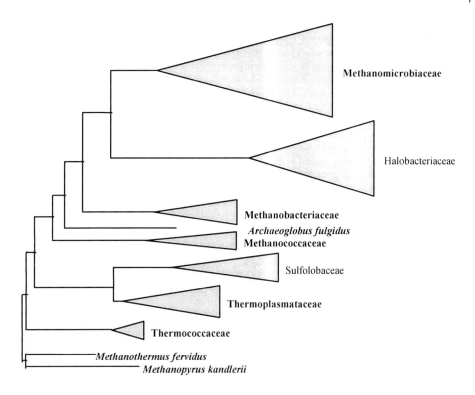

FIG. 2. Phylogenetic tree of Domain *Archaea*. Extremophilic anaerobes are distributed in all

2.1. THE THERMOPHILES

In this section we consider thermophiles *stricto sensu* and extreme thermophiles (Table 2). During the last ten years extensive isolation and characterization studies of thermophilic anaerobes from thermophilic and mesophilic ecosystems has led to the identification of several new heterotrophic clostridial species (Canganella and Wiegel 1993). Proteolytic anaerobic rods affiliated to the genus *Coprothermobacter* and *Thermobrachium* have been isolated from digestors (Ollivier et al. 1985) and from thermal springs (Engle et al. 1996), respectively. *Caloramator indicus* (Chrisostomos et al. 1996) is the most frequently species isolated from subsurface geothermal aquifers (Patel unpublished). *Clostridium thermocellum* is the most intensely studied species with respect to its cellulolytic activities at high temperatures.

The homoacetogen *Thermoanaerobacter* (formerly *Acetogenium*) *kivui* (Lowe et al. 1993), some members of the genera *Desulfotomaculum*, and *Methanobacterium* are regarded as thermophiles. *Desulfotomaculum kuznetsovii* (Nazina et al. 1989) and *D. thermoacetoxidans* (Min and Zinder 1990), which completely oxidize organic compounds in the presence of sulfate were isolated from thermal mineral waters and thermophilic anaerobic bioreactors, respectively. *Methanobacterium thermoautotrophicum* was found in ecosystems ranging from hydrothermal areas to sewage sludge (Lowe et al. 1993).

TABLE 2 . Comparative properties of genera of thermophilic anaerobes

Genera	Temperature range °C	optimum temperature °C	Habitat	Number of[a] species
Autotrophic *Bacteria*				
Desulfurobacterium	40-75	70	hydrothermal vent	1
Heterotrophic *Bacteria*				
Thermoanaerobacter	35-85	55-70	hot springs, oil well, volcanic lake, compost	10
Dictyoglomus	-	78	hot springs	2
Thermoanaerobacterium	35-75	60-65	hot springs	6
Thermotoga	46-90	65-70	oil wells, hot springs	4
Geotoga	30-60	45-50	oil wells	2
Petrotoga	35-65	55-60	oil wells	2
Fervidobacterium	41-80	65-70	hot springs, artesian basin	5
Thermosipho	35-80	70-75	hydrothermal vent	2
Anaerobaculum	28-60	55	oil well	1
Anaerobranca	30-66	57	hot spring	1
Caldicellulosiruptor	45-80	68-75	hot springs, lake	3
Caloramator	30-80	55-68	hot springs, artesian basin	3
Clostridium	26-66	48-65	compost, digesters	12
Carboxydothermus	40-78	70-72	hot spring	1
Coprothermobacter	35-70	55-63	digesters	2
Deferribacter	50-65	60	oil well	1
Moorella	43-65	58	hot springs	3
Thermoanaerobium	50-78	65-68	hot spring, artesian basin	1
Thermobrachium	35-75	62-66	hot springs, compost	1
Thermohydrogenium	45-75	65	industrial yeast biomass	1
Thermoterrabacterium	50-74	65	hot spring	1
Sulfate-reducing *Bacteria*				
Desulfotomaculum	22-74	54-68	hot springs, artesian basin digesters, deep subsurface	8
Desulfacinum	40-65	60	oil well	1
Thermodesulfobacterium	45-85	65-70	oil well	2
Thermodesulfovibrio	40-70	65	hot springs	1
Thermodesulforhabdus	44-74	60	oil well	1
Syntrophic *Bacteria*				
Thermosyntropha	52-70	60-66	hot spring	1
Methanogenic *Archaea*				
Methanobacterium	38-75	55-65	digesters, compost	4
"Methanothermobacter"	35-80	55-70	digesters, compost	2
Methanosarcina	30-69	50-55	digester	1
Methanosaeta	nd	60	digester	1

[a] **The number of species** validly described at this Table and the folloing ones, is up dated to March 1999.

Thermoanaerobacter, *Thermoanaerobacterium*, and *Dictyoglomus* are members of the "*Thermoanaerobiaceae*". The members of the genera *Thermoanaerobacter* and *Thermoanaerobacterium* can be routinely isolated from a broad range of habitats such as soils, thermal volcanic hot springs, oil-producing wells, sugar beet extraction juices, and sugar cane juices (Patel et al. 1991, Fardeau et al. 1993, Lee et al. 1993, Wynter et al. 1996). All isolates are reported to be saccharolytic heterotrophs and have the ability to use thiosulfate as an electron acceptor.

The order *Thermotogales* contains five genera, namely *Fervidobacterium*, *Thermotoga*, *Geotoga*, *Petrotoga*, and *Thermosipho* (Patel et al. 1985, Stetter et al. 1990, Davey et al. 1993, Lowe et al. 1993, Jeanthon et al. 1995, Ravot et al. 1995, Andrews and Patel, 1996). They have been isolated from hot springs, hot marine sediments, oil reservoirs, and subsurface artesian aquifers and are saccharolytic. Several members of this order produce L-alanine from glucose fermentation (Ravot et al. 1996) and utilize S° as electron acceptor to produce sulfide (Adams 1990, Lowe et al. 1993). *Thermotoga maritima* uses S° to oxidize H_2, which inhibits growth, so that H_2 oxidation appears to be a detoxification process.

2.2. THE HYPERTHERMOPHILES

Hyperthermophiles are fascinating microorganisms for microbiologists, physiologists, biochemists, and molecular biologists. The interest in these organisms further increased with the report of Stetter (1986) on the isolation of organisms that could grow above 100°C, outpassing temperature previously considered as the upper limit for life. Anaerobic hyperthermophiles, with the exception of *T. maritima*, *T. neapolitana*, and *T. hypogea*, are members of the domain *Archaea* (Stetter et al. 1990, Lowe et al. 1993) (Table 3). Anaerobic hyperthermophilic *Archaea* include heterotrophs, sulfate-reducing, and methanogenic bacteria, and are divided into seven orders named *Thermoproteales*, *Pyrodictiales*, *Thermococcales*, *Archaeoglobales*, *Igneococcales*, *Methanobacteriales*, and *Methanococcales*.

The environmental distribution of these organisms depends on their salt tolerance and therefore, some genera are adapted to the low-salt continental solfataric springs while others inhabit marine hydrothermal systems. *Desulfurococcus* and *Pyrobaculum* species have been isolated from terrestrial, solfataric muds, or thermal springs (Zillig et al. 1982, Huber et al. 1987) whereas *Pyrococcus* and *Pyrodictium* species have been isolated from marine solfataric mud (Stetter et al. 1983, Fiala and Stetter 1986). The two last species exhibit, together with *Hyperthermus butylicus*, the upper limit temperature for growth (105-110°C) ever reported for any anaerobic microorganism.

Methanogenic rod-shaped archaea, e.g. *Methanopyrus* and *Methanothermus*, have also been isolated from terrestrial, solfataric muds (Stetter et al. 1981, Huber et al. 1989), while *Methanococcus* species are more widespread in hydrothermal, submarine or deep-sea vents (Jones et al. 1983, Burggraf et al. 1990). Deep-sea hydrothermal vents are also inhabited by the sulfate-reducing *Archaeoglobus* spp. (Stetter et al. 1990, Lowe et al. 1993). Stetter et al. (1993) reported on presence of hyperthermophiles in the North Sea and Alaskan oil reservoirs, extending the range of non-volcanic ecosystems in which these bacteria were found. *Pyrococcus*, *Thermococcus*, and *Archaeoglobus* species are found in oil reservoirs along with representatives of domain *Bacteria*, e.g. *Thermotoga*

TABLE 3 . Comparative properties of genera of hyperthermophilic anaerobes

Genera	Temp. range °C	optimum temp. °C	optimum pH	Habitat	Number of species
AUTOTROPHS					
Acidophilic					
Stygiolobus	nd[a]-90	80	2.5-3	volcano	1
HETEROTROPHS					
Acidophilic					
Thermococcus	55-100	85-88	5.8-6	hydrothermal vents, solfataras	4
Pyrodictium	80-110	97-105	4.2-7.2	deep sea sediments.	3
Pyrolobus	90-113	106	5.5	hydrothermal vents	1
Caldococcus	55-100	88	6.4	hydrothermal vents	1
Thermodiscus	nd	88	5	Solfataras	1
Desulfurococcus	76-97	85-90	6-6.4	Solfataras	4
Stetteria	68-102	95	6	hydrothermal vents	1
Staphylothermus	62-98	92	4.5-8.5	hydrothermal vents	1
Thermoproteus	78-97	87-88	5-5.6	hot springs	2
Thermophilum	nd-100	85-90	5	solfataras	2
Alkaliphilic					
Thermococcus	56-103	75-85	8-9	hydrothermal vents	2
Neutrophilic					
Thermococcus	50-98	75-88	6.5-7.5	deep sea sediments hydrothermal vents	7
Pyrococcus	67-104	96-103	7-7.5	deep sea sediments	3
Hyperthermus	nd	95-107	7	hydrothermal vents	1
Thermoproteus	nd-100	75-88	6.5-7	hot springs	1
Pyrobaculum	75-104	100	7	solfataras	3
Ferroglobus	65-95	85	7	hydrothermal vents	1
Thermotoga[b]	55-90	80	7	hot springs	3
Sulfate-reducers					
Archaeoglobus	60-90	75-82	7	hydrothermal vents	3
Methanogens					
Methanothermus	55-97	83-88	6.5	solfataras	2
"*Methano-thermococcus*"	30-91	65-85	6.5-7.5	hydrothermal vent	2
"*Methano-caldococcus*"	50-85	85	6	deep sea sediment	1
"*Methanoignis*"	45-91	88	5.7	hydrothermal vent	1
Methanopyrus	84-110	98	6.5	hydrothermal vent	1

[a]nd = not determined ; [b]Domain *Bacteria*

hypogea (Fardeau et al. 1997). Hyperthermophiles, with the exception of the methanogens *Methanopyrus* and *Methanothermus* and the heterotrophic members of the *Thermoproteales*, have similar morphological features. Most of them are cocci or disc-shaped microorganisms (Stetter et al. 1990, Lowe et al. 1993) able to reduce S° to sulfide; in some case this reaction is energy-yielding (Adams 1990).

2.3. THE PSYCHROPHILES

Psychrophilic anaerobic microrganisms have been isolated from low temperature environments such as Antacrtica saline ice lakes, saline ponds, estuary sediments, deep granitic rock aquifers, or spoiled vacuum-packed refrigerated beef (Table 4). Two fermentative saccharolytic members of the genus *Clostridium*, *C. vincentii* (Mountfort et al. 1997) and *C. estertheticum* (Collins et al. 1992), grew at temperatures ranging from 1 to 15°C and 0 to 24°C, respectively, and produced acetate and butyrate as the major volatile fatty acids. *Clostridium vincentii* was isolated from the sediment below a cyanobacterial mat in a low salinity pond in Antarctica. It used N-acetyl glucosamine, a cyanobacterial wall constituent, and might therefore be of ecological importance in its environment for decomposing organic matter.

The homoacetogenic psychrophilic bacteria studied to date are members of the genus *Acetobacterium* (Zhilina et al. 1995). Beside their ability to ferment sugars and lactate to acetate, they also have the ability to oxidize H_2 with concomitant reduction of CO_2 to acetate. A gas-vacuolated, sulfate-reducing bacterium, *Desulforhopalus vacuolatus* isolated from a temperate estuary grew optimally at 18-19°C. It incompletely oxidized propionate and lactate to acetate (Isaksen and Teske 1996).

Methanogenic degradation of organic matter has been found to occur under psychrophilic conditions in deep lakes or deep granitic rock aquifers at temperatures lower than 20°C, but no psychrophilic methanogens have been phenotypically characterized so far from these environments. Two methanoarchaea have been isolated from Ace Lake in Antarctica. One is a methylotroph, *Methanococcoides burtonii*, which grows optimally at 23.4°C (Franzmann et al. 1992), while the second, *Methanogenium frigidum*, is a hydrogenotroph and exhibits its optimum growth at 15°C (Franzmann et al. 1997).

TABLE 4. Comparative properties of species of psychrophilic anaerobes

Species	temperature range °C	optimum temperature °C	Habitat
Heterotrophs			
Clostridium estertheticum	1-15	nd	refrigerated beef
Clostridium vincentii	0-24	12	saline pond
Acetobacterium bakii	1-30	20	low temp. environ.
Acetobacterium paludosum	1-30	20	low temp. environ.
Acetobacterium finetarium	1-35	30	low temp. environ.
Sulfate-reducers			
Desulforhopalus vacuolatus	0-24	18-19	estuary sediment
Methanoarchaea			
Methanogenium frigidum	0-17	15	saline ice lake
Methanococcoides burtonii	0-29	23	saline ice lake

3. Extreme pH

3.1. THE ALKALIPHILES

Alkalophilic anaerobic microorganisms have been isolated from a great variety of environments including soda and salt lake sediments, river and marine sediments, sewage sludge, microbially heated compost, hot springs, and subterrestrial environments (Table 5). Soda and salt lake sediments represent so far the major source for isolation of mesophilic alkaliphilic anaerobes. Studies on the microbial communities from such environments have received more attention during the past decade. These communities include key trophic groups involved in the decomposition of organic matter mainly derived from cyanobacteria (Zhilina and Zavarzin, 1994). Fermentative, homoacetogenic, methanogenic, and sulfate-reducing microorganisms have been recovered from these ecosystems. Interestingly, no anaerobic acetate oxidizers have been isolated from soda lakes, indicating that the mechanism of organic matter mineralization in such ecosystems is still unclear.

Amongst the alkaliphilic heterotrophs, *Spirochaeta* species produces acetate, ethanol, lactate, and hydrogen from sugar metabolism (Zhilina et al. 1996b), whereas *Natroniella acetigena,* which grows optimally at pH 9.7-10, ferments a limited range of substrates such as lactate and ethanol to acetate (Zhilina et al. 1996a). Hydrogen and formate are the only substrates oxidized by the sulfate-reducing bacterium *Desulfonatronovibrio hydrogenovorans* (Zhilina et al. 1997), whereas *Desulfonatronum palustre* uses ethanol (Pikuta et al. 1998). Hydrogen has also been reported as an electron donnor for methanogenesis by *Methanobacterium alcaliphilum*. Methylotrophic methanogens such as *Methanosalsus zhilinae* (Mathrani et al. 1988) and *Methanohalophilus oregonense* (Liu et al. 1990), which grow optimally around pH 9.0, were isolated from alkaline lakes. In these environments, sulfide produced by the sulfate-reducing bacteria can be oxidized by the phototrophic anoxygenic *Halorhodospira* species. Alkalitolerant heterotrophs belonging to the genera *Haloanaerobium* and *Acetohalobium* have been isolated from neutrophilic salt lakes (Zhilina and Zavarzin 1990, Tsai et al. 1995).

Anaerobic alkalithermophilic bacteria are comprised of physiologically different genera and species, but all are members of the Gram-positive low G+C DNA containing phylum. The most alkaliphilic anaerobic thermophiles are the spore-former *Clostridium paradoxum* (Li et al. 1993) and the non-spore-former *Clostridium thermoalcaliphilum* (Li et al. 1994). They are saccharolytic, grow optimally at pH from 9.2 to 10.1 and have been isolated from various sewage facilities. Another saccharolytic alkalitolerant anaerobe (pH optimum 8.1), *Caloramator indicus,* isolated from the Indian Artesian Basin (Chrisostomos et al. 1996) is a very close phylogenetic relative of *Thermobrachium celere. Thermobrachium* and *Caloramator* isolates appear to dominate deep subsurface thermal aquifers (Patel, unpublished).

Proteolytic activities were reported from the alkalitolerant *Anaerobranca horikoshi* isolated from water and soil in the Yellowstone National Park (Engle et al. 1995) and *Thermobrachium celere* isolated from a great variety of environments including hot springs (Engle et al. 1996). Svetlitshnyi et al. (1996) reported on *Thermosyntropha lipolytica* which grew optimally between pH 8 and 9 on volatile fatty acids from C4 to

C18 when associated with an hydrogenotrophic partner. The most thermophilic (hyperthermophilic) alkalitolerant (optimum pH: 8.0-9.0) anaerobes belong to the genus *Thermococcus*, domain *Archaea*. They were isolated from hydrothermal vents, use proteinaceous compounds, and are able to reduce elemental sulfur to sulfide. We recently isolated from an oil reservoir an alkalitolerant (optimum pH: 8.0) hyperthermophilic species of the genus *Thermotoga*, domain *Bacteria*, phylogenetically closely related to *T. hypogea* (Fardeau et al. 1997). Similarly to *T. hypogea*, this isolate reduced thiosulfate to sulfide.

Among the other anaerobic thermophilic *Archaea*, the most alkaliphilic methanogens known so far are the methylotrophic *Methanosalsus zhilinae* (Mathrani et al. 1988), the hydrogenotrophic *Methanobacterium thermoflexum* (Kotelnikova et al. 1993), *Methanobacterium wolfei* (Winter et al. 1984), and some *Methanobacterium thermoautotrophicum* strains previously described as *M. thermoalcaliphilum* (Zeikus and Wolfe 1972, Blotevogel and Fischer, 1985, Blotevogel et al. 1985).

TABLE 5 . Comparative properties of genera of alkaliphilic anaerobes

Genera	pH range	pH optimum	Habitat	Number of species
MESOPHILIC				
Heterotrophs				
Natrionella	8.1-10.7	9.7-10	soda lakes	1
Acetohalobium	nd	7.4-8	saline lakes	1
Haloanaerobium	5.8-10	6.7-7.4	saline lakes	2
Spirochaeta	7.9-10.7	8.7-9.7	alkaline lakes	3
Sulfate-reducers				
Desulfonatronovibrio	8-10.2	9.5-9.7	soda lakes	1
Desulfonatronum	7.5-10	9.5	soda lakes	1
Methanoarchaea				
Methanobacterium	6.5-9.2	7.8-9.1	soda lake, subterrestrial rock	2
Methanosalsus	nd	9.2	saline lakes	1
Methanolobus	7.6-9.4	7.2-9.2	marine sediments saline aquifer	2
Phototrophs				
Halorhodospira	nd	7.5-9.5	soda lakes	5
THERMOPHILIC				
Heterotrophs				
Clostridium	7-11	9.2-10.1	sewage plants	2
Anaerobranca	6.9-10.3	8.5	hot springs	1
Thermosyntropha	7.1-9.5	7.6-8.9	soda lake	1
Thermobrachium	5.4-9.5	8.2	ubiquitous	1
Caloramator	6.2-8.9	8.1	deep aquifer	1
Thermoanaerobacter	4.5-9.8	5.5-8.5	hot springs	2
Methanoarchaea				
Methanobacterium	6-10	7.5-8.5	digestor	2

3.2. THE ACIDOPHILES

Most of the anaerobic acidophilic microorganisms are heterotrophic hyperthermophiles and are members of domain *Archaea* (Table 3). The exception is *Stygioglobus azoricus* (Segerer et al. 1991), which is the most acidophilic anaerobic microorganism known to date (optimum growth at pH 2.5-3.0). It uses elemental sulfur to oxidize only H_2 and is therefore considered to be a chemolithotroph. The other hyperthermophilic acidophiles are considered moderate to slightly acidophiles as they have an optimum pH for growth from 4.2 to 6.5. They generally grow on proteinaceous compounds such as peptides and reduce elemental sulfur to sulfide, with some also being able to oxidize hydrogen. *Thermoproteus tenax* is the only thermophilic acidophile reported that uses a wide range of substrates including sugars (glucose, starch), alcohols (ethanol, methanol), and organic acids (malate, fumarate) (Zillig et al. 1981).

4. Extreme salinity: the halophiles

The inland salt lakes of the world such as the Dead Sea or the Great Salt Lake are inhabited by halophilic microorganisms which require NaCl for growth. Such environments are exposed to intensive evaporation and can therefore become extremely saline. They show a great variability in ionic composition, total salt concentration, and pH. Human activity also creates highly saline habitats such as the solar salterns, which can be saturated in NaCl.

Sulfate is an important electron acceptor involved in the mineralization of organic matter in hypersaline ecosystems. However, at salinities higher than 15%, this mineralization is limited partly by poor rates of sulfate reduction and the absence of methanogenesis from H_2 and acetate (Oren 1988, Ollivier et al. 1994). Accordingly the accumulation of H_2 and volatile fatty acids is observed, thus indicating that catabolism via interspecies H_2 transfer hardly occurs in hypersaline environments.

Halophilic organisms can be divided into three groups on the basis of their growth responses to salt concentration (Larsen 1962): (i) the slight halophiles (optimum growth at 2-5% NaCl, 0.34-0.85 M), (ii) the moderate halophiles (optimum growth at 5-20% NaCl, 0.85-3.4 M), and (iii) the extreme halophiles (optimum growth at 20-30% NaCl, 3.4-5.1 M).

Fermentative halophilic anaerobes belong mainly to the order *Haloanaerobiales*, which include two families, the *Haloanaerobiaceae* and the *Halobacteroidaceae* (Oren 1992, Patel et al. 1995). All species ferment carbohydrates except the homoacetogen *Acetohalobium arabaticum* (Zhilina and Zavarzin 1990), which reduces CO_2 to acetate and grows on betaine and trimethylamine. *Halocella cellulolytica* is the only cellulolytic microorganism so far described within the *Haloanaerobiales* (Simankova et al. 1992). *Sporohalobacter* and *Orenia* species differ from all other species in being sporogenous. Most of bacterial species are considered as moderate halophiles with optimum growth occurring from 3 to 15% NaCl. *Halobacteroides lacunaris* (Zhilina et al. 1991), *Haloanaerobacter chitinovorans* (Liaw and Mah 1992), *Acetohalobium arabaticum,* and *Haloanaerobium lacusroseus* (Cayol et al. 1995) can be considered as extreme halophilic bacteria, since they grow optimally at 18% NaCl. *Halothermothrix orenii* is the only strict anaerobe described (Cayol et al. 1994). A second thermophilic halophilic fermentative species *"Thermohalobacter berrensis"* was isolated from a solar saltern in France and was phylogenetically placed in the order *Clostridiales* (Cayol et al. 1998). Beside the terrestrial saline environments such as solar saltern and salt lakes, haloanaerobes also inhabit the subterrestrial subsurface. They were isolated from oil field brines, are members of the genus *Haloanaerobium*, order *Haloanaerobiales*, and include *H. salsugensis* (Bhupathiraju et al. 1994), *H. congolense* (Ravot et al. 1997), and *H. acetoethylicum* (Rengpipat et al. 1988). A free-living moderately halophilic spirochete, *Spirochaeta smaragdinae*, was isolated from an African oil field in Congo (Magot et al. 1997).

Bacterial sulfate reduction is an important process of mineralization of organic matter in anoxic saline environments. Biological sulfate reduction was observed in hypersaline ecosystems containing large amounts of sulfate (Nissenbaum and Kaplan 1976). However, only a limited range of substrates, that included H_2, formate, and lactate, were involved in this process. Evidence of oxidation of volatile fatty acids such as acetate or butyrate by sulfate reducers has so far been demonstrated at salinities lower

TABLE 6. Comparative properties of genera of halophilic anaerobes

Genera	NaCl range %	NaCl opt. %	Habitat	Number of species
Heterotrophs				
Haloanaerobium	2 - 30	7.5 - 20	saline lakes, oil well	8
Halocella	5-20	15	mats, salted lakes	1
Halothermothrix	4-20	5-10	saline lakes	1
Halobacteroides	2 - 30	9 - 18	saline lakes. oil well, solar saltern	3
Haloanaerobacter	5-30	14-15	solar saltern	2
Orenia	3 - 18	3 - 12	saline lakes.	1
Natroniella	10 - 26	12 - 15	soda lakes	1
Sporohalobacter	4 - 15	8 - 9	saline lakes	1
Acetohalobium	10 - 25	15 - 18	saline lakes	1
Sulfate-reducers				
Desulfovibrio	0,2 - 18	5 - 8	saline lakes, oil well deep sea sediments	3
Desulfohalobium	3 - 25	10	saline lakes	1
Desulfocella	2-19	4-5	saline lakes	1
Methanoarchaea				
Methanohalophilus	1-21	3 - 12	saline lakes, mats salinarium	4
Methanosalsus	1-12	4	saline lakes	1
Methanohalobium	15-30	24	saline lagoons	1
Methanocalculus	0-12,5	5	oil well	1
Phototrophs				
Rhodovibrio	5 - 25	7 - 14	solar salterns	2
Halorhodospira	3 - 31	7 - 21	solar salterns. saline lakes	4
Thiohalocapsa	4 - 20	8	solar salterns	1
Halochromatium	4 - 20	10	solar salterns	1

than 13% (Brandt et al. 1999) and 19% (Brandt and Ingvorsen 1997), indicating that at higher salinities the resident strains are mainly incomplete oxidizers. The process of incomplete oxidation favors the accumulation of acetate in sediments of saline environments. Currently no extreme halophilic sulfate reducer capable of acetate oxidation has so far been described. Moderate halophiles (optimum growth at 5-10% NaCl), which have been isolated from marine salterns, saline lakes, deep sea sediments, and oil field brines, belong to four genera: *Desulfovibrio, Desulfotomaculum, Desulfocella,* and *Desulfohalobium. Desulfohalobium retbaense* (Ollivier at al. 1991), isolated from the hypersaline Retba Lake in Senegal has an upper limit of 25% NaCl for growth which is the highest amongst sulfate reducers.

Few purple anoxygenic phototrophic bacteria were isolated from hypersaline habitats. Beside sunlight, they need H_2S and/or hydrogen, and organic compounds as electron donnors for anaerobic growth. Most of the isolates from hypersaline ponds in marine salterns are moderately halophiles, requiring 6-11 % NaCl for optimal growth. They belong to the genera *Rhodovibrio, Halochromatium, Thiohalocapsa* and *Halorhodospira.* Extremely halophilic purple bacteria have most commonly been isolated from alkaline brines in athalassohaline environments such as desert lakes (Imhoff 1992). They require 20- 25 % NaCl for optimal growth (Ollivier et al 1994) and belong to the family *Ectothiorhodospiraceae.*

Microbiological studies in hypersaline ecosystems confirmed that the use of methylated compounds by methanogens predominate over hydrogen and acetate utilization (Oren 1988). To our knowledge, the upper NaCl concentration so far reported for the methanogens using H_2 or formate is 12-13% for *Methanocalculus halotolerans,* a strain recently isolated from an oil well (Ollivier et al. 1998). In this respect, most methanogens that inhabit hypersaline environments are strict methylotrophs. They belong to three genera, and with the exception of the extreme halophile *Methanohalobium evestigatum,* are moderate halophilic cocci. All were isolated from saline lagoons (Zhilina and Zavarzin, 1987). Marked differences in temperature for growth has been observed. Although most methanogenic species are mesophilic, *Methanosalsus zhilinae* and *Methanohalobium evestigatum* are moderately thermophilic, with optimum temperature of 45°C and 50°C, respectively.

5. High Pressure: the Barophiles

Beside low temperature, the deep-sea biosphere is influenced by high hydrostatic pressure. While the variability of pressure adaptation in deep sea bacteria has been well studied amongst strains growing under aerobic conditions (Yayanos 1986), the response of deep sea anaerobic bacteria to hydrostatic pressure is poorly documented. Barotolerant and barosensitive hyperthermophilic archaeal strains have been reported (Jannasch and Wirsen 1984). Experiments in which these microorganisms were exposed to elevated pressure have indicated that growth was sometimes stimulated by the pressure and maximal temperature for growth extended by a few degrees (Holden and Baross 1995). *Thermococcus barophilus* is a typical barophilic hyperthermophilic archaeon. It exhibited improved growth and growth rates under increasing pressure at all temperatures tested. In addition, it behaved as an obligate barophile between 95°C and

100°C (Marteinsson et al. 1999). A mesophilic barophilic sulfate reducing-bacterium, *Desulfovibrio profundus*, isolated from deep marine sediments obtained from the Japan Sea, was shown to grow optimally at 10-15 MPa (Bale et al. 1997). Sulfate-reducing bacteria phylogenetically related to *D. profundus* were isolated from Cascadia Margin sediments in the Pacific Ocean.

6. Conclusion

Anaerobic extremophiles belong to domains *Bacteria* and the *Archaea*. Our knowledge on their biodiversity has significantly increased in the past two last decades, particularly that on thermophilic-hyperthermophilic, alkaliphilic, and halophilic microorganisms. All extreme environments harbor fermentative bacteria, sulfate-reducers, and methanogens, but the complete anaerobic oxidation of organic matter to CO_2 and CH_4 or CO_2 and H_2S has not been documented thus far in most of them. The discovery of biological mechanisms involved in the anaerobic oxidation of acetate is thus a challenge in particular for alkaline and hypersaline environments. The complete biodiversity of extremophilic anaerobes is yet to be elucidated. Molecular biology tools to study total biodiversity and/or uncultured biodiversity (e.g rRNA sequencing, reverse probing, denaturing gradient gel electrophoresis, whole cell hybridization) has improved our knowledge on biodiversity. Real Time PCR is a relatively new but powerful tool, which allows rapid and simultaneous detection, identification, and quantification of microorganisms. It is yet to be used in microbial ecology. Once implemented, it will provide an added dimension to studies on microbial diversity. These tools are currently being developed in our laboratory (Woo et al. 1997a,b, 1998, 1999).

The increase in interest on enzymes that have unique stability and kinetic properties under extreme conditions and their potential application in industrial processes (α-amylases and pectinases in the food industry, lipases and proteases in the detergent industry, xylanases for bleaching up paper, etc...) has extended our about the habitats of extremophilic microorganisms.

7. Acknowledgements

We acknowledge the contribution of P. A. Roger in improving the manuscript.

8. References

Adams, M. W. W. (1990) FEMS Microbiol. Rev. 75, 219-238.
Adams, M. W. W. (1993) Annu. Rev. Microbiol. 47, 627-658.
Andrews, K. A. and Patel, B.K.C. (1996) Int. J. Syst. Bacteriol. 46, 265-269.
Bale, S. J., Goodman, K., Rochelle, P. A., Marchesi, J. R., Fry, J. C., Weightman, A. J. and Parkes, R. J. (1997) Int. J. Syst. Bacteriol. 47, 515-521.
Bhupathiraju, V. K., Oren, A., Sharma, P. K., Tanner, R. S., Woese, C. and McInerney, M. J. (1994) Int. J. Syst. Bacteriol. 44, 565-572.
Blotevogel, K. H. and Fischer, U. (1985) Arch. Microbiol. 142, 218-222.
Blotevogel, K. H., Fisher, U., Mocha, M. and Jannsen, S. (1985) Arch. Microbiol. 142, 211-217.

88

Brandt, K. K. and Ingvorsen, K. (1997) Syst. Appl. Microbiol. 20, 366-373.

Brandt, K. K., Patel, B. K. C. and Ingvorsen, K. (1999) Int. J. Syst. Bacteriol. 49, 193-200.

Burggraf, S., Fricke, H., Neuner, A., Kristjansson, J., Rouviere, P., Mandelco, L., Woese, C. R. and Stetter, K. O. (1990) Syst. Appl. Microbiol. 13, 263-269.

Canganella, F. and Wiegel, J. (1993) In: Clostridria and Biotechnology. pp. 393-429.

Cayol, J.-L., Ollivier, B., Patel, B. K. C., Prensier, G., Guezennec, J. and Garcia, J.-L. (1994) Int. J. Syst. Bacteriol. 44, 534-540.

Cayol, J.-L., Ollivier, B., Patel, B. K. C., Ageron, E., Grimont, P. A. D., Prensier, G. and Garcia, J.-L. (1995) Int. J. Syst. Bacteriol. 45, 790-797.

Cayol, J.-L., Ducerf, S., Garcia, J.-L, Patel, B. K. C., Thomas, P. and Ollivier, B. (1998) In: Thermophiles 98. Int. Conf., Brest.

Chrisostomos, S., Patel, B. K. C., Dwivedi, P. P. and Denman, S. E. (1996) Int. J. Syst. Bacteriol. 46, 497-501.

Collins, M. D., Rodrigues, U. M., Dainty, R. H., Edwards, R. A. and Roberts, T. A. (1992) FEMS Microbiol. Lett. 96, 235-240.

Davey, M. E., Wood, W. A., Key, R., Nakamura, K. and Stahl, D. A. (1993) Syst. Appl. Microbiol. 16, 191-200.

Engle, M., Li, Y., Rainey, F. A., DeBlois, S., Mai, V., Reichert, A., Mayer, F., Mesmer, P. and Wiegel, J. (1996) Int. J. Syst. Bacteriol. 46, 1025-1033.

Engle, M., Li, Y., Woese, C. and Wiegel, J. (1995) Int. J. Syst. Bacteriol. 45, 454-461.

Fardeau, M.-L., Cayol, J.-L. , Magot, M. and Ollivier, B. (1993) FEMS Microbiol. Lett. 113, 327-332.

Fardeau M.-L., Ollivier, B., Patel, B. K. C., Magot, M., Thomas, P., Rimbault, A., Rocchiccioli, F. and Garcia, J.-L. (1997) Int. J. Syst. Bacteriol. 47, 1013-1019.

Fardeau, M.-L, Ollivier, B., Patel, B. K. C., Thomas, P., Magot, M. and Garcia, J.-L. (1997) In: 97th General Meeting of the American Society for Microbiology, Miami Beach, USA. Poster R-20, p. 528.

Fiala, G. and Stetter, K. O. (1986) Arch. Microbiol.145, 56-61.

Franzmann, P. D., Liu, Y., Balkwill, D. L., Aldrich, H. C., Conway de Macario, E. and Boone, D. R. (1997) Int. J. Syst. Bacteriol. 47, 1068-1072.

Franzmann, P. D., Springer, N., Ludwig, W., Conway de Macario, E. and Rhode, M. (1992) Syst. Appl. Microbiol. 15, 573-581.

Holden, J. F. and Baross, J. (1995) FEMS Microbiol. Ecol. 18, 27-34.

Huber, R., Kristjansson, J. K. and Stetter, K. O. (1987) Arch. Microbiol. 149, 95-101.

Huber, R., Kurr, M. , Jannasch, H. W. and Stetter, K. O. (1989) Nature 342, 833-834.

Imhoff, J. F. (1992) In: A. Balows, H. G. Trüper, M. Dworkin, W. Harder and K. H. Schleifer (eds.). The Prokaryotes, Vol.4. Springer Verlag, New York pp. 3222-3229.

Isaksen, M. F. and Teske, A. (1996) Arch. Microbiol. 166, 160-168.

Jannasch, H. W. and Wirsen, C. O. (1984) Arch. Microbiol. 139, 281-288.

Jeanthon, C., Reysenbach, A.-L., L'Haridon, S., Gambacorta, A., Pace, N. R., Glénat, P. and Prieur, D. (1995) Arch. Microbiol. 164, 91-97.

Jones, W. J., Leigh, J. A., Mayer, F., Woese, C. R. and Wolfe, R. S. (1983) Arch. Microbiol. 136, 254-261.

Kotelnikova, S. V., Obraztsova, A. Y., Gongadza, G. M. and Laurinavichius, K. S. (1993) Syst. Appl. Microbiol. 16, 427-434.

Larsen, H. (1962) In: I. C. Gunsalus and R. Y. Stanier (eds.). The Bacteria, Vol. 4. Academic Press, N. Y., London pp. 297-342.

Lee Y. E., Jain, M. K. , Lee, C., Lowe, S. E. and Zeikus, J. G. (1993) Int. J. Syst. Bacteriol. 43, 311-316.

Li, Y., Engle, M., Mandelco, L. and Wiegel, J. (1994) Int. J. Syst. Bacteriol. 44, 111-118.

Li, Y., Mandelco, L. and Wiegel, J. (1993) Int. J. Syst. Bacteriol. 43, 450-460.

Liaw, H. and Mah, R. A. (1992) Appl. Environ. Microbiol. 58, 260-266.

Liu, Y., Boone, D. R. and Choy, C. (1990) Int. J. Syst. Bacteriol. 40, 111-116.

Lowe, S. E., Jain, M. K. and Zeikus, J. G. (1993) Microbiol. Rev. 57, 451-509.

Magot, M., Fardeau, M.-L., Arnauld, O., Lanau, C., Ollivier, B., Thomas, P. and Patel, B. K. C. (1997) FEMS Microbiol. Lett. 155, 185-191.

Marteinsson, V. T., Birrien, J.-L., Reysenbach, A.-L., Vernet, M., Marie, D., Gambacorta, A., Messner, P., Sleytr, U. B. and Prieur, D. (1999) Int. J. Syst. Bacteriol. 49, 351-359.

Mathrani, I. M., Boone, D. R., Mah, R. A., Fox, G. E. and Lau, P. P. (1988) Int. J. Syst. Bacteriol. 38, 139-142.

Min, H. and Zinder, S. H. (1990) Arch. Microbiol. 153, 399-404.

Mountfort, D. O., Rainey, F. A., Burghardt, J., Kaspar, H. F. and Stackebrandt, E. (1997) Arch. Microbiol. 167, 54-60.

Nazina. T. N.. Ivanova, A. E. , Kanchaveli, L. P. and Rozanova, E. P. (1989) Microbiology (Engl. Transl. Mikrobiologiya) 57, 659-663.

Nissenbaum, A. and Kaplan, I. R. (1976) In: J. O. Nriagu (ed.), Environmental Biogeochemistry, Vol.1. Ann Arbor Science Publishers, Ann Arbor, Michigan. pp. 309-325.

Ollivier, B., Caumette, P., Garcia, J.-L. and Mah, R. A. (1994) Microbiol. Rev. 58, 27-38.

Ollivier, B., Fardeau, M.-L., Cayol, J.-L., Magot, M., Patel. B. K. C., Prensier, G. and Garcia, J.-L. (1998) Int. J. Syst. Bacteriol. 48, 821-828.

Ollivier, B., Hatchikian, C. E,. Prensier, G., Guezennec, J. and Garcia, J.-L. (1991) Int. J. Syst. Bacteriol. 41, 74-81.

Ollivier, B., Mah, R. A., Ferguson, T. J., Boone, D. R., Garcia, J.-L. and Robinson, R. (1985) Int. J. Syst. Bacteriol. 35, 425-428.

Oren, A. (1988) Anton. Leeuwenhoek 54, 267-277.

Oren, A. (1992) In: A. Balows, H. G. Trüper, M. Dworkin, W. Harder and K. -H. Schleifer (ed.), The Prokaryotes, Vol.2. Springer-Verlag, New York pp. 1893-1900.

Patel. B.K.C., Morgan, H.W. and Daniel, R.M. (1985) Arch. Microbiol. 141, 63-69.

Patel, B. K. C., Skerratt, J. H. and Nichols, P. D. (1991) Syst. Appl. Microbiol. 14, 311-316.

Patel, B.K.C., Andrews, K.T., Ollivier, B., Mah, R.A. and Garcia, J. -L. (1995) FEMS Microbiol. Lett. 134, 115-119.

Pikuta, E. V., Zhilina, T. N., Zavarzin, G. A., Kostrikina, N. A., Osipov, G. A. and Rainey, F. A. (1998) Microbiology (Engl. Transl. Mikrobiologiya) 67, 105-113.

Ravot, G., Magot, M., Fardeau, M.-L., Patel, B. K. C., Prensier, G., Egan, A., Garcia, J.-L. and Ollivier, B. (1995) Int. J. Syst. Bacteriol. 45, 308-314.

Ravot. G.. Magot, M., Ollivier, B., Patel, B. K. C., Ageron, E., Grimont, P. A. D., Thomas, P. and Garcia, J.-L. (1997) FEMS Microbiol. Lett. 147, 81-88.

Ravot, G., Ollivier, B., Fardeau, M.-L., Patel, B. K. C., Andrews, K., Magot, M. and Garcia, J.-L. (1996) Appl. Environ. Microbiol. 62, 2657-2659.

Rengpipat, S., Langworthy, T. A. and Zeikus, J. G. (1988) Syst. Appl. Microbiol. 11, 28-35.

Segerer, A. S., Trincone, A., Gahrtz, M. and Stetter, K. O. (1991) Int. J. Syst. Bacteriol. 41, 495-501.

Simankova, M. V., Chernych, N. A., Osipov, G. A. and Zavarzin, G. A. (1992) Syst. Appl. Microbiol. 16, 385-389.

Stetter, K. O. (1986) Nature 300, 258-260

Stetter, K. O., Fiala, G., Huber, G., Huber, R. and Segerer, A. (1990) FEMS Microbiol. Rev. 75, 117-124.

Stetter, K. O., Huber, R., Blöchl, E., Kurr, M., Eden, R. D.. Fielder, M., Cash, H. and Vance, I. (1993) Nature 365, 743-745.

Stetter, K. O., König, H. and Stackebrandt, E. (1983) Syst. Appl. Microbiol. 4, 535-551.

Stetter, K. O., Thomm, M., Winter, J, Wildgruber, G., Huber, H., Zillig, W., Janecovic, D.. König, H.. Palm. P. and Wunderl, S. (1981) Zb. Bakteriol. Mikrobiol. Hyg. 1 Abt. Orig. C2, 166-178.

Svetlitshnyi, V., Rainey, F. A. and Wiegel, J. (1996) Int. J. Syst. Bacteriol. 46, 1131-1137.

Tsai, C.-R., Garcia, J.-L., Patel, B. K. C., Cayol, J.-L., Baresi, L. and Mah, R. A. (1995) Int. J. Syst. Bacteriol. 45, 301-307.

Wiegel, J. (1998) Extremophiles 2, 257-267.

Winter, J., Lerp, C., Zabel, H. P., Wildenauer, F. X., König. H. and Schindler, F. (1984) Syst. Appl. Microbiol. 5, 457-466.

Woese, C. R., Kandler, O. and Wheelis, M. L. (1990). Proc. Natl. Acad. Sci. USA 87, 4576-4579.

Woo, T. H. S., Patel, B. K. C., Symonds, M. L., Norris, M. A., Dohnt, M. F. and Smythe, L. D. (1997a) J. Clin. Microbiol. 35, 3140-3146.

Woo, T. H S., Patel, B. K. C., Symonds, M. L., Norris, M. A.. Dohnt, M. F. and Smythe. L. D. (1997b) Anal. Biochem. 256, 132-134.

Woo, T. H. S., Patel, B. K. C., Cinco, M., Smythe, L. D.. Symonds, M. L., Norris, M. A.. and Dohnt, M. F. (1998) Syst. Appl. Microbiol. 21, 89-96.

Woo, T. H. S., Patel, B. K. C., Cinco, M., Smythe, L. D.. Norris, M. A., Symonds, M. L., Dohnt, F. and Piispanen, J. (1999) J. Microbiol. Meth. (in press).

Wynter, C., Patel, B. K. C., Bain, P., De Jersey, J., Hamilton, S. and Inkerman, P. A. (1996) FEMS Microbiol. Lett. 140, 271-276.

Yayanos, A. A. (1986) Proc. Nat. Acad. Sci. USA 83, 9542-9546.

Zeikus, J. G. and Wolfe, R. S. (1972) J. Bacteriol. 109, 707-713.

90

Zhilina, T. N. and Zavarzin, G. A. (1987) Dokl. Akad. Nauk. SSSR 293, 464-468 (Rus.).

Zhilina, T. N. and Zavarzin, G. A. (1990) Dokl. Akad. Nauk. SSSR 311, 745-747 (Rus.).

Zhilina, T. N. and Zavarzin, G. A. (1994) Curr. Microbiol. 29, 109-112.

Zhilina, T. N., Bolotina, N. P., Lysenko, A. M. and Osipov. G. A. (1995) Arch. Microbiol. 163, 29-34.

Zhilina, T. N., Miroshnikova, L. V., Osipov, G. A. and Zavarzin, G. A. (1991) Microbiology (Engl. Transl. Mikrobiologiya) 60, 714-724.

Zhilina, T. N., Zavarzin, G. A., Detkova, E. N. and Rainey. F. A. (1996a) Curr. Microbiol. 32, 320-326.

Zhilina, T. N., Zavarzin, G. A., Rainey, F., Kevbrin, V. V., Kostrikina, N. A. and Lysenko, A. M (1996b) Int. J. Syst. Bacteriol. 46, 305-312.

Zhilina, T. N., Zavarzin, G. A., Rainey, F. A., Pikuta, E. N., Osipov, G. A. and Kostrikina, N. A. (1997) Int. J. Syst. Bacteriol. 47, 144-149.

Zillig, W., Stetter, K. O., Prangishvilli, D., Schäfer, H. , Wunderl, S., Janekovic, D., Holz, I. and Palm, P. (1982) Zbl. Bakt. Hyg., I. Abt. Orig. C3, 304-317.

Zillig, W., Stetter, K. O., Schäfer, H. , Janekovic, D., Wunderl, S., Holz, I. and Palm, P. (1981) Zbl. Bakt. Hyg. C2, 205-227.

III

THERMOPHILES AND ACIDOPHILES

Biodata of Dr. **Kenneth M. Noll**, contributor (with co-author Susan E. Childers) of
"Sulfur Metabolism Among Hyperthermophiles"

Dr. Kenneth M. Noll is an Associated Professor in the Department of Molecular and
Cell Biology at the University of Connecticut, Storrs, CT. He gained his Ph.D. at the
University of Illinois, Urbana, IL. in 1987. Dr. Noll's main interests are in microbial
physiology (thermophilic bacteria and Archaea, biochemical basis of thermophily);
microbial evolution and the evolution of metabolic pathways; chromosome structure and
evolution in Prokaryotes.
E-Mail: noll@uvonnvm.uconn.edu

Biodata of Dr. **Susan E. Childers**, co-author (with K.M. Noll) of *"Sulfur Metabolism
among Hyperthermophiles"*

Dr. Susan Elizabeth Childers is a Postdoctoral Research Associate at the Department
of Microbiology, University of Massachusetts, Amherst, MA. She received her Ph.D. in
Microbiology from the University of Connecticut, Storrs in 1997. Her main interests are
in the following areas: Iron reduction in hyperthermophiles; physiological and molecular
studies of biofilm cells of *Pseudomonas aeruginosa* and sulfur reduction in the
hyperthermophilic bacterium *Thermotoga neapolitana*.
E-Mail: childers@microbio.umass.edu

J. Seckbach (ed.), Journey to Diverse Microbial Worlds, 93-105.
© 2000 *Kluwer Academic Publishers. Printed in the Netherlands.*

SULFUR METABOLISM AMONG HYPERTHERMOPHILES

KENNETH M. NOLL[1] and SUSAN E. CHILDERS[2]
[1]*Department of Molecular and Cell Biology*
University of Connecticut
Storrs, Connecticut 06269-3125 USA
[2]*Department of Microbiology*
University of Massachusetts
Amherst, Massachusetts 01003-5720 USA

1. Introduction

Sulfur biochemistry figures prominently in the bioenergetics of many strictly anaerobic hyperthermophiles (organisms capable of growth to at least 90^0C). Sulfidogenesis by the reduction of elemental sulfur to sulfide is common among the deepest phylogenetic branches of extant microbes (Figure 1). These sulfidogens are all hyperthermophiles and most are Archaea. The capacity to carry out elemental sulfur reduction is found in both archaeal lineages. Two groups of Bacteria, the Thermotogales and Aquifecales, are also hyperthermophilic sulfidogens. Other forms of sulfur metabolism are also found among hyperthermophiles including sulfate reduction by the archaeon *Archaeoglobus* and sulfur oxidation by *Sulfolobus* and its relatives. Examples of each of these will be considered here. For many of these organisms, we do not know if the organism derives any energy from carrying out sulfur reduction. This review will compare mechanisms of sulfidogenesis in selected hyperthermophiles to highlight features that might shed light on the mechanisms and evolution of this metabolism.

2. Geothermal Habitats

Hydrothermal systems are characterized by the mixing of hot, reduced geothermal fluid with cold, oxidized seawater or air. The contrasting oxidative states of a variety of chemical species provides a rich source of energy for microbes able to tolerate the extreme conditions. Geothermal heating of water is found in surface terrestrial environments, shallow marine sediments, and deep-sea vents. In terrestrial systems hydrothermal fluids come into direct contact with air while in marine systems hydrothermal fluid contacts sulfate-rich seawater. Oxygen provides a strong oxidant in terrestrial hot springs allowing microbial oxidation of elemental sulfur to sulfuric acid. Since these springs are only weakly buffered, highly acidic ecosystems unique to these geothermal sites are created (Brock 1978). In seawater, abundant sulfate provides an oxidant for marine sulfate reducing organisms at moderate pH. Terrestrial hot springs cannot heat above 100^0C while marine systems, particularly the deep-sea systems, can attain temperatures ≤350°C. The following discussion of the sulfur chemistry of hydrothermal systems will primarily make reference to deep-sea systems. Shallow marine sediments and the anoxic portions of terrestrial hot springs will have similar chemistries

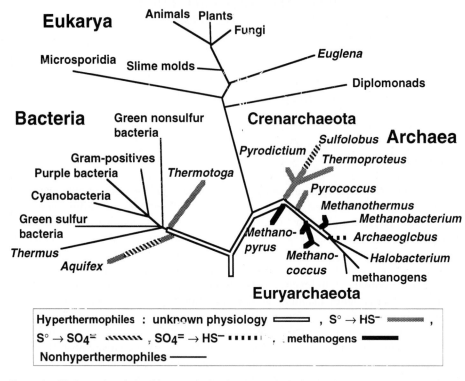

Figure 1. Phylogenetic relationships among the three Domains as determined by comparisons of small subunit rRNA sequences. The distribution of sulfur metabolisms among the hyperthermophiles is depicted. *Aquifex* is capable of both elemental sulfur reduction and oxidation. Tree is modified from (Stetter 1996).

Sulfide emitted in hydrothermal fluid is primarily derived from the underlying heated basalt while a smaller portion is derived from reduction of seawater sulfate during subduction (Von Damm 1990) (Figure 2). Some of the sulfide reacts with metals and precipitates above, below or within the vent chimneys as iron, zinc, copper, or lead sulfides depending upon the metals present in the fluid (Von Damm 1990). Upon contact with oxidizing seawater, sulfide can also oxidize to elemental sulfur which is sometimes found within chimney walls (McCollom and Shock, 1997). The sulfur compounds available to hyperthermophiles in hydrothermal environments are not known. Polysulfides are likely present and have been shown to accumulate to physiologically significant levels under temperature and pH conditions found in many geothermal locations (Schauder and Müller 1993). Although polysulfides rather than elemental sulfur are likely the sulfur substrate for many hyperthermophiles capable of "sulfur" reduction, this has only been demonstrated experimentally with

Pyrococcusfuriosus and *Pyrodictium brockii* (Blumentals *et al.* 1990; Pihl *et al.* 1990). Growth of *Thermococcus* species with elemental sulfur results in the intracellular accumulation of a variety of cyclic polysulfides with aliphatic and aromatic side chains (Ritzau *et al.* 1993). Their physiological roles are not known, however.

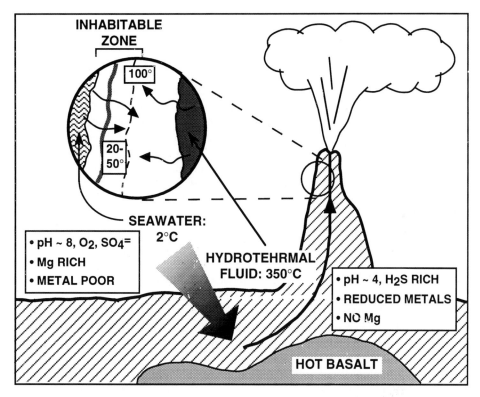

Figure 2. Schematic diagram of a deep-sea hydrothermal vent. Inset is a cross-section of the chimney wall showing the 20-50°C and 100°C isotherms caused by the outflow of hydrothermal fluid mixing with the inflowing seawater. Drawing is modified from, Kennish *et al.* (1992) and McCollom and Shock (1997).

In marine environments, the majority of hyperthermophiles live in the area of mixing of the reduced hydrothermal fluid and oxidized seawater. Most organisms are found stratified in the outer layers of hydrothermal vent chimneys and flanges (Harmsen *et al.* 1997a; Harmsen*et al.* 1997b; Hedrick *et al.* 1992). The distribution of organisms suggests they respond to the thermal and chemical gradients established within these structures and is consistent with a model distribution based upon free energies available from chemical species within a vent system (McCollom and Shock 1997). In that model, sulfur-metabolizers are distributed in two regions, a 20-50°C isotherm and a 100°C isotherm (Figure 2). In the former, mesophilic sulfide-oxidizers take advantage of the sulfide emissions and the dissolved oxygen in the inflowing seawater to drive chemolithoautotrophic growth. In the latter, hyperthermophilic sulfate reducers (*Archaeoglobus*) may be exposed to sufficient sulfate from seawater along with organic compounds from other microbes or vent fluid to drive chemoheterotrophic growth. Although not indicated in the model, polysulfide-reducing hyperthermophiles would likely be found near the 100^0c isotherm as well.

3. Hyperthermophilic Organisms

3.1. ARCHAEA: EURYARCHAEOTES

3.1.1. *Pyrococcus furiosus*
Pryrococcusfuriosus is the type species of a genus of heterotrophic archaea which grow optimally above 90^0c. Numerous oxidoreductases involved in its central catabolism have been purified (Adams and Kletzin 1996). It is an obligate organotroph catabolizing peptides and oligosaccharides with production of hydrogen, carbon dioxide, acetate, and alanine as major products by a novel glycolytic pathway (Kengen and Stams, 1994). Hydrogen accumulation in batch cultures inhibits growth, but can be prevented by sulfidogenesis from elemental sulfur or polysulfides. Sulfidogenesis is catalyzed by at least two sulfur reductase activities in *P. furiosus*: a hydrogenase and a sulfide dehydrogenase.

   P. furiosus hydrogenase is a highly thermostable, soluble nickel-iron hydrogenase (Bryant and Adams 1989). Reduced *P.. furiosus* ferredoxin stimulates hydrogenase-catalyzed hydrogen evolution *in vitro*, but apparently NADPH serves as its electron donor (Ma *et al.*, 1994). When a sulfur reductase activity found in *P. furiosus* cell extracts was purified later, it was found to be this hydrogenase (Ma *et al.*, 1993). Polysulfides are the primary substrates for this activity. Given its dual catalytic activity, the enzyme was called a sulfhydrogenase. Two sulfhydrogenase subunits' genes (β and γ) have sequence similarities with subunits of a dissimilatory *Salmonella* sulfite reductase (Pedroni *et al.* 1995). It was speculated that perhaps these subunits may catalyze the reduction of polysulfide separate from the hydrogen evolution activity of the other subunits (δ and α) in a manner similar to a hydrogenase from *Methanobacterium thermoautotrophicum* in which two factor F_{420}-reducing hydrogenase catalytic subunits are distinct from two subunits that catalyze reduction of the CoM-SS-HTP heterodisulfide (see Section 4).

   A second sulfur reductase activity in *P. furiosus* was found in a soluble enzyme originally named sulfide dehydrogenase (Ma and Adams 1994). Sulfide dehydrogenase catalyzes the reduction of polysulfide to H_2S using NADPH as electron donor. It also functions as a reduced ferredoxin:NADP oxidoreductase (FNOR). The FNOR activity is probably its major role *in vivo*. An *in vitro* system using purified enzymes was shown to mediate electron flow from pyruvate oxidative decarboxylation catalyzed by purified pyruvate:ferredoxin oxidoreductase (POR) to hydrogenogenesis catalyzed by sulfhydrogenase (Figure 3A). Recently the POR was shown to catalyze pyruvate decarboxylase activity (Ma *et al.* 1997).

   Measurements of growth yields on maltose with and without elemental sulfurindicate that energy may be conserved by sulfur reduction in *P. furiosus* (Schicho *et al.* 1993). Sulfur reduction had been postulated to serve as an alternative means to dispose of reducing equivalents and prevent the accumulation of growth inhibiting concentrations of hydrogen (Fiala and Stetter, 1986). As such, sulfur reduction would serve to enhance fermentation, but not generate energy *per se*. Studies indicated, however, that maintaining low hydrogen concentrations during cultivation does not provide growth stimulation like that observed when sulfur is provided (Schicho *et al.* 1993). Thus sulfidogenesis does not seem to enhance growth by alleviating end product inhibition of fermentation. Since none of the enzymes involved in sulfidogenesis are membrane-bound, a respiratory mechanism of energy conservation is unlikely. How sulfidogenesis may provide extra energy to *P. furiosus* is still unknown.

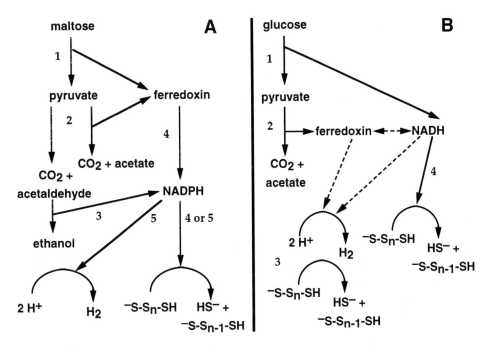

Figure3. Glycolysis and sulfidogenesis in A. *Pyrococcus furiosus* and B. *Thermotoga maritima.* Arrows in the glycolytic pathways represent several steps. Numbers refer to the following enzymes: A. 1. glyceraldehyde:ferredoxin oxidoreductase,2. pyruvate:ferredoxin oxidoreductase. 3. alcohol dehydrogenase. 4. ferredoxin:NADP oxidoreductase (sulfide dehydrogenase). and 5. sulfhydrogenase. B. 1. glyceraldehyde dehydrogenase. 2. pyruvate:ferredoxin oxidoreductase. 3. sulfhydrogenase. 4. NADH:polysulfide oxidoreductase.

3.1.2. *Archaeoglobus*
Archaeoglobus is a sulfate-reducing archaeon found in geothermally-heated sediments and oil deposits (Stetter *et al.* 1993; Stetter *et al.*1987). It grows heterotrophically by dissimilatory sulfate reduction and chemolithoautotrophically by thiosulfate reduction using hydrogen as reductant. Elemental sulfur is not reduced. Electrons derived from lactate oxidation are carried by a variety of carriers including ferredoxin, a menaquinone, and the deazaflavin cofactor, coenzymeF_{420} (Kunow *et al.* 1993). Genes encoding cytochromes were found, so a complete membrane-bound respiratory pathway

is possible (Klenk *et al.*, 1997). The pathway by which electrons from organic substrate or hydrogen oxidation are used to reduce sulfate or thiosulfate is unknown. It is reasonable to assume that electron transport is coupled to generation of a transmembrane ion gradient, but how this is linked to sulfate reduction is unclear. The adenylsulfate reductase which catalyzes the initial reductive step has been purified from the soluble fraction of cell extracts (Speich *et al.* 1994; Speich and Trüper 1988). Associated activities catalyzed by bisulfite reductase and APS reductase have also been reported. *Archaeoglobus* is closely related to methanogenic archaea and it shares many physiological traits with methanogens including a common carbon dioxide fixation pathway (Klein *et al.* 1993; Schwörer *et al.*1993). It also shares with methanogens the heterodisulfide reductase, a key enzyme in methanogen bioenergetics (see Section 4). The role of this putative membrane-bound enzyme remains to be elucidated.

3.2. ARCHAEA: CRENARCHAEOTES

3.2.1. *Pyrodictium*
Organisms of the genus *Pyrodictium* are among the most thermophilic organisms with optimal growth temperatures above 100^{0}C (Pley *et al.* 1991; Stetter *et al.* 1983). They grow as a network of disc-shaped cells connected by cytoplasmic filaments. All grow chemolithoautotrophically by hydrogen oxidation and sulfur or, in some species, thiosulfate reduction (Dirmeier *et al.* 1998). Cells do not grow attached to elemental sulfur, but as a thin web above the sulfur (Stetter *et al.* 1983). If cells are separated from the sulfur by a dialysis membrane, growth still occurs suggesting the actual substrates for sulfur reduction are soluble compounds such as polysulfides (Pihl et al. 1990).

Two species have been examined with respect to the mechanism of sulfidogenesis: *P. brockii* and *P. abysii*. In *P. brockii*, hydrogen-linked sulfidogenesis is catalyzed by a membrane-bound complex consisting of a hydrogenase, a quinone, a cytochrome, and sulfur reductase. The membrane-bound hydrogenase of *P. brockii* was purified and found to be a NiFe hydrogenase able to use a variety of artificial electron acceptors including those of positive potential (Pihl and Maier 1991). A quinone apparently mediates electrontransfer between hydrogenase and a c-type cytochrome (Pihl *et al.* 1992). Both the quinone and cytochrome were purified, but the structure of the quinone was not determined. Membranes of *P. brockii* catalyze hydrogen oxidation coupled with sulfide production (Pihl *et al.* 1992), but the sulfur reductase has not been purified from this species. A sulfur-reducing complex has been purified from the membrane of *P.abysii* (Dirmeier *et al.* 1998). This complex consists of nine major polypeptides including cytochromes *b* and *c* and a hydrogenase. No evidence for quinones could be found, so the mechanism of electron transfer in these two species may be rather different. In neither case is there any evidence regarding how (or if) energy is conserved in electron transfer between hydrogen and the sulfur substrate.

3.2.2. *Sulfolobus*
Members of the genus *Sulfolobus* are capable of sulfide or elemental sulfur oxidation to sulfate typically with oxygen as electron acceptor (Shivvers and Brock 1973). Species of the closely related genera *Desulfolobus* and *Acidianus* are additionally able to grow anaerobically by hydrogen-coupled sulfur reduction (Segerer *etal.* 1985; Zillig *et*

al. 1985). An extensive electron transport pathway is used when cells grow heterotrophically by oxygen respiration (Schäfer 1996). However, the mechanism by which cells grow autotrophically by sulfur oxidation is unknown. The sulfur oxidizing enzyme has been purified from *S. brierleyi* and *D. ambivalens.* The *S. brierleyi* enzyme is an oxygenase-type oxidase that catalyzes elemental sulfur oxidation to sulfite (Emmel *et al.* 1986). The *D. ambivalens* enzyme is a soluble enzyme that produces sulfite and hydrogen sulfide from elemental sulfur under an oxygen atmosphere (Kletzin 1989). The enzyme is only produced when cells are grown aerobically and it does not catalyze the reduction of sulfur to sulfide under a hydrogen atmosphere (Kletzin 1989; Kletzin 1992). Thus the sulfur reductase activity of this organism resides in a different enzyme.

3.3. BACTERIA

3.3.1. *Thermotoga*
Thermotoga species are members of the Thermotogales, the first heterotrophic lineage to evolve among the Bacteria (Achenbach-Richter *et al.* 1987; Huber *et al.* 1986). They are hyperthermophilic strict anaerobes that grow by fermenting sugars to acetate, hydrogen, carbon dioxide, and a small amount of lactate. Like *Pyrococcus*, some species also form alanine as an end product (Ravot *et al.* 1996). Of the glucose substrate, 85% passes through the Embden-Meyerhof-Parnas pathway while the remaining 15% is catabolized through the Entner-Doudoroff pathway (Schönheit and Schäfer 1995). Elemental sulfur is not required for growth of *Thermotoga*, but itstimulates growth (Belkin et al. 1986; Huber et al. 1986). Growth enhancement is observed with elemental sulfur, polysulfides, cystine, dimethyl disulfide, and thiosulfate (Childers *et al.* 1992; Ravot *et al.* 1995). Growth with sulfane sulfur prevents accumulation of toxic hydrogen and sulfide concentrations up to 10 mM are tolerated (Belkin *et al.* 1986). The enzymes involved in sulfidogenesis have not been as thoroughly characterized as those of *P. furiosus,* but some interesting differences are emerging from early studies (Figure 3B).

Purified hydrogenase from *T. maritima* also catalyzes polysulfide reduction to sulfide in the presence of hydrogen but at a rate much lower than the *P. furiosus* sulfhydrogenase (Juszczak *etal.* 1991; Ma *et al.* 1993). The enzyme was reported to be a soluble iron hydrogenase and is highly oxygen sensitive in its pure form. Although *T. maritima* evolves hydrogen during metabolism, the purified hydrogenase had low hydrogen evolution activity. Reduced ferredoxin isproduced as a result of oxidative decarboxylation of pyruvate by pyruvate:ferredoxin oxidoreductase (POR), but *T. maritima* ferredoxin could not serve as an electron donor to purified hydrogenase.

A second sulfur reductase activity with 12- to 48-fold higher activity in cell extracts than sulfhydrogenase is due to an NADH:polysulfide oxidoreductase (NPOR) (Childers and Noll 1994). NADH, but not NADPH, can provide electrons for polysulfide reduction *in vitro.* Elemental sulfur harvested from cultures or polysulfide can serve as electron acceptor. Neither cystine nor thiosulfate were reduced *in vitro.* The NPOR has been partially purified and the activity resides in a 90 kDa flavoprotein apparently composed of two 43 kDa subunits (Childers 1997). Its activity is oxygen labile. The NPOR can be separated chromatographically from the activity usually ascribed to the

NADH:ferredoxin oxidoreductase, thus, NPOR does not seem to be necessary to mediate electron transfer between ferredoxin andnicotinamide cofactors (Childers and Noll 1995).

Electrons derived from the POR reaction must be converted to NADH to allow polysulfide reduction by sulfur-grown cells. In other strict anaerobes, this interconversion is mediated by an NADH:ferredoxin oxidoreductase which is commonly measured by reduction of the dye methyl viologen by NADH (an NADH:methyl viologen oxidoreductase or NMOR) (Jungermann *et al.* 1973). NMOR activity was reported in a *Thermotoga* species (Janssen and Morgan, 1992), but we discovered this activity is associated with the particulate fraction of cell extracts (Käslin *et al.* 1998). As expected, it appears to be exposed on the cytoplasmic face of the cell membrane. We also found that *Thermotoga* ferredoxin cannot serve as a substrate for this enzyme *in vitro*. POR activity can be coupled with NAD^+ reduction *in vitro* only in the presence of the membrane fraction (Käslin 1997), suggesting there is no soluble reduced ferredoxin:NAD^+ oxidoreductase. We prepared membrane vesicles of *T. neapolitana* and found they contained NMOR activity and also catalyzed hydrogen oxidation (Käslin *et al.* 1998). The relationship between the membrane-associated hydrogenase and the soluble hydrogenase has yet to be established. Typically, membrane-bound hydrogenases catalyze hydrogen oxidation, but this activity has not been demonstrated in *Thermotoga*, although growth attributed to hydrogen-linked ferric iron reduction has been reported (Vargas *et al.* 1998).

3.3.2. *Aquifex*

Species of *Aquifex* are microaerophilic hyperthermophiles able to use a variety of electron donors and acceptors for chemolithoautotrophic growth. The type species, *A. pyrophilus*, couples oxidation of hydrogen, elemental sulfur or thiosulfate with reduction of oxygen or nitrate (Huber *et al.* 1992). When thiosulfate or elemental sulfur serve as the electron source, sulfuric acid is produced. However it is not an acidophile like *Sulfolobus*. If both oxygen and sulfur are available during cultivation under a hydrogen atmosphere, sulfide is produced during the later stages of growth. Although sulfide can apparently be generated by sulfur reduction under microaerophilic conditions, growth under strictly anaerobic conditions by hydrogen/sulfur chemolithotrophy has not been reported. The genes encoding putative sulfide dehydrogenase subunits were found in the genome sequence of another species, *A. aeolicus* (Deckert *et al.* 1998). No sulfur reductase activity has been examined in any *Aquifex* species, so the relationship of these traits to those of other sulfur-reducing hyperthermophiles is unknown. A membrane-bound respiratory system consisting of a hydrogenase, cytochrome *c*, and a novel quinone is present in the closely-related thermophile, *Hydrogenobacter* (Igarashi and Kodama, 1990) and genes encoding similar components were identified in the genome sequence of *A. aeolicus* (Deckert *et al.* 1998). Conceivably, these organisms could link hydrogen oxidation to sulfurreduction *via* a respiratory chain similar to that found in *Pyrodictium*.

4. Conclusions

Whether sulfur compounds are used as substrates for true respiratory systems or as electron sinks to enhance fermentation, they are clearly important to many hyperthermophiles. These compounds could have served as the first external electron acceptors and so played an important role in the evolution of energy metabolisms. With the exception of the acidophiles, polysulfides are readily available to modern hyperthermophiles and these (and not elemental sulfur) probably serve as electron acceptors for the "sulfur-reducing" hyperthermophiles (Schauder and Müller 1993).

The mechanisms by which modern hyperthermophiles reduce polysulfides are only partially understood, so the evolutionary relationships among these mechanisms can only be guessed at this time. Given the phylogenetic distribution of polysulfide reduction (Figure 1), hydrogen/polysulfide chemolithoautotrophy may have been the first metabolic system to exploit electron flow to sulfidogenesis. Among the crenarchaeotes, *Pyrodictium* still retains this metabolism. The euryarchaeotal lineage may have been founded by a proto-methanogen that fixed carbon dioxide by the methanogen variation of the reductive acetyl-CoA pathway now unique to the euryarchaeotal autotrophs. Energy might have been provided to this pathway by hydrogen/polysulfide chemolithotrophy. The evolution of methanogens might have come with the development of a means to link these two metabolisms *via* the development of the methylcoenzyme M methyltransferase, methylcoenzyme M methylreductase, and heterodisulfide reductase activities. Modern methanogens depend upon sulfur biochemistry for energy generation. The final step of methanogenesis generates a heterodisulfide of two cofactors:2-mercaptoethane sulfonate (HS-CoM) and 7-mercaptothreonine phosphate (HS-HTP) (Bobik *et al.* 1987). The reduction of this heterodisulfide is catalyzed by a membrane-associated heterodisulfide oxidoreductase complex using electrons derived from hydrogen oxidation (Heiden *etal.*, 1993, Setzke *et al.* 1994). This reaction generates a transmembrane proton by which the organisms conserve the energy of the final methanogenic step (Peinemann *et al.* 1990). Coenzyme M and HS-HTP, serve as electron donors in fumarate reduction in modern methanogens (Bobik and Wolfe 1989), suggesting they could have served other roles in the proto-methanogen prior to the evolution of methanogenesis. Modern methanogens can reduce elemental sulfur in anon-growth dependent manner by an unknown mechanism (Stetter and Gaag, 1983). It is tempting to speculate that the heterodisulfide reductase might have originally been used to mediate electron transfer from hydrogenase to polysulfides, but there is no evidence that it is capable of using polysulfides as substrates. The purified enzyme is specific for CoM-SS-HTP and does not reduce cystine, oxidized lipoamide, or oxidized glutathione (Hedderich *et al.* 1989).

Ultimately comparisons of sequences of the polysulfide reductases from a number of hyperthermophiles will be necessary to answer questions about their evolutionary relatedness and the possibility that this biochemical activity was present in the common ancestor. It may be that different enzymes were used to shunt excess electrons to polysulfides in different lineages, so this activity will have had multiple origins. Given the increasing accumulation of genome sequence information from a number of hyperthermophiles, these questions may soon be amenable to such investigations.

5. Acknowledgments

Work on *Thermotoga* physiology in the author's laboratory was supported by grants from the National Science Foundation (MCB-9418197), the US Department of Energy(DE-FG02-93ER20122) and the University of Connecticut Research Foundation.

6. References

Achenbach-Richter, L., Gupta, R., Stetter, K. O. and Woese. C. R. (1987) *System. Appl. Microbiol.* .9, 34-9.
Adams, M. W. W., and Kletzin, A. (1996) *Advances in Protein Chemistry* **48**, 101-180.
Belkin, S., Wirsen, C. O., and Jannasch, H. W. (1986) *Appl. Environ. Microbiol.* **51**, 1180-1185.
Blumentals, I. I., Itoh, M., Olson, G. J., and Kelly, R. M. (1990) *Appl. Environ. Microbiol.* **56**, 1255-1262.
Bobik, T. A., Olson, K. D., Noll, K. M., and Wolfe, R. S. (1987) *Biochem. Biophys. Res. Commun.* **149**, 455-460.
Bobik, T. A., and Wolfe, R. S. (1989) *J. Biol.Chem.* **264**, 18714-18718.
Brock, T. D. (1978) *Thermophilic Microorganisms and Life at High Temperatures*, Springer-Verlag, NY.
Bryant, F. O., and Adams, M. W. W. (1989) *J. Biol.Chem.* **264**, 5070-5079.
Childers, S. E. (1997) Ph.D. Dissertation, University of Connecticut, Storrs, p 138.
Childers, S. E., and Noll, K. M. (1994) *Appl. Environ.Microbiol.* **60**, 2622-2626.
Childers, S. E., and Noll, K. M. (1995) In: *Abstr. Ann. Meet. Amer. Soc. Microbiol.*, Washington, DC, p 558.
Childers, S. E., Vargas, M., and Noll, K. M. (1992) *Appl. Environ. Microbiol.* **58**, 3949-3953.
Deckert, G., *et al.* (1998) *Nature* **392**, 353-358.
Dirmeier, R., Keller, M., Frey, G., Huber, H., and Stetter, K. O. (1998) *Eur. J. Biochem.* **252**, 486-491.
Emmel, T., Sand, W., König, W. A., and Bock, E. (1986) *J. Gen. Microbiol.* **132**, 3415-3420.
Fiala, G., and Stetter, K. O. (1986) *Arch.Microbiol.* **145**, 56-61.
Harmsen, H. J. M., Prieur, D., and Jeanthon, C. (1997a) *Appl. Environ. Microbiol.* **63**, 2876-2883.
Harmsen, H. J. M., Prieur, D., and Jeanthon, C. (1997b) *Appl. Environ. Microbiol.* **63**, 4061-4068.
Hedderich, R., Berkessel, A., and Thauer, R. K. (1989) *FEBS Lett.* **255**, 67-71.
Hedrick, D. B., Pledger, R. D., White, D. C., and Baross, J. A. (1992) *FEMS Microbiol. Ecol.* **101**, 1-10.
Heiden, S., Hedderich, R., Setzke, E., and Thauer, R. K. (1993) *Eur. J. Biochem.* **213**, 529-535.
Huber, R., Langworthy, T. A., König, H., Thomm, M., Woese, C. R., Sleytr, U. B., and Stetter, K. O. (1986) *Arch. Microbiol*.**144**, 324-333.
Huber, R., Wilharm, T., Huber, D., Trincone, A., Burggraf, S., König, H.,Rachel, R., Rockinger, I., Fricke, H., and Stetter, K. O. (1992) *Syst. Appl. Microbiol.* **15**, 340-351.
Igarashi, Y., and Kodama, T. (1990) *FEMS Microbiol. Rev.* **87**, 403-406.
Janssen, P. H., and Morgan, H. W. (1992) *FEMS Microbiol. Lett.* **96**, 213-217.
Jungermann, K., Thauer, R. K., Leimenstoll, G., and Decker, K. (1973) *Biochim. Biophys. Acta* **305**,268-280.
Juszczak, A., Aono, S., and Adams, M. W. (1991) *J. Biol. Chem.* **266**, 13834-13841.
Käslin, S. (1997) Diplom, University of Zürich, Zürich, p 52.
Käslin, S., Childers, S. E., and Noll, K. M. (1998) *Arch. Microbiol.* **170**, 297-303.
Kengen, S. W. M., and Stams, A. J. M. (1994) *Arch. Microbiol.* **161**, 168-175.
Kennish, M. J., Lutz, R. A., and Simoneit, B. R. T. (1992) *Rev. Aquatic Sci.* **6**, 467-477.
Klein, A. R., Breitung, J., Linder, D., Stetter, K. O., and Thauer,R. K. (1993) *Arch. Microbiol.* **159**, 213-219.
Klenk, H. P., *et al.*, (1997) *Nature* **390**, 364-370.
Kletzin, A. (1989) *J. Bacteriol.* **171**, 1638-1643.
Kletzin, A. (1992) *J. Bacteriol.* **174**, 5854-5859.
Kunow, J., Schwörer, B., Stetter, K. O., and Thauer, R. K. (1993) *Arch. Microbiol.* **160**,199-205.
Ma, K., and Adams, M. W. W. (1994) *J. Bacteriol.* **176**, 6509-6517.
Ma, K., Hutchins, A., Sung, S. J. S., and Adams, M. W. W. (1997) *Proc. Natl. Acad. Sci. U S A* **94**, 9608-9613.
Ma, K., Schicho, R. N., Kelly, R. M., and Adams, M. W. W. (1993) *Proc. Natl. Acad. Sci. U S A* **90**,5341-5344.
Ma, K., Zhou, Z. H., and Adams, M. W. W. (1994) *FEMS Microbiol. Lett.* **122**, 245-250.

McCollom T. M., and Shock, E. L. (1997) *Geochlm. Cosmochim. Acta* **61**, 4375-4391.

Pedroni, P., Della Volpe, A., Galli, G., Mura, G. M., Pratesi, C., and Grandi, G. (1995) *Microbiology* **141**, 449-458.

Peinemann, S., Hedderich, R., Blaut, M., Thauer, R. K., and Gottschalk, G. (1990) *FEBS Lett.* **263**, 57-60.

Pihl, T. D., Black, L. K., Schulman, B. A., and Maier, R. J. (1992) *J. Bacteriol.* **174**, 137-143.

Pihl, T. D., and Maier, R. J. (1991) *J. Bacteriol.* **173**, 1839-1844.

Pihl, T. D., Schicho, R. N., Black, L. K., Schulman, B. A., Maier, R. J., and Kelly, R. M. (1990) *Genet. Eng. Rev.* **8**, 345-377.

Pley, U., Seger, J., Woese, C. R., Gambacorta, A., Jannasch, H. W., Fricke, H., Rachel, R., and Stetter, K. O. (1991) *System. Appl. Microbiol.* **14**, 245-253.

Ravot, G., Ollivier, B., Fardeau, M. L., Patel, B. K., Andrews, K. T., Magot, M., and Garcia, J. L. (1996) *Appl. Environ. Microbiol.* **62**, 2657-2659.

Ravot, G., Ollivier, B., Magot, M., Patel, B. K. C., Crolet, J.-L.,Fardeau, M.-L., and Garcia, J.-L. (1995) *Appl. Environ.Microbiol.* **61**, 2053-2055.

Ritzau, M., Keller, M., Wessels, P., Stetter, K. O., and Zeeck, A. (1993) *Leibigs. Ann. Chem.*, 871-876.

Schäfer, G. (1996) *Biochim. Biophys. Acta* **1277**, 163-200.

Schauder, R., and Müller, E. (1993) *Arch. Microbiol.* **160**, 377-382.

Schicho, R. N., Ma, K., Adams, M. W. W., and Kelly, R. M. (1993) *J. Bacteriol.* **175**,1823-1830.

Schönheit, P., and Schäfer, T. (1995) *World J. Microbiol. Biotechnol.* **11**, 26-57.

Schwörer, B., Breitung, J., Klein, A. R., Stetter, K. O., and Thauer,R. K. (1993) *Arch. Microbiol.* **159**, 225-232.

Segerer, A., Stetter, K. O., and Klink, F. (1985) *Nature* **313**, 787-789.

Setzke, E., Hedderich, R., Heiden, S., and Thauer, R. K. (1994) *Eur. J. Biochem.* **220**, 139-148.

Shivvers, D. W., and Brock, T. D. (1973) *J. Bacteriol.* **114**, 706-710.

Speich, N., Dahl, C., Heisig, P., Klein, A., Lottspeich. F., Stetter, K. O., and Trüper, H. G. (1994) *Microbiology* **140**, 1273-1284.

Speich, N., and Trüper, H. G. (1988) *J. Gen. Microbiol.* **134**, 1419-1425.

Stetter, K. O. (1996) *FEMS Microbiol. Rev.* **18**, 149-158.

Stetter, K. O., and Gaag, G. (1983) *Nature* **305**, 309-311.

Stetter, K. O., Huber, R., Blöchl, E., Kurr, M., Eden, R. D.,Fielder, M., Cash, H., and Vance, I. (1993) *Nature* **365**, 743-745.

Stetter, K. O., König, H., and Stackebrandt, E. (1983) *System. Appl. Microbiol.* **4**, 535-551.

Stetter, K. O., Lauerer, G., Thomm, M., and Neuner, A. (1987) *Science* **236**, 822-824.

Vargas, M., Kashefi, K., Blunt-Harris, E. L., and Lovley, D. R.(1998) *Nature* **395**, 65-67

Von Damm, K. L. (1990) *Annu. Rev. Earth Planet. Sci.* **18**. 173-204.

Zillig, W., Yeats, S., Holz, I., Böck, A., Gropp, F., Rettenberger, M., and Lutz. S. (1985) *Nature* **313**, 789-791.

Biodata of **Joseph Seckbach**, editor of this volume, and the chief editor of the book-series of **COLE** (Cellular Origin and Life in Extreme Habitats). He is the author of the chapters "*Acidophilic Microorganisms,*" "*A Vista into the Diverse Microbial World: An Introduction to Microbes at the Edge of Life*" (with co-author A. Oren), and "*Introduction to Astronomy; Origin, Evolution, Distribution and Destiny of Life in the Universe*" (with co-authors F. Westall and J. Chela-Flores) in this current volume.

Dr. Joseph Seckbach edited (and contributed two chapters to) *Enigmatic Microorganisms and Life in Extreme Environments* (Kluwer Academic Publishers, The Netherlands, 1999). See: **http://www.wkap.nl/bookcc.htm/0-7923-5492-3.** Likewise, he organized and contributed to the "*Cyanidium book*" entitled *Evolutionary Pathways and Enigmatic Algae: Cyanidium caldarium (Rhodophyta) and Related Cells* [Kluwer, 1994, see: http://www.wkap.nl/bookcc.htm/0-7923-2635-0]. He is the co-author (with author R. Ikan) of the Hebrew-language publication *Chemistry Lexicon* (1991, 1999) and co-editor of *From Symbiosis to Eukaryotism: Endocytobiology VII* (E.Wagner et al., eds.) published by the University of Freiburg and Geneva (1999).

Dr. Seckbach earned his Ph.D. from the University of Chicago (1965) and spent his postdoctoral years in the Division of Biology at Caltech (Pasadena, CA). Then he headed a team at the University of California, Los Angeles (UCLA) searching for extraterrestrial life. Dr. Seckbach has been with the Hebrew University of Jerusalem since 1970 and performed algal research and taught Biological courses. He spent sabbatical periods in Tübingen (Germany), UCLA and Harvard University. At Louisiana State University (LSU, Baton Rouge), he served (1997/1998) as the first selected occupant of the John P. Laborde endowed Chair for the Louisiana Sea Grant and Technology Transfer, and as a visiting Professor in the Department of Life Sciences.

Among his publications are books, scientific articles concerning plant ferritin (phytoferritin), cellular ultrastructure, evolution, acido-thermophilic algae, and life in extreme environments. He has also edited and translated books of popular science. Dr. Seckbach's recent interest is in the field of enigmatic microorganisms and life in extreme environments.

E-Mail: **seckbach@cc.huji.ac.il**

J. Seckbach (ed.), Journey to Diverse Microbial Worlds, 107-116.
© 2000 *Kluwer Academic Publishers. Printed in the Netherlands.*

ACIDOPHILIC MICROORGANISMS

JOSEPH SECKBACH
Hebrew University of Jerusalem
Jerusalem, 91904, Israel

1. Introduction

In the last decade great interest has developed concerning microorganisms that live and thrive in extreme environments (Seckbach 1999). These life forms are mainly microbes that survive at extreme high or low levels of temperature (thermophiles-psychrophiles), hypersalinity (halophiles), pressure (barophiles), dryness and desiccation, and exceptional ranges of pH (acidity and alkalinity) see Seckbach (1997, 1999), Horikoshi and Grant (1998) and Madigan and Marrs (1997). Further, knowledge of these extremophiles may lead to practical applications for extracting enzymes and isolating chemicals for scientific, medical and industrial usage (Pick 1999). For industrial applications, extremophile cultures have less trouble with contamination, and processing time is shorter.

It has been suggested that the Earth's primordial atmosphere higher in carbon dioxide content, and warmer than today. If the primeval atmosphere has been reduced than perhaps it contained also more ammonia. The microorganisms (mainly Prokaryotes) that currently thrive in hot temperatures and at lower pH levels have been viewed as ancient cells and perhaps the progenitors of the rest of the biological world. Others consider the modern thermoacidophiles as secondarily reduced from the compared ancient common ancestors, so they are not the same and not progenitors (D. Searcy, personal communication).

Microbial responses to various pH ranges are important to many areas of ecology. For example, the mechanism of the passage of resistance to acidic pH is important in understanding the passage of human pathogens through the acid of the stomach. Likewise, the microbial degradation of industrial waste that takes place at higher pH ranges is important for the environment.

Organisms that live in low levels of pH are termed "acidophiles." We are aware of moderate (growing in pH levels under 6.0) and extreme (thriving at pH ranges 0–4) acidophiles. Among them one can find members from all three domains of the living organisms (Archaea, Bacteria, Eukarya), see also chapters by Seckbach and Oren, Weiss Bizzoco, in this volume.

2. Ecology and Chemistry of Acid Environments

Acid hot springs are those in which sulfide derived from magma or wall rock is oxidized by O_2 or NO_3^- to sulfuric acid; free hydrochloric acid may also be involved. Springs biologically labeled "acid" are those with pH values below 4 (Castenholz 1979). Solfatara fields are usually acidic of (pH 5–6) and rich in sulfur at the surface (soil, mud, holes or in surface waters) as in Pozzuoli near Naples (Italy), in Iceland, and in Yellowstone National Park (USA). Marine hydrothermal systems are situated in shallow and abyssal depths consisting of hot fumaroles, springs, and deep sea vents ("black smokers") with a temperature range up to 400°C. Some black smokers' chimney walls are strongly acidic (pH 3). Within terrestrial solfatara soils the upper acidic layer may contain some O_2 and appear ochre-colored due to Fe^{+3}. Solfatara fields abound in sulfur that is formed by oxidation of H_2S at the surface, and sulfate and sulfite are present. Similarly, in seawater sulfate, may exist in high amounts. In most acid environments associated with thermal hot springs oxidation of hydrogen sulfide (H_2S) occurs when it comes out from the volcanic gas and reacts with the air to yield sulfuric acid (H_2SO_4) or hydrogen chloride (HCl). *Thiobacillus thiooxidans* generates sulfuric acid from the oxidation of sulfide or elemental sulfur ($HS^- + 2O_2 \rightarrow SO_4^= + H^+$).

The acidophiles can thrive in such a sour world owing to protective mechanisms. Table 1 shows the lower pH limit of several organisms; cyanobacteria, Bacteria, protists, plants, lower and higher animals. From this table one sees that Bacteria, Protista (algae and fungi) and invertebrates have the lowest values of pH, while the cyanobacteria and the vertebrates as well as the higher plants show higher values of external pH tolerance.

Table 1. Lower pH limits for different Groups of Organisms

Group of Organisms	Lower pH Limit
Archaea	- 0.06 – 5
Bacteria	0 – 1
Cyanobacteria	4 – 5
Algae	~ 0
Fungi	~ 0
Protozoa	< 2
Higher Plants	2 – 4
Invertebrates	< 2
Vertebrates	3.5 – 4

Data adapted from Brock (1971), Schleper et al. (1995), Stetter (1998), Seckbach and Oren (in this volume).

3. Internal pH of Acidophiles

For the integrity of acid–labile molecules, the internal pH of the acidophiles has to be nearer neutral than the external one. For example, compounds like chlorophylls, DNA, and ATP are unstable in acid, break down, and may be destroyed at low internal pH levels. Those microorganisms, which grow at low external pH, have to develop a reliable mechanism to prevent the entrance of hydrogen ions or to exclude them so as to protect the sensitive internal substances. They have to use strong proton pump activity to reject H^+ ions or develop selective permeability properties of the membrane for preventing the uncontrolled entrance of protons into the cells. For further discussion, see Seckbach (1994), Pick (1999) and Seckbach and Oren in this volume.

In the past a technical problem prevented the measurement of the internal pH of many cells (their size measures in microns or even less in diameter). It was predicted and later proven that several extremophilic acidophiles, such as *Cyanidium caldarium, Cyanidioschyzon merolae* and *Galdieria sulphuraria* (Rhodophyta), do not possess an internal extreme lower pH level (Brock 1978, Beardall and Entwisle 1984; Seckbach 1994, Roberts 1999). Those unicellular algae have a maximum growth temperature of 57°C and thrive at very low pH levels (even in 1N sulfuric acid); see Allen (1959), Brock (1978), and Seckbach (1994). The heavy cell wall of *C. caldarium* is composed of a high percentage of proteinaceous molecules with a low percentage of cellulose. Maintenance of the internal pH of *Cyanidium* appears to result perhaps by chloroplast that is making ATP and that powers a H^+-ATPase on the cell membrane. Table 2 represents the ratios between the external and the internal milieu of selected microorganisms. The existence of a steep proton gradient across the plasmalemma of *Cyanidium* may have an important implication for nutrient transport, (symport with H^+), and this merits further investigation.

Sasaki et al. (1999) observed accumulation of sulfuric acid in four species of Dictyotales (Brown Algae, Phaeophyceae). Their vacuoles were estimated to be from 0.5 to 0.9 by pH measurements of their cell extracts in distilled water. It is interesting that other species of these genera do not show such low vacuolar pH levels. The authors cite additional data on sulfuric acid and low pH ranges in other genera of brown algae (see references of Sasaki et al. 1999). Perhaps these algae prevent internal acidity damage by draining all the internal harmful acid into their "safety deposit box" within the vacuoles. The acid tolerance of some green algae (Chlorophyta) and both thermoacidophiles, *Cyanidium* (Rhodophyta) and the archaeal *Thermoplasma acidophilum* are notable in that they maintain a remarkably stable internal pH over an external range of approximately 4 pH units (Table 2). However, at environmental pH values greater than 6.0, the ability of the cells to regulate their internal pH is lost, and there is a marked increase in internal pH as the external pH rises

In the green halophilic alga *Dunaliella acidophila* a combination of factors contributes to reduce the influx of protons and to facilitate extrusion of protons from the cell (Pick 1999). They include positive-inside transmembrane potential, positive surface charge, and existence of a potent plasma membrane H^+-ATPase. This pH homeostatic mechanism in *D. acidophila* allows an internal cytosolic solution of pH 7 while outside it may be at pH 1 (see Figure 1 in Pick, 1999). The data for the ΔpH across the cell membrane in *D. acidophila* are 7.2 (internal) vs. pH 0.5 (external) and also 6.2 (internal) vs. pH 3.0 (external). The average pH gradient (for these two determinations of *D. acidophila*) is ~5 pH units across the plasma membrane (Table 2).

Table 2. Some values of internal vs. external pH values for Prokaryotes and Eukaryotes.

	External pH (A)	internal pH (B)
Archaea and Bacteria*		
Thermoplasma acidophilum	2.0	6.0
Sulfolobus acidocaldarius	2.5	6.0
Bacillus acidocaldarius	3.0	6.3
Algae		
Cyanidium caldarium	2.1	6.6
Chlorella saccharophila	4.0	7.1
Chlorella vulgaris Beij	5.3	6.6
Chlorella pyrenoidosa Chick	3.1	7.0
Chara corallina Klein ex Willd.	4.5	7.3
Scenedesmus quadricauda (Turp.) Breb.	3.1	6.9
Euglena mutabilis Schmitz	2.8	5.7
*Dunaliella acidophila****	(0.5)-3.0	(7.2)-6.2

Adapted from data of Langworthy (1979)*, Beardall and Entwisle (1984)** and Pick (1999)***

4. Moderate Acidophiles

Many soils and lakes are mildly acidic, and several algae (Table 2) such as the green algae *Chlorella* and *Chara* have been found in these habitats. Among the moderate acidophiles are the protists and fungi as well as the rumen microbes within the ruminate animals. For example, some acidophiles live in the digestive track of termites or in other insects. There are yeast and fungi that are able to grow at ~pH 2, although they usually show optimal growth around pH 5.5 to 6.0.

5. Hyperacidophiles

In general, Prokaryotes have adapted more readily to extreme conditions than have Eukaryotes. There are several microbes in the animals' stomachs where the pH is 1–2. In other acidic niches, hydrogen sulfide is chemically oxidized to sulfate ($SO_4^=$) resulting in low pH (H_2S is probably oxidized to H_2SO_4, and Na_2S to sulfate).

Extreme acidophilic hyperthermophilic organisms were found within terrestrial solfatara fields. These microbial cells are coccoid-shaped aerobes, facultative anaerobes, or obligatory anaerobes. They are acidophilic due to their need for acidic pH ~3, and all belong to *Sulfolobales* branch. Members of *Sulfolobus* are strict aerobes growing autotrophically by either oxidizing S to $SO_4^=$ or reducing it to H_2S. (Segerer et al. 1985). Weiss Bizzoco (1999) studies new prokaryotic Archaea and Bacteria in Yellowstone National Park. He has identified a half dozen new thermoacidophilic cells and presented their fine structures (see chapters by Weiss Bizzoco, Seckbach and Oren, in this volume). *D. acidophila* and *C. caldarium* are among the most extreme acidophilic algae which flourish at pH 0–3 with a optimum at pH of 1 and 4, respectively. *D. acidophila* is a mesophile that grows at a temperature of 18–26°C, while *C. caldarium* is a thermophile and is found at a maximum temperature of >55°C. Schelper et al. (1995) reported that *Picrophilus oshimae* and *P. torridus* have isolated from extremely acidic places and thrive at ~ 60°C, these Archaea have the lowest known acidity, with an optimum of pH 0.7, while the lowest pH was at −0.06±0.01 measured at 20°C. The authors questioned the physiological aspects as how these hyperacidophiles cope with their severe environment, which is either with a strong proton pump or a low proton membrane permeability (Schelper et al. 1995).

5.1. FURTHER ON ACIDOPHILIC PROKARYOTES

The best-known caldoactive acidophilic organism is the facultative chemolithotroph *Sulfolobus acidocaldarius*. It is widespread in acid hot springs and hot volcanic soils, where it oxidizes (with O_2) elemental sulfur to sulfuric acid (or Fe^{2+} to Fe^{3+}), using CO_2 as its carbon source. It does so optimally at 70–75°C and at pH 2–3 but extends its range to waters that are constant at 90°C (Castenholz 1979). The archaeol *Stygioglobus azoicus* grows in a medium of pH 2.5–3 (Roberts 1999, Seckbach and Oren, in this volume) while the thermophilic archaeol cells, *Picrophilus* from Japanese solfataras are hyperacidic and grows at slightly below pH 0 (Schleper et al. 1995). Weiss Bizzoco (1999) has recently reported the features and ultrastructures of five Archaea-like and Bacteria-like new organisms living in high temperature (76 to 89°C) and low pH levels of 1.5 to 2.3. The hyperthermophilic archaean *Pyrolobus fumarii* (its maximum growth occurs at 113°C) lives at pH range of 4 to 6.5 (Stetter 1998, Madigan and Oren 1999).

The archaeal thermoacidophilic *Thermoplasma acidophilum* is related in several of its features to eukaryotic cells (Searcy et al. 1978). Furthermore, certain thermophilic characteristics could be preadaptations for the eukaryotic condition. Seckbach (1994, 1996, 1997) and Seckbach et al. (1993) suggested that such an archaean cell might serve as the host for an endosymbiotic cyanobacteria which perhaps lead to the establishment of the Cyanidiacean chloroplast. *T. acidophilum* lacks a cell wall and its cytoplasmic membrane has to be the protective border to avoid the entrance of H^+ ions. This archaean cell is restricted to hot, acidic environments, it was initially discovered in a burning refuse pile outside of an abandoned coal mine at pH 1–2 and a temperature of 59°C (Searcy et al. 1978). Since then it has been isolated worldwide, wherever there are acidic soils containing sulfur and a temperature of at least 32°C. Thermal acidic environments were once more common on Earth than they are today, and their current inhabitants might be progeny of the ancient primordial cells. Thus, conditions similar to those required by *T. acidophilum* apparently were common on the Earth in Precambrian times, and thermophilic organisms were probably much more prevalent than they are today. Recently archaean *Thermoplasma*-like rRNA has been purified from cold seawater and a variety of non-extreme environments (DeLong et al. 1994, Hinrichs et al. 1999).

5.2. "SOUR" ALGAE

A few green algae like *Chlamydomonas acidophila* (Chlorophyta) grow under pH 1–2 and have been observed in Japanese and Czechoslovakian volcanic waters. The chlorophytan *Euglena mutabilis* was found at pH levels of 1 to 5 (Gessner, 1959). Seckbach (1997) lists additional species of algae known to be acidophiles. The ubiquitous and obligatory thermoacidophilic red alga *Cyanidium caldarium* (and its cohorts) thrive under very acidic conditions of pH 0 to 4 (Seckbach 1992, 1994, 1999). Allen (1959) reported that *C. caldarium* also grows in a solution of 1 N H_2SO_4 (placing this algal culture in such a strong acid solution can be a useful method for purifying these cells of contaminants). *C. caldarium*, is considered a primitive eukaryote, its members may serve as "Bridge" algae (between the prokaryotic cyanobacteria and the primitive Rhodophytes) see Seckbach et al. (1983). Seckbach and coworkers proposed an interspecies evolutionary line within the Cyanidiaceae from *Cyanidioschyzon*, the most primitive member of this group ⇨ *Cyanidium*, the medium alga and best known member ⇨ *Galdieria*, the more advanced cell resembles the lower rhodophytes (Seckbach et al. 1983, Seckbach 1992, 1994, 1999). It is interesting that some species of this genus have been observed inside caves, growing at ambient conditions of temperature and neutral pH levels (Hoffmann 1994, Seckbach 1994, 1999). Tables 1 and 2 lists additional species of acidophilic algae.

Carbon dioxide streaming into the cultural media will lower the pH of these solutions to about pH 5 (since $CO_2 + H_2O$ ⇨ $HCO_3^- + H^+$). Many algae can not survive such pH lowering of their media. Seckbach and Libby (1970, 1971) treated several algal cultures with a bubbling stream of pure CO_2 into their media. Most of these algal genera could not survive this treatment because of the pH decline in their culturing media. It is interesting that one culture[1] of *Scenedesmus sp.* flourished with the increase of CO_2. Furthermore, these cells showed morphological structural changes from thin crescent-like (in air) to more spherical cells in the CO_2-enriched cultures (Seckbach and Libby 1970, 1971). Likewise *C. caldarium* shows a strong "love" for pure CO_2 and indicated a larger cell count and higher rate of photosynthesis under such semi-anaerobic treatments (Seckbach et al. 1970). At this strong acidic solution only CO_2 species is dissolved and absorbed, while HCO_3^- is not present or taken up by the algae. For further data see Seckbach and Walsh (1999) and Seckbach and Oren (this volume).

5.3. TART YEAST AND FUNGI

The acidophilic thermophilic yeast *Saccharomycopsis guttulanta* grows in the stomach of rabbit (Brock 1969). Roberts (1999) reported on the Eukaryote acidophiles and listed additional acidophilic fungi like *Acontium cylatium, Cephalosporium sp.,* or *Trichospore cerebriae* (see, Schleper et al. 1995, Seckbach and Oren, in this volume). There are many other yeast and fungi that are able to grow at pH values down to 2.

---

[1] We termed this species "ZAV" since, firstly, at the beginning we did not identify its taxon and secondly, because we obtained it from the Russian microbiologist **George Zavarzin**.

6. Summary

Acidophilic microorganisms are found in all three domains of life. These acido-extremophiles grow at pH levels of 0 to 4. Among the acidophiles are sulfur bacteria, Archaea and phototrophic hot spring protists like the thermoacidophilic alga *Cyanidium caldarium* (Seckbach 1994, 1997), *Dunaliella acidophilum* (Pick 1999), and diatoms. Acidothermophilic Archaea include *Sulfolobus*, a chemoautotroph living in hot-springs as well as *Thermoplasma*, a heterotroph in hot coal tailings; both thrive in hot acidic habitats. The hyperthermophilic archaean *Pyrolobus fumarii* (max. growth at $113^{\circ}C$) lives in a pH range of 4 to 6.5 (Stetter 1998, Madigan and Oren 1999).

The extreme acidity surrounding of these microorganisms is mainly external, the internal cellular medium has been found to be around neutral. It is known that high intracellular acidity may impair the function of several organelles and cell constituents. DNA is especially sensitive to damage by depurination. Acidophiles may use a proton pump for selective proton membrane permeability; such a defense mechanism would assist them to maintain their intracellular pH at a neutral value. Recently the accumulation of sulfuric acid has been reported in the vacuoles of some brown algae (Sasaki et al. 1999).

7. Acknowledgement

I thank Professors Dennis Searcy (University of Massachusetts at Amherst) and Rick Weiss Bizzoco (San Diego State University) for critically reading the manuscript and pointing out their constructive comments.

7. References

Allen, M.B. (1959) Arch. *Mikrobiol.* 32: 270-277.

Beardall, J. and Entwisle, L. (1984) Phycologia 23: 397-399.

Brock, T.D. (1969) in: P. Meadow and S.J. Pirt (eds.) *Symposia Soc. Gen. Microbiol.* XIX. *Microbial Growth.* Great Britain. pp. 15-41.

Brock, T.D. (1971) in: I. A. Bernstein (ed.) *Bichemical Responses to Environmental Stress* Plenium Press, N.Y. pp. 32-37.

Brock, T.D. (author) (1978) *Thermophilic Microorganisms and Life at High Temperatures.* Springer Verlag, New York-Heidelberg-Berlin. pp. 255-302.

Castenholz, R.W. (1979) in: M. Shilo (ed.) *Strategies of Microbial Life in Extreme Environments.* Life Sciences Research Report 13. Berlin: Dahlem Konferenzen. pp. 373-392.

DeLong, E.F., Wu, K.Y., Prézelin, B.B. and Jovine, R.V.M. (1994) Natrure 371: 695-697.

Gessner, F. (1959) Hydro Botanik, Deutscher Verlag der Wissen. Berlin. II: 280-284.

Hinrichs, K-U, Hayes, J.M., Sylva, S.P., Brewer, P.G. and DeLong, E.F. (1999) Nature 398: 802-805.

Hoffmann, L. (1994) in: J. Seckbach (ed.) *Evolutionary Pathways and EnigmaticAlgae: Cyanidium caldarium (Rhodophyta) and Related Cells,* Kluwer Academic Publishers, Dordrecht, The Netherlands. pp. 175-182.

Horikoshi, K. and Grant, W.D. (eds.) (1998) *Extremophiles: Microbial Life in Extreme Environments,* Wiley-Liss Publication, New York.

Langworthy, T.A. (1979) in: M. Shilo (ed.) *Strategies of Microbial Life in Extreme Environments.* Life Sciences Research Report 13. Berlin: Dahlem Konferenzen. pp. 417-432.

Madigan, M.T. and Marrs, B. L. (1997) Sci. Amer. 276 (4): 66-71.

Madigan, M.T. and Oren, A. (1999) Curr. Opinion in Microbiol. 2: 265-269.

116

Pick, U. (1999). in: J. Seckbach (ed.) *Enigmatic Microorganisms and Life in Extreme Environments.* Kluwer Academic Publishers, Dordrecht, The Netherlands, pp. 465-478.

Roberts, D. (1999) in: J. Seckbach (ed.) Enigmatic Microorganisms and Life in Extreme Environments, Kluwer Academic Publishers, Dordrecht, The Netherlands. pp 163-173.

Sasaki, H., Kataoka, H., Kamiya, M. and Kawai, H. (1999) J. Phycol. 35 (4):732-739.

Searcy, D.G., Stein, D.B. and Green, G.R. (1978) BioSys. 10: 19-28.

Seckbach, J. (1992) in: W. Reisser (ed.) *Algae and Symbioses.* Biopress Limited.Bristol, UK. pp. 399-426.

Seckbach, J. (ed.) (1994) *Evolutionary Pathways and EnigmaticAlgae: Cyanidium caldarium (Rhodophyta) and Related Cells,* Kluwer Academic Publishers, Dordrecht, The Netherlands.

Seckbach, J. (1996) in: J. Chela-Flores and F. Raulin (eds.) *Chemical Evolution: Physics of the Origin and evolution of Life.* Kluwer Academic Publishers, Dordrecht, The Nethelands. pp. 197-213.

Seckbach, J. (1997) in: C.B. Cosmovici, S. Bowyer and D.Werthimer (eds.) *Astronomical and Biochemical Origins and the Search for Life in the Universe.* Proceeding of the 5th International Conference on Bioastronomy IAU Colloquium No. 161 (Capri, July 1-5, 1996) Editrice Compositori, Italy. pp 511-523.

Seckbach, J. (1999) In: J. Seckbach (ed.) *Enigmatic Microorganisms and Life in Extreme Environments.* Kluwer Academic Publishers, Dordrecht, The Netherlands, pp. 425-435.

Seckbach, J. and Libby, W.F. (1970) Space Life Sci. 2: 121-143.

Seckbach, J. and Libby, W.F. (1971) in: C. Sagan, T.C. Owen and H.J. Smith (eds.) *Planetary Atmospheres,* IAU, Symposium No. 40. D. Reidel Publishing Company, Dordrecht-Holland. pp. 62-83.

Seckbach, J. and Walsh, M.M. (1999) in: E. Wagner, J. Normann, H. Greppin, J.H.P. Hackstein, R.G. Herrmann, K.V. Kowallik, H.E.A. Schenk and J. Seckbach (eds.). *From Symbiosis to Eukaryotism, Endocytobiology VII.* University of Geneva, pp. 85-104..

Seckbach, J., Baker, F.A. and Shugarman, P.M. (1970) Nature 227: 744-745.

Seckbach, J., Fredrick, J.F. and Garbary, D.J. (1983) in: H. Schenk and W. Schwemmler (eds.) *Endocytobiology II.* Walter de Gruyter and Comp. Berlin. pp. 947-962.

Seckbach, J. Ikan, R., Nagashima, H. and Fukuda, I. (1993) in: S. Sato, M. Ishida and H. Ishikawa (eds.) *Encdocytobilogy V.* Tübingen University Press. Tübingen, Germany. pp. 241-254.

Segerer, A., Stetter, K.O. and Klink F. (1985) Nature 313: 787-789.

Schleper, C., Pühler, G. , Kühlmorgen, B. and Zillig, W. (1995) Nature 375: 741-742.

Stetter, K.O. (1998) in: K. Horikoshi and W.D. Grant (eds.) *Extremophiles: Microbial Life in Extreme Environments.* Wiley-Liss, New York. pp. 1-24.

Weiss Bizzoco, R.L. (1999) in: J. Seckbach (ed.) *Enigmatic Microorganisms and Life in Extreme Environments.* Kluwer Academic Publishers, Dordrecht, The Netherlands. pp. 305-314.

Biodata of **R. L. Weiss Bizzoco** author of *"New Acidophilic Thermophilic Microbes."*

Dr. **Richard L. Weiss Bizzoco** is a Professor of Biology at San Diego State University. He earned his Ph.D. from Indiana University (1972) and spent his postdoctoral years in Berkeley, Harvard, and at UC Irvine Medical School. His current interest is characterization of new hyperthermal acidophiles growing at 700C to 900C at pH 2-3.
E-mail: <**rbizzoco@sunstroke.sdsu.ed**>.

J. Seckbach (ed.), Journey to Diverse Microbial Worlds, 117-128.
© 2000 *Kluwer Academic Publishers. Printed in the Netherlands.*

NEW ACIDOPHILIC THERMOPHILIC MICROBES

R.L. WEISS BIZZOCO, N. BANISH, M. LU, and S. SAAVEDRA
Department of Biology, San Diego State University
San Diego, CA 92182

1. Low Ph High Temperature Environments

1.1. NEW ORGANISMS

We have examined samples from low pH high temperature habitats and found an unusual variety of new microbes within both mixing pools and flowing springs. These new microbes were all discovered in the acid hot springs of Yellowstone National Park, WY, USA (YNP). The mixing pools are turbid habitats with free floating microorganisms. In contrast, the organisms discovered in the channels of flowing hot springs were attached to sulfur crystals like *Sulfolobus*-type cells in nearby springs, but they are rod shaped rather than coccoid like *Sulfolobus* (Brock *et al.*, 1972). In the sections that follow we present information on the variety of new rod shaped microbes growing below pH 3 and at 70°C or higher.

2. New Microbes From A Mixing Pool

2.1. A ROD SHAPED ORGANISM WITH FLAGELLA

Whole mounts of samples were examined by electron microscopy of negatively stained cells and by thin section electron microscopy. On one rod shaped cell prepared by negative staining of whole mount samples (*Fig. 1*) we observed long wavy filaments which were easily identified as flagella. On the cell shown here the polar location of flagella can be seen. They usually originate from the end of the cell. On other organisms from acid hot springs such structures are made of protein or glycoprotein (Jarrell *et al.*, 1996) and function as organelles of motility, and that is their likely function here as well. Cells we observed had from 1 to 3 flagella at one pole. Some cells had bipolar flagella. As shown here, these cells usually had two flagella at one end and one fragmented flagellum at the opposite end. The flagella were either wavy or straight (*Fig. 2*). This observation is identical with our unpublished electron microscopic observations of *Thermoplasma* flagella. When observed at higher magnification these filaments show a distinct linear substructure. A typical example is shown in *Fig. 1*, inset. Near the insertion point the flagellum appears as thin linear elements (Fig 1, arrow). These are shown to be continuous as parallel lines running along the long axis of the flagellum. Pili were occasionally observed at the pole opposite the flagellum. These were thin short pili with a diameter of ~4 nm and a length of about 3 μm. They appeared to extend directly away from the pole of the cell and were somewhat irregular in appearance. They were not seen along the sides or over the surface of the cell. The cell shape was that of a short wide rod. Staining with the DNA stain, DAPI (not shown), revealed a condensed nucleoid, indicating that these cells were most likely in an active phase of cell division. *Sulfolobus* cells that actively divide in culture had a similar arrangement of DNA condensed into a nucleoid (Poplawski and Bernander, 1997). In stationary phase *Sulfolobus* cells the staining was more or less uniform over the volume of the cell. The wide cell morphology of this short rod readily permits their identification by means of phase contrast microscopy, particularly when it is coupled with the characteristic DNA staining pattern. These cells were commonly seen in samples from Frying Pan Spring and comprised the second most abundant cell type observed in this habitat. The cell wall of this organism lacked any obvious subunit structure such as that displayed by other rod shaped microbes that have a *Sulfolobus* like cell wall. Thus, this wide short rod most

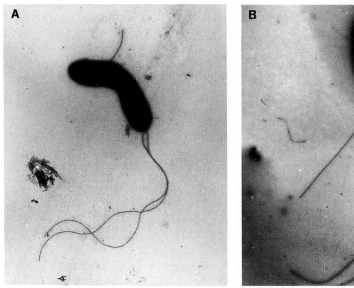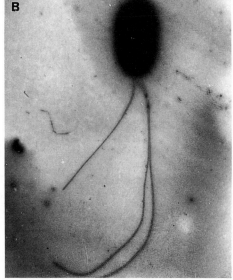

Figure 1. *A.* Wide short rod with two flagella originating from one end and a single flagellum arising from the opposite pole. The flagella are intertwined and have smooth undulations. Inset, Enlargement showing flagellar insertion point and linear substructure of flagellum (arrow). *B.* Rod shaped cell with 3 flagella arising from one end of the cell. One flagellum is straight while the other to are curved. Compare with the wavy flagella in *A.*. Frying Pan Spring, YNP: 75.5°C; pH 2.7. A., B. X 8, 500.

2.2. A ROD SHAPED ORGANISM WITH A SURFACE ARRAY

In our examination of whole mounts after negative staining we encountered a second new organism growing in Frying Pan Spring. This organism, was filamentous and was characterized by a distinctive surface layer of subunits. This is shown in *Fig. 2*. The cell is a regular rod with rounded ends. The cell did not show surface wrinkling characteristic of microbes with a layered cell envelope structure. Other rod-shaped cells with a surface array have been recognized in YNP (see Section 7). The organism shown in *Fig.* 2 has a very distinct two dimensional wall lattice and is filamentous. It is not a thin filament as its mean diameter is about 0.45 μm and it is approximately 5 μm long. The cell wall subunits appear to form a hexagonal array. Each unit cell presents a solid center that seems to limit the penetration of the aqueous uranyl acetate negative stain. This response differs from the staining character of *Sulfolobus* cell walls which showed penetration of stain within a hole in the center of the subunit. The wall array in *Fig.* 2 seems to be robust in arrangement as there is almost no obvious collapse of the cell envelope as with other filamentous cells observed in samples taken from this same extreme habitat.

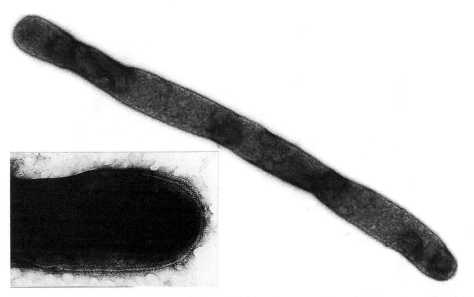

Figure 2. Elongated filamentous organism with a regular array over the cell surface. The two dimensional lattice of the cell surface layer is ordered as a hexagonal array. Inset shows details of the cell wall on a different cell. Frying Pan Spring, YNP: 75.5°C; pH 2.7. X 17, 000. Inset. X 67, 000

2.3. TWO THIN FILAMENTOUS ORGANISMS

When we examined the mixture of microbial types, we were able to discern subtle differences among the numerous rod shaped cells in our samples from Frying Pan Spring. The main cell type was a thin filamentous organism (*Fig. 3*) with a diameter of 0.30 μm and a variable length measuring from 4 to 35 μm in whole mounts or in samples observed by phase contrast microscopy. Our DAPI staining showed extended lengths of DNA in cells. These cells were neither flagellated nor piliated. Their cell surface as shown here was layered. The cells had no bulbous outpocketings or spheres at the ends of cells or branching described on other mildly acidic thermophiles like *Thermocladium* (Itoh *et al.*, 1998). The resemblance to *Thermophilum* and

Thermoproteus (Thermoproteales) relates mainly to the filamentous character of the cell, since the cell surface (*Fig. 2*) seems unlike the archaeal type of subunit cell wall described for the above organisms in the Thermoproteales (Stetter and Zillig, 1985).

A second cell type with a diameter of 0.4 μm was also present (*Fig. 2*). It had a length of about 8 μm and a layered cell wall. Like the other cell it had no pili or flagella or other structures arising from the cell surface. No array or pattern appeared on the cell surface.

Figure 3. Long thin filaments of two different cell diameters, 0.3 μm and 0.4 μm. Frying Pan Spring, YNP: 75.5°C; pH 2.7. X 87, 000.

3. Preservation of Organisms Living at Low pH and High Temperature

3.1. COMPARISON OF PRESERVATION METHODS

The features of these new microorganisms include both flagella and pili. The basic cell wall structure of cells with an S-layer can also be resolved. Thin section electron microscopy was used to reveal the other features of the cell wall. All of the organisms we observed from Frying Pan Spring were rod shaped. They had either a layered cell wall structure or an arrayed cell wall when observed after negative staining. We observed the other features of the cell wall in sections by comparing their structure using three different preparative methods, 1) vapor fixation in spring water, 2) conventional fixation, and 3) freeze substitution. We used this approach because the cells grow in nature at pH 2.7, but it is recognized that the chemical fixatives are most effective at near neutral pH. This leaves a dilemma as to the most appropriate way to preserve cells from acid habitats.

3.2 VAPOR FIXATION AT pH 2.0

Samples from Frying Pan Spring were fixed by exposing the cells in spring water to osmium tetroxide vapors in a closed chamber at 23°C. After the cells were fixed ultrathin sections were examined in the electron microscope. A representative sample is shown in Figure 4. The cell envelope layers appear in section along with short lengths of sectioned flagella (arrow). The cell carries open areas representing lost or extracted cytoplasm. The cells range in diameter from 0.20-0.25 μm X ~6 μm in length.

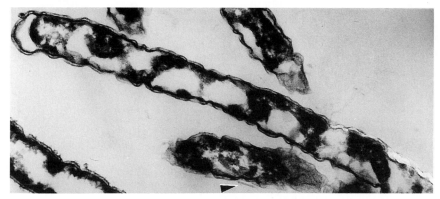

Figure 4. Vapor fixation in osmium tetroxide in spring water. Cell filament with layered cell envelope. The cytoplasm contains clear partly extracted or empty areas, but the envelope structure and flagella (arrow) are preserved. Frying Pan Spring, YNP: 75.5°C; pH 2.7. X 47, 500.

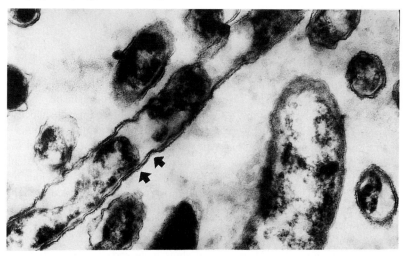

Figure 5. Thin and thick cells. Cells fixed at pH 7, embedded and sectioned. Thin filaments have a cell envelope bearing subunits (arrows). Frying Pan Spring, YNP: 75.5°C; pH 2.7. X 42, 000.

3.3 FIXATION BY CONVENTIONAL METHODS AT pH 7.0

Samples from Frying Pan Spring were cooled to room temperature and fixed in glutaraldehyde and osmium tetroxide by conventional methods using HEPES buffer at neutral pH. The results of this method are shown in *Figure 5*.

3.4 FREEZE SUBSTITUTION AT pH 2.0

Samples from Frying Pan Spring were concentrated by centrifugation and a pellet of cells was rapidly frozen in spring water using liquid propane at -195°C. Frozen samples were substituted (-90°C) in osmium tetroxide-acetone until suitably preserved. The resulting cells are shown in *Fig. 6*.

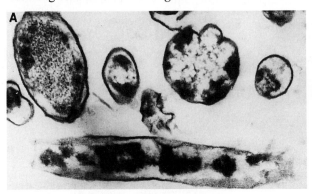 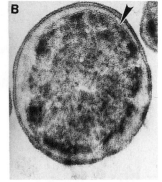

Figure 6. Freeze substitution. *A*. Cell filaments. *B*. Cross sectioned cells with smooth wall and membrane (arrowhead) Frying Pan Spring, YNP: 75.5°C; pH 2.7. A. X 39, 000. B. X 56, 500.

4. New Microbes from a Flowing Spring

4.1. EXAMINATION OF A THERMAL GRADIENT

Amphitheater Springs is a flowing sulfur spring. We sampled elemental sulfur crystals at temperatures from 70 to ~80°C. *Figure 7A* shows the microorganisms on sulfur at a site near the origin of the spring at 78°C. At this temperature both rod-shaped (R) and spherical *Sulfolobus* type (S) cells were present in about equal numbers. In other springs in this same solfatara basin, *Sulfolobus* type cells are usually more abundant than rod-shaped cells. At 78°C in this spring many rod shaped organisms are present. At a slightly lower temperature of 74°C (*Fig. 7B*) thin rods were the most obvious microbes on the sulfur crystal. There was one unusual cell type with a rounded end (Fig. 7B, inset). This feature could represent a site of active growth for these cells. Lobed spherical cells were also seen at 74°C. At 70°C rod-shaped cells attached to sulfur were far more abundant (Fig 7C) than the *Sulfolobus* type cells.

We found both long thin filaments and short wide rods in this particular spring. The filaments were of moderate length 2-10 μm, whereas in nearby springs at the same temperature they were extremely long, up to 100 μm in length. The diameter of filaments varied from a low value of 0.13 μm to a value of 0.80 μm. A common feature of the samples we examined was a 'carved out' sulfur crystal surface near the attached cell as illustrated in *Fig. 7B*.

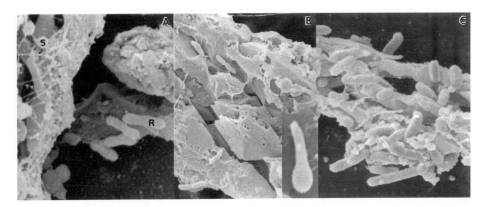

Figure 7. New microbes along a thermal gradient. *A.* Rod-shaped cells appear at 78°C along with lobed spheres (S). *B.* The rods increase in number at 74°C and align along the sulfur crystal surface. Inset. Cell with one end enlarged. *C.* At 70°C rod shaped cells attach to sulfur crystals and form large cell masses. Amphitheater Springs, YNP: 80°C; pH 2.0. A. X 2,600. B. X 1,200. Inset. X 4, 500. C. X 1,600.

4.2. X-RAY ANALYSIS OF SULFUR CRYSTALS

To aid in the confirmation that cells were in fact attached to sulfur rather than some other crystalline material, we examined the distribution of sulfur in crystals using digital X-ray mapping. *Fig. 8 A,B* shows a sulfur-only digital map of the crystal structure. We found that most of the sulfur mapped to the crystals to which cells in our samples attached. We also found that Si may be a minor component that co-localizes with sulfur crystals in the flowing hot spring habitat. We next examined the composition of the observed crystals (*Fig. 7*) using X-ray microanalysis. In the resulting X-ray spectrum (*Fig. 8*) a major peak for sulfur was seen at 2.307 keV. This peak represents the major line spectrum (Kα) for this element. The other peaks in the X-ray spectrum can be accounted for by the gold-palladium (Au, Pd) coating, the adhesive (C), the aluminum (Al) stub, and other elelments (Si) in the sample.

4.3. ATTACHMENT TO SULFUR CRYSTALS

Sulfolobus attaches to sulfur by pili (Brock., 1978). To determine how rod-shaped cells attach, we examined our samples by negative staining of whole mounts. shows On one cell with a surface array (*Fig. 9A*) we found no pili or flagella. Many of the cells, including some with a cell wall array had 1 or 2 pili. Our microscopic observations show that rod-shaped cells attach to sulfur by their poles or along the length of the cell. Pili were found at both sties on many of the cells we examined. As shown in *Figure 9B* some cells had both flagella and pili. The cell illustrated here has a portion of the flagellum and 3 polar pili. The flagellum on this cell has thin lines along its length. In this sense it resembles the flagellum presented in *Fig. 1.* An unusual feature of this particular flagellum is the presence of a feather-like fringe (*Fig. 9, inset*) that extends at a trailing angle from the flagellar surface. The most likely function of such an architectural feature is to increase the hydrodynamic efficiency of the flagellar beat. In

126

contrast, pili most likely function in the attachment of these new rod-shaped microbes to sulfur, as with *Sulfolobus*.

A major difference was observed both in the attachment of rod-shaped cells and in the presence or absence of pili on these cells. With *Sulfolobus* as large numbers of cells accumulate the pili often form rope-like aggregates as the cells layer on the sulfur crystals. This gives rise to bundles of 50 or more pili from several different cells. These 'ropes' were just visible at the level of the scanning electron microscope (see *Fig. 7A* {S}) as thin stranded material. In whole mounts observed after negative staining pili in these rope-like structures are particularly prominent. With rod-shaped cells as the numbers of organisms increase, the stability of attachment to sulfur relies not on the presence of large numbers of pili, but instead is increased by direct cell-cell connections (*Fig. 7C*) between organisms already attached to the sulfur crystal surface.

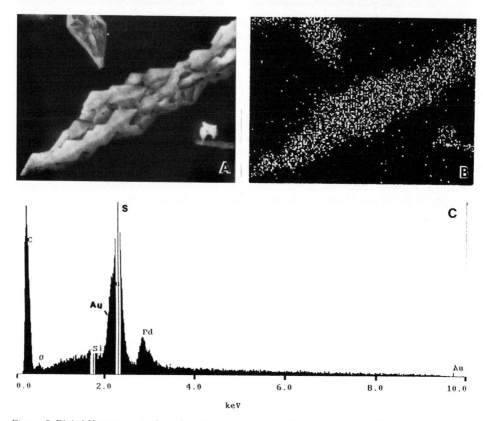

Figure. 8. Digital X-ray map. *A.* Scanning electron micrograph of sulfur crystals. *B.* Digital X-ray map of sulfur in *Fig. 8A*. Sulfur localizes mainly to the crystal structures seen in *Fig. 8A*. *C.* X-ray spectrum from *Fig. 8A*. Sulfur (S) is the main element. Amphitheater Springs, YNP: 70°C; pH 2.0. A, B. X 450.

The properties of new microbes described in our study are summarized in Table 1., Here we identify the significant features of each new organism observed at a specific sampling site along with the sample temperature and pH. We have limited this summary to those organisms that we have been able to recognize repeatedly in our analysis.

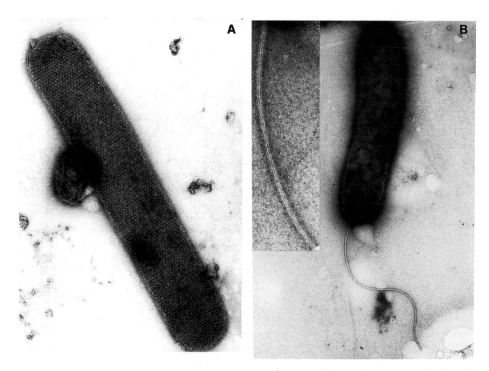

Figure 9. Features of cells attached to sulfur. *A.* Cell wall array on non-piliated cell. *B.* Flagellated cell with 3 polar pili. Inset shows fringe along flagellar surface and thin lines within flagellum. Amphitheater Spring 2, YNP: 74°C; pH 2.0. A. X 36, 000. B. X 18,500. Inset. X 93, 500.

TABLE 1. New Microbes in acid thermal habitats

Location	(°C)	pH	Diameter (μm)	Properties
Amphitheater Spring 2	74	2.0	0.4-0.45	subunit wall
	74	2.0	0.7-0.8	subunit wall flagella, pili
Frying Pan Spring	75.5	2.4	0.3	layered wall thin filament
	75.5	2.4	0.4	layered wall thin filament
	75.5	2.4	0.4	moderate size filament subunit wall
	75.5	2.4	0.7-0.8	curved rod; short and wide flagella, pili

128

4. Summary

We have identified several new microbes in acid hot springs. We examined two different habitats, a mixing pool and a flowing spring with sulfur deposits. New microbes were found in both habitats. The organisms in the mixing pool were examined by thin section electron microscopy and found to have a layered cell wall structure resembling that of Gram negative bacteria. Other organisms in the pool displayed a cell wall array. The archaea-like organism in the flowing sulfur springs are characterized by a subunit wall resembling that of *Sulfolobus*. Negative staining revealed the presence of both flagella and pili on cells from both habitats.

In our early studies we found these new organisms were readily isolated and easy to maintain in culture. The only organism so far reported at temperatures above 70°C at low pH is *Sulfolobus*. Taken together this collection of new organisms from low pH high temperature environments represents a new area of investigation that is worthy of further study.

5. Acknowledgments

I wish to acknowledge John D. Varley, the Director of the Yellowstone Center for Resources and the National Park Service for providing the necessary resources and access to the hot springs of Yellowstone National Park, WY, USA Steven B. Barlow provided facilities and training for electron microscopy and X-ray analysis.

6. References

Brock, T.D., Brock, K.M., Belly, R.T., and Weiss, R.L. (1972) *Sulfolobus*: a new genus of sulfur-oxidizing bacteria living at low pH and high emperature, Arch. Mikrobiol. **84**, 54-68.

Brock, T.D. (1978) Thermophilic microorganisms and life at high temperature. Springer-Verlag, New York.

Itoh, T., Suzuki, K-i., and Nakase, T. (1998) *Thermocladium modestius* gen. nov., sp. nov., a new genus of rod-shaped extremely thermophilic crenarchaeote. Int. J. Syst. Bacteriol. **48**, 879-887.

Jarrell, K.F., Bayley, D.P., and Kostyukova, A.S. (1996) The Archael Flagellum: a Unique Motility Structure, J. Bacteriol. **178**, 5057-5064.

Poplawski, A. and Bernander, R. (1997) Nucleoid structure and distribution in thermophilic archaea. J. Bacteriol. **179**, 7625-7630.

Stetter, K.O. and Zillig, W. (1985) *Thermoplasma* and the thermophilic sulfur dependent Archaebacteria, The Bacteria **8**, 85-170.

IV

PSYCHROPHILES AND BAROPHILES

Biodata of **Dr. Ronald W. Hoham** author (with co-author H.U. Ling) of *"Snow algae: The effects of chemical and physical factors on their life cycles and populations."*

Dr. Ronald W. Hoham is a Professor in the Department of Biology, Colgate University, Hamilton, NY. He received his Ph.D. from the University of Washington, Seattle, WA. in Botany (Phycology) in 1971. His main research interests are: Cryophilic algae, (snow algae) in temperate North America with emphasis on evolution, speciation, life histories, physiological ecology, nutrition, cultures and distribution. Dr. Hoham has published numerous articles in scientific journals; written book chapters; acted as a book editor. He also served as, an editor for Phycological journals, as a reviewer for articles submitted to various journals, and reviewed grant proposals as well as text books in Biology, Botany and Phycology.

E-mail: rhoham@mail.colgate.edu

J. Seckbach (ed.), Journey to Diverse Microbial Worlds, 131-145.
© 2000 *Kluwer Academic Publishers. Printed in the Netherlands.*

Biodata of **Dr. Hau Ling** co-author (with Dr. R.W.Hoham). of *Snow Algae: The effects of chemical and physical factors on their life cycles and populations."*

Dr. Hau Ling is currently a Phycologist at the Australian Antarctic Division, Kingston, Tasmania, Australia. He received his Ph.D. degree at the University of Tasmania in the field of desmid reproduction and cytogenetics in 1978. Dr. Ling is an algal taxonomist with extensive experience in the isolation, culturing, life histories and identification of freshwater algae. His recently completed volume on *The Freshwater Algae of Australia* is being published in the *Bibliotheca Phycologica series*. He is currently working on the non-marine algae of the Windmill Islands region, Antarctica, concentrating mainly on the snow algae.

E-mail: **hau_lin@antdiv.gov.au**

SNOW ALGAE:
The Effects of Chemical and Physical Factors on Their Life Cycles and Populations

RONALD W. HOHAM[1] and H. U. LING[2]

[1]*Department of Biology, Colgate University, Hamilton, NY 13346, USA and* [2]*Antarctic Division, Kingston, Tasmania, Australia 7050.*

1. Introduction

Algae, fungi and bacteria are microorganisms commonly found in snow and ice habitats (Kol 1968, Horner 1985, Hoham and Duval 1999). These microbes may encounter conditions of extreme temperature, acidity, irradiation levels, minimal nutrients and desiccation when liquid water is no longer available (Hoham 1992). When liquid water is present, it allows for their growth and reproduction (Pollock 1970), and special adaptations and mechanisms for surviving in cold temperatures occur (Bidigare et al. 1993, Hoham and Duval 1999). Microorganisms play a fundamental role in the biogeochemistry of snow and ice (Jones 1991) and are involved in the primary production, respiration, nutrient cycling, decomposition, metal accumulation and food webs associated with these habitats (Fjerdingstad et al. 1978, Hoham 1980, Jones 1991, Hoham et al. 1993, Hoham and Duval 1999).

Comprehensive lists of snow algal taxa were previously reported from Japan (Fukushima 1963) and worldwide (Kol 1968). Besides the saccoderm desmid, *Mesotaenium berggrenii*, the majority of the dominant snow algae belong to the green algal flagellate class, Chlamydomonadaceae (Ettl 1983), although research on Antarctic snow algae has added a species of *Chlorosarcina* and two species of *Desmotetra* (Ling 1996). Historically, a lack of information on the life cycles of the various species has made the taxonomy of the snow algae difficult and confusing. Recent studies, however, have not increased the number of snow algal species substantially because research has concentrated on the life cycles of individual species rather than on floristics. On the contrary from these studies, several previously described "taxa" were reduced to synonymy when they were found to be nothing more than stages in the life cycle of a single species (Hoham et al. 1979, Hoham 1980).

This chapter focuses on the life cycles and populations of snow algae and how they are influenced by physical and chemical factors in this freshwater habitat. Subjects on snow algae not included here are population ecology, cell physiology and special adaptations, cell structure, interactions with other snow microbes, human interests and future research directions, all of which are covered in an extensive recent review by Hoham and Duval (1999). It also does not include discussions on sea ice that is predominantly a marine environment (See reviews on this subject by Horner 1985, Palmisano and Garrison 1993, Kirst and Wiencke 1995 and Horner 1996) or biological ice nucleation (See the comprehensive bibliography on this subject by Warren 1994).

2. Life Cycles and Laboratory Mating Experiments

2.1. LIFE CYCLES

The life cycles of the green algal flagellates, *Chloromonas* and *Chlamydomonas*, have been studied extensively in snow (Hoham 1980, Ling and Seppelt 1998, Hoham and Duval 1999), and the phases of these life cycles correlate with physical and chemical factors at the time of snowmelt (Hoham 1980, Hoham et al. 1989, Jones 1991, Ling and Seppelt 1993, Hoham and Duval 1999). Active metabolic phases occur in spring or summer when the snow melts, nutrients and gases are available, and light penetrates through the snowpack (Hoham 1980). The process begins with germination of resting spores at the snow-soil interface [old snow - new snow interface in persistent snowfields] producing biflagellate zoospores (Hoham 1980). These cells swim in the liquid meltwater surrounding the snow crystals towards the upper part of the snowpack, and their position in the snowpack is determined by irradiance levels and spectral composition (Hoham et al. 1998). Visible blooms of snow algae occur a few days after germination. Both asexual and sexual biflagellates develop in some species. The sexual cells (gametes) fuse to form resting zygotes, and in other species, asexual resting spores develop directly from asexual biflagellate vegetative cells called zoospores (Hoham 1980, Hoham et al. 1993, Ling 1996). The resting spores eventually adhere to the soil or debris over the soil when the snowpack has melted or remain on old snow in persistent snowfields. From year to year, populations of snow algae stay in approximately the same localities. The resting spores remain dormant during summer, may form daughter cells through cell division after the first freezes in autumn, are covered with new snow in fall, winter and spring, and do not germinate again until the factors repeat themselves as given above (Hoham 1980). The entire life cycle process is illustrated in Figure 1.

Since snow algal flagellates are typically restricted to the snow environment, the species selected for in snow in temperate regions are those requiring minimal nutrients. It is important that snow algae produce some type of resistant spore prior to complete snowmelt, and nutrient depletion in snow appears to correlate with the transition from vegetative phase to resistant spore (Czygan 1970, Hoham et al. 1989, Jones 1991). The selection for phototactic species allows populations to migrate for optimal photosynthesis and to locate new nutrient sources (Hoham 1975b). In the Antarctic, it is different because penguin rookeries provide the snow environment with abundant nutrients for growth of some snow algae.

Studies of snow algal life histories have emphasized the green algae in the order Volvocales, and these life histories were reviewed by Hoham and Duval (1999). One such life history is that of *Chloromonas brevispina* (Hoham et al. 1979). This shows how biflagellate vegetative cells increase the populations size through the formation of additional biflagellates (zoospores) or are transformed into the sexual gametes that unite producing zygotes that become resting spores (Figure 2).

Asexual reproduction would be the fastest route to produce resting spores because more time is needed to complete the sexual phase of the snow algal life cycle (Hoham 1992), and asexually produced resting spores occur in certain species of *Chloromonas* (Hoham et al. 1993, Ling 1996). Thus in localities where the snowpack is inconsistent

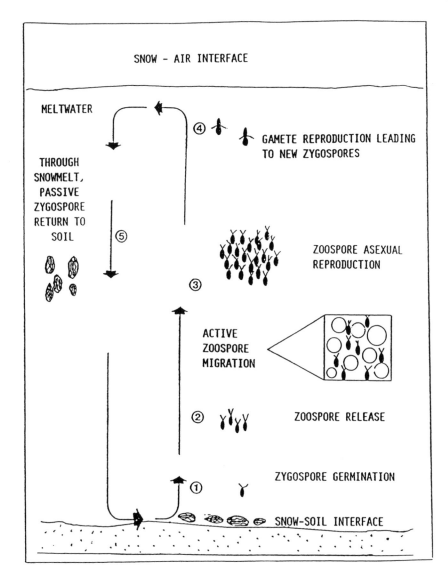

Figure 1. Life cycle of snow algal flagellates such as *Chloromonas* and *Chlamydomonas*. Zygospores germinate releasing zoospores (1 & 2), zoospores become active vegetative cells (3), the latter produce gametes when nutrients become limiting (4) and new zygospores (5) do not germinate until the following year (modified after Gamache, 1990).

from year to year, asexual reproduction may be favored through natural selection. However, some of these populations of snow algae lacking in genetic diversity and living in an inconsistent habitat may be headed for evolutionary extinction.

2.2. LABORATORY MATING EXPERIMENTS.

Using modifications of Hoshaw (1961), laboratory mating experiments revealed that Colgate University strains 593A and 593C were normal + and - mating strains from

136

Québec that produced normal mating pairs and zygospores (Hoham et al. 1993). These were the first normal mating strains of *Chloromonas* isolated from snow in North America, and zygospores with markings similar to those in *C. brevispina* developed in the laboratory from these strains (Hoham et al. 1979, Hoham and Clive, unpubl.). Other

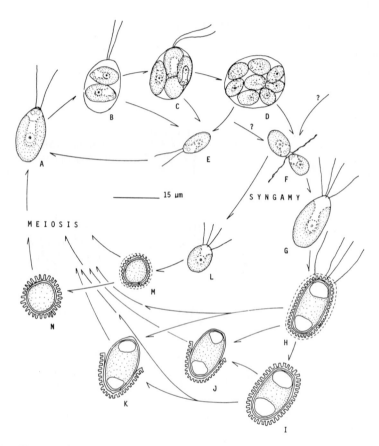

Figure 2. Life cycle of *Chloromonas brevispina*: vegetative cell (A), division of vegetative cells (B-D), zoospore (E), fusion of gametes (F), larger quadriflagellate zygote formed early when nutrient supply is higher (G), zygote identical to *Cryocystis brevispina* (H), zygote identical to *Cryocystis japonica* (I), zygote identical to *Cryodactylon glaciale* (J), zygote identical to *Oocystis lacustris f. nivalis* (K), smaller quadriflagellate zygote formed later when nutrient supply is lower (L), zygote identical to *Trochiscia rubra* (if pink or red) or *Troschiscia cryophila* (if green) (N) (after Hoham et al. 1979. Phycologia. 18, 68).

strains of *C. brevispina* from Québec and Vermont were crossed with strains 593A and 593C, some of which are self-mating and others appear to have no mating potential (Yoder et al. 1995, unpubl.). Similar variation in genetic compatibility within a species complex was also reported for the green algal desmid, *Micrasterias thomasiana* (Blackburn and Tyler 1987).

In the normal sexual mating *Chloromonas* sp.-D from the Tughill Plateau, New York, three types of mating pairs were observed (oblong-oblong, oblong-sphere and sphere-sphere) (Figures 3-5), and each pair results in a quadriflagellate zygote (Figure 6). The oblong-oblong mating pairs diminished through the 8-hr time duration of the experiments, and the spherical-type mating pairs peaked at 6-7 hrs after the experiments began (Hoham et al. 1997, 1998). Another *Chloromonas* from widely separate geographical areas (Adirondack Mtns, New York, White Mtns, Arizona and Jay Pk, Vermont) is asexual and belongs to *Chloromonas* sp.-A (Hoham et al. 1993), and not to *Chloromonas polyptera* as suggested earlier by Hoham (1992). Even though *Chloromonas* sp.-A produces rare abnormal mating configurations, resting zygotes did not develop. The abnormal mating pairs observed in the laboratory for *Chloromonas* sp.-A were similar to those observed in the field for *Chlamydomonas nivalis* (Kawecka and Drake 1978) and for *Chloromonas rostafinskii* (Kawecka 1983/1984). In very rare

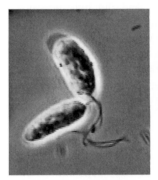
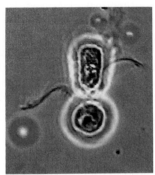
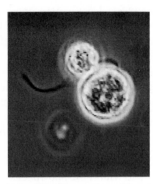

Figures 3-5. Mating pair types found in *Chloromonas* sp.-D: oblong-oblong (3), oblong-sphere (4) and sphere-sphere (5). Scale bar (-----------------) = 15 μm for Figures 3-8.

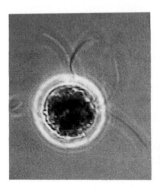
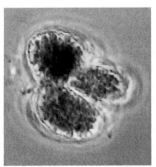

Figures 6-8. Quadriflagellate zygote (6), abnormal mating with cytoplasmic bridges between three cells without flagella (7) and fusion between three flagellate cells (8) in *Chloromonas* sp.-D.

cases, triple fusions were observed (Hoham 1992), and these were mostly between geographically isolated strains of *Chloromonas* sp.-A. Triple fusions were noted in *Chlamydomonas nivalis* in field samples from red snow in Poland (Kawecka and Drake 1978) and in laboratory mating experiments in *Chloromonas* sp.-D from the Tughill

Plateau, New York (Hoham et al. 1997, 1998) (Figures 7- 8). Populations of snow algae that reproduce asexually rather than sexually would result in subsequent generations lacking in genetic diversity. Abnormal matings would promote asexual reproduction if non-viable zygotes are the end product.

3. Interrelationships with Physical Factors

3.1. TEMPERATURE

Temperature optima for algae living in snow were reviewed by Hoham (1975a, 1980), and it was suggested that true snow algae have optimal growth at temperatures below 10°C. The snow alga, *Chloromonas pichinchae*, was designated as an obligate cryophile (psychrophile) because the motile vegetative cells grew at temperatures between 1-5°C, developed into abnormal clumps at 10°C, and did not survive at temperatures above 10°C (Hoham 1975a, 1975b). Working primarily with bacteria, however, Morita (1975) defined a true psychrophile with an optimal temperature for growth below 16°C, an upper limit for growth at 20°C and a minimal temperature for growth at 0°C or lower. Psychrotrophic bacteria were defined as organisms that can grow at 0°C, but optimally at 20-25°C (Morita 1975). Other examples of obligate and non-obligate snow algae were discussed by Hoham and Duval (1999).

3.2. MELTWATER FLOW AND WATER CONTENT

Meltwater flow, horizontal ice layers and vertical ice fingers affect the position of the algae within the snowpacks (Hoham 1975b, Hoham et al. 1979, 1983), and water flows either gravitationally or laterally and often follows the horizontal ice layers and vertical ice fingers. The range of water flow varied between 3-150 cm hr^{-1}, and maximum flow occurred during mid-afternoon when air temperatures were highest, there was a direct sun exposure and when snowbanks were connected (Hoham 1975b). Minimum water flow occurred during late evening and early morning hours when air temperatures were lowest, there was no sun exposure and in isolated snowbanks. This meltwater flow was correlated with developmental stages and potential sources of nutrients for the snow alga, *Chloromonas pichinchae* (Hoham 1975b). Nutrients in the meltwater would be available at the snow-soil interface for algal growth at the time of resting spore germination and throughout the snowpack for growth of vegetative cells (Hoham et al. 1983).

Percent water content (number of ml H_2O melted from 100 cm^3 core samples of snow) was reviewed by Hoham et al. (1983) from snowbanks in Washington [44-69%], Arizona [43-63%] and Montana [52-72%], and these percentages were correlated with algal life cycle phases in different species of *Chloromonas*. The highest percentages [57-63%] associated with the snow alga, *Chloromonas pichinchae*, occurred in early to mid afternoon in Washington snow when air temperatures and light intensities were highest (Hoham 1975b). At this time, the alga was in the swimming asexual vegetative phase. Gamete production, release and fusions [sexual phase] occurred during lower percentage readings [47-54%]. Water content percentages of 42-52% and 60-96% were recorded in association with the snow alga, *Mesotaenium berggrenii*, and the red ice alga, *Chloromonas rubroleosa*, respectively, from Antarctica (Ling and Seppelt 1990, 1993).

3.3. LIGHT

The amount of light that penetrates snow is a function of wavelength, snow density and depth (Richardson and Salisbury 1977, Warren 1982). Of importance to subniveal phototrophs is the amount of photosynthetically active radiation (PAR) received at a particular depth. This region of the light spectrum includes visible or white light (400-700 nm) with blue light penetrating the deepest (Sze 1993). Curl et al. (1972) reported that the major portion of spectral energy in snow is between 450-600 nm with a peak at 475 nm. Also, blue and green light at 455 and 526 nm (Hoham et al. 1983) and at 454 and 530 nm, respectively (Hoham 1975b), penetrated the deepest in snow for most time periods sampled. While most light is attenuated by snow, 1% of incident PAR was measured at a depth of 1 m in wet snow, and this promoted photosynthesis and algal germination (Curl et al. 1972). Light was measured penetrating through 2 m of wet snow, suggesting that photoactive responses can occur under minimal light conditions (Richardson and Salisbury 1977). In laboratory mating experiments using *Chloromonas* sp.-D isolated from Tughill Plateau snow, New York State, it was found that maximum mating occurred at a photon irradiance of 95 μmol m^{-2}s^{-1} (PAR of 400-700 nm) under both wide-spectrum and cool-white regimes, and blue light regimes produced more matings than green or red light regimes (Hoham et al. 1997, 1998). Current investigations with *Chloromonas* sp.-D imply that there is no optimal photoperiod for maximum mating in this species, but additional experiments are being conducted to verify this (Hoham et al. 1999, unpubl.).

In polar regions [Antarctica], red snow is characteristic of low nutrient areas whereas green snow is most often associated with higher nutrients near seabird rookeries (Bidigare et al. 1993, Broady 1996). In temperate regions, however, the distribution of red, green and orange snow, colored by populations of algae, is related to irradiance levels of sunlight received by the populations with green cells receiving the least and red cells the most (Fukushima 1963, Kol 1968, Pollock 1970, Hoham 1971). In horizontal ice layers, snow algae are often found concentrated several centimeters below the snowpack surface (Hoham 1975b, 1980), adjust their vertical photo-position (Hoham 1975b) and support populations of up to 1 x 10^6 cells ml^{-1} (Hoham 1987). Snow algal populations are also associated with vertical ice fingers (Hoham et al. 1979).

Using field experiments in Norwegian snow, the vertical distribution of *Chlamydomonas nivalis* associated with the water surface surrounding the snow crystals, and its appearance in the snow layers was related to the melting process rather than to phototaxis (Grinde 1983). It also appears that Grinde used resting spores stages of *C. nivalis* in the experiments, and these cells would not reform flagella even if wetted. Thus it is difficult to conclude that *C. nivalis* is not phototactic from this study. Kessler et al. (1992) suggested that biflagellate green cells of *Chlamydomonas nivalis* were strongly oriented by gravity, but direction of gravitaxis was degraded by collisions between cells.

4. Productivity and Biogeochemical Cycles

4.1 PRIMARY PRODUCTIVITY AND RESPIRATION

Snow algae are often subjected to overnight freezes or repeated freeze-thaw events, and photosynthesis was reported from frozen samples of snow algae that were later thawed (Hoham 1975a, 1975b, Mosser et al. 1977). Optimum temperatures for photosynthesis in certain strains of snow algae were reported at -3 to 4°C (Hoham 1975a, Mosser et al. 1977).

Chlorophyll a concentrations that were very heterogeneous in snow were attributed to populations of snow algae (Thomas 1972). Using ^{14}C (^{14}C-HCO$_3^-$ and ^{14}CO$_2$), Mosser et al. (1977) reported that *Chlamydomonas nivalis* photosynthesized optimally in the field at 10 or 20°C, but retained substantial activity at temperatures as low as 0 or -3°C. Using ^{14}C, similar quantities of µg C fixed mm^{-3} cell volume hr^{-1} were recorded for *Chlamydomonas nivalis* (0.05-0.97, Mosser *et al.* 1977), mixed populations of snow algae including *Chlamydomonas* sp. (0.04-1.85, Komárek et al. 1973) and *Chlamydomonas nivalis* (0.002-0.86, Fogg 1967). Higher amounts of fixed carbon (5.7-34.2) were reported for *Chlamydomonas nivalis* (Thomas 1972), but he questioned these values because of a high assay of CO$_2$ concentration in the snow meltwater.

Carbon production in algal photosynthesis ranged from 1.2×10^{-2} - 12.3×10^{-2} µg C (ml snow)$^{-1}$ hr^{-1} in red snow containing *Chlamydomonas nivalis* and *Trochiscia americana* (Thomas 1994). In adjacent white snow samples, photosynthesis values ranged from $0.00-0.16 \times 10^{-2}$ µg C (ml snow)$^{-1}$ hr^{-1}. Even though algal photosynthesis occurred in white snow, ratios of photosynthesis in red snow to white snow ranged from 27-79. Thomas (1994) also converted the algal photosynthesis values to units used by Mosser et al. (1977) for red snow comparisons and found that the values reported from both studies fell into the same range.

The effects of total UV on photosynthetic uptake of radioactive carbon in green and red snow from the Sierra Nevada Mtns, California, were studied by Thomas and Duval (1995). They found UV inhibited uptake by 85% in green snow containing, *Chloromonas*, but inhibited uptake by only 25% in red snow containing spores of *Chlamydomonas nivalis* [1994 field data]. They concluded that red snow found in open, sunlit areas was better adapted to UV than green snow found in forested, shaded locations. Thomas (pers. comm., 1995) found that *Chlamydomonas nivalis* photosynthesis was not inhibited by UV as was the case in 1994 (Thomas and Duval 1995). Thomas attributed these differences in UV inhibition to a snowpack 4X greater in 1995 than in 1994 that allowed the cells in 1995 to become better adapted to UV because of the longer growth season.

4.2. DISSOLVED GASES AND pH

The dissolved gases, CO$_2$ and O$_2$, are important criteria concerning microbial populations in the snowpack. The concentration of dissolved CO$_2$ in snow ranged from 2.5-5.0 mg L^{-1}, and 9-13 mg L^{-1} for dissolved O$_2$ [D.O.], in snow containing populations of snow algae (Hoham 1975b, Hoham and Mullet 1977). Brooks et al. (1993) reported a CO_2 flux at the snow/air interface of 320-360 mg C m^{-2} day^{-1} and suggested a minor source within the snowpack. They also indicated that most snow is probably oxygenated enough to support metabolic activity for aerobic microbes.

Snow algae affect pH values in snow (Hoham et al. 1989). Green snow with algae from Whiteface Mtn, New York, had higher pH values than snow without algae (5.87 vs 5.63 at 1341 m and 5.17 vs 4.98 at 1265 m), reputedly the result of CO_2 uptake during photosynthesis (Hoham et al. 1989). Similar increases in pH values were reported in green snow from Svalbard where biflagellate cells were photosynthetically active (Müller et al. 1998). Newton (1982), however, reported a lower pH (6.2) in regions of Svalbard snow colonized by *Chlamydomonas nivalis* compared to areas that were not colonized (pH of 7.0-7.6) suggesting that the alga excreted organic materials (acids and polysaccharides) into the snow lowering the pH. However, Newton's study was conducted when the asexual spore (cyst) stage was present, a phase that would have reduced metabolic activity. Interestingly, Müller et al. (1998) found a similar situation in Svalbard where snow with red-orange resting spores of algae were found in lower pH (up to 0.4-0.7 pH unit lower) than in control samples without algae. It appears that the relationship between pH and algae in snow depends upon the metabolic state and phase of the snow algal life cycle (Hoham et al. 1989, 1993).

In temperate regions snow microbes are subjected to high acidity. Observations of pH in snow ranged from 4.0-6.3 in western North America (Hoham et al. 1983), as low as 3.4 for meltwater in the Adirondacks, New York (Schofield and Trojnar 1980), and from 3.5 to 5.4 in south central Ontario with the lowest pH during the initial snowmelt (Goodison et al. 1986). Lukavský (1993) reported a pH range of 4.6-4.9 from snowfields in the High Tatra Mtns, Europe, and a higher pH in old snow from the Bohemian Forest Mtns [no values given], and both habitats were associated with snow algae. A pH range of 4.4-6.2 was reported from Svalbard snow algal samples (Müller et al. 1998).

Within a relatively small area of the Windmill Islands region, Antarctica, different pH ranges in snow populated by algae were observed. These included 4.5-5.7 for the reddish-brown, green algal desmid, *Mesotaenium* (Ling and Seppelt 1990), 4.6-6.2 for snow colored red by the green alga, *Chloromonas rubroleosa* (Ling and Seppelt 1993), and for green snow, 6.3-6.9 for *Chlorosarcina* sp., 6.8-7.8 for *Desmotetra* sp. and 6.7-8.1 for *Chloromonas polyptera* (Ling and Seppelt 1999). The green snow habitats were receiving snowmelt from active Adelie penguin colonies where the soil pH was 7.18 ± 0.33 (Roser et al. 1993).

The effects of acidic snow may result in the natural selection of snow microbes with greater tolerance to acidity. Hoham and Mohn (1985) reported that strains of the snow alga, *Chloromonas* [currently thought to belong to *Chloromonas* sp.-A], from the Adirondack Mtns, New York, had growth optima between pH 4.0-5.0 compared to other strains isolated from the White Mtns, Arizona, with pH optima of 4.5-5.0. Since there was a significant difference in growth between these geographically isolated strains at pH 4.0 ($P < 0.05$), this suggests that snow algae were adapting to the more acidic precipitation found in eastern North America.

4.3. NUTRIENTS, NUTRIENT CYCLING AND CONDUCTIVITY

Snow is an environment that is limiting in nutrients (Hoham 1989, Jones 1991), and nutrient loads in the snowpack are spatially distributed (Tranter et al. 1987, Davies et al.

1989). This spatial variability may correlate with the spatial distributions of microbes such as the snow algae (Hoham 1980, Tranter, 1993, pers. comm.). Nutrient depletion (particularly NO_3^-) coincided with shifts in phases of the life cycle of snow algae such as *Chloromonas* (Hoham et al. 1989). The importance of NO_3^- on the forest floor such as in the Adirondacks, New York (Rascher et al. 1987), may play an important role in the life cycles of some snow algal species at the time of their germination. The snowmelt waters concentrated in sulfuric acid and nitrogenous anions probably affect the snow microbiota (Hoham and Mohn 1985, Bartuma et al. 1990, Williams 1993), and microbial processes in surface soils beneath the snow may also contribute to this acidity (Arthur and Fahey 1993). Many of the processes and features occurring in snow such as sublimation, melt and meltwater channels determine the concentration of nutrients that affect the snow community (Goodison et al. 1986).

Nutrients are deposited on snow by wind, precipitation, weathering of rock and by animals (Jones 1991, Hoham and Duval 1999). High nutrient loads of nitrogen and phosphorus were reported in eastern European snow comparable to eutrophic waters (Komárek et al. 1973); however, the source of these nutrients was not given. Cell surfaces of the snow alga, *Chlamydomonas nivalis*, showed prolific accumulation of aerosol debris both locally and globally derived (Tazaki et al. 1994a, 1994b). Yellow snow over the European Alps and the Subarctic was derived from a Saharan dust storm in Africa in March 1991 (Franzén et al. 1994a, 1994b). Weathering of rock may also add to the nutrient composition of snow (Kawecka 1986). The desmid, *Mesotaenium*, was observed in snowmelt downslope from moraine and rock aggregations, and this observation was probably a result of enhanced melting or mineral leachate from the upgradient rock (Ling and Seppelt 1990). Most snow algae from Antarctica were from snow samples located at seabird rookeries where there was an ample supply of nitrogen (Bidigare et al. 1993), and vegetation, birds and small mammals contribute to the patchiness of nutrients described in snow (Jones 1991).

Snow algae deplete nutrients for their growth and development in their life cycles. This subject was reviewed by Hoham and Duval (1999), and their review included studies from Whiteface Mtn, New York (Hoham et al. 1989, Jones 1991), Lac Laflamme, Québec, Canada (Gamache 1991, Germain 1991, Jones 1991), the Himalayas (Yoshimura et al. 1997) and Ontario, Canada (Gerrath and Nichols 1974). High levels of Si, P, S and organics were found in red and green cells of *Chlamydomonas nivalis* from Cornwallis Island, Canada, and high Ca content in the green cells only (Tazaki et al. 1994a). They suggested that both P and S were of vital importance to the algae under conditions of such extreme low temperature. In Svalbard snow, high levels of Fe, Ca, Mg, K, P and Al were found in algal cells despite the very low concentration of these ions in the extra-cellular meltwater (Müller et al. 1998).

Coniferous litter, dust and debris are important sources of nutrients for microbes living in snow (Hoham and Duval 1999). The effects that extracts from coniferous litter and different snow meltwaters had on growth in the snow algae, *Raphidonema nivale* and *Chloromonas pichinchae*, were investigated by Hoham (1976). The coniferous leachate experiments indicated increased growth for *C. pichinchae* in extracts from the five species of conifers used. *Raphidonema nivale*, however, responded very differently to the same extracts, and at the lowest concentrations used, growth was inhibited. Even some morphological malformations occurred in *R. nivale* grown in higher extract concentrations. The results of these experiments correlated with field observations that *C. pichinchae* was more abundant in the snowfields surrounded by conifers than was *R.*

nivale. Phenolic compounds have been identified from coniferous spruce leachates at high altitudes in snow that are potentially responsible for allelopathic interferences (Gallet and Pellissier 1997). Vitamins such as B_1 and B_{12} are needed by certain microbes that live in snowpacks beneath coniferous tree canopies (Hoham et al. 1989). It is not known, however, if the vitamins are derived directly from the canopy or from other microbes in the snowpack such as bacteria, fungi or lichen pieces (Hoham and Duval 1999).

The interactions between snow algae and snow chemistry affect conductivity readings in snow (Hoham et al. 1989). From Whiteface Mtn, New York, conductivity values were lower in snow samples containing algae compared to snow samples without algae (13.1 vs 19.5 µS cm^{-1} at 1341 m and 9.6 vs 16.4 µS cm^{-1} at 1265 m). These differences were due to nutrient uptake and algal metabolism. However, conductivity in Svalbard snow was higher in regions of algal colonization compared to regions without algae (12 vs 4-7 µS cm^{-1}) (Newton 1982), and it was suggested that algal activity results in an increase of ionic concentrations as well as preferring regions that receive more wind-blown materials (the latter probably raised the conductivity values here). Other conductivity values reported from snow algal studies include 8-15 (Arizona, Hoham et al. 1983), 4-6 and 18-20 (Montana, Hoham et al. 1983), and 0.3-17 µS cm^{-1} (Svalbard, Müller et al. 1998). In the Windmill Islands region, Antarctica, snow algal species were found with different conductivity values for *Mesotaenium berggrenii* (6-33), *Chloromonas rubroleosa* (25-85), *Chlorosarcina* sp. (39-44), *Chloromonas polyptera* (56-950) and *Desmotetra* sp. (279-426 µS cm^{-1}) (Ling and Seppelt 1999). The relationship between conductivity values and snow algae may depend upon the species, the metabolic state and phase of the snow algal life cycle, the metabolic state of other microbes such as bacteria and fungi, and the degree to which the snow has been leached by meltwater.

4.4. BIOACCUMULATION OF HEAVY METALS

Metals enter the niveal food chain and accumulate in algal cells thousands of times their concentration in surrounding snow (Fjerdingstad 1973, Fjerdingstad et al. 1974, Hoham et al. 1977, Fjerdingstad et al. 1978, Hoham 1980). In *Mesotaenium berggrenii*, the cells accumulate significant quantities of iron in the form of an iron tannin complex which apparently acts as an UV shield (Ling and Seppelt 1999).

The concentrations of trace metals in green snow caused by the alga, *Chloromonas pichinchae*, from Washington, U.S.A., was compared to red snow samples caused by the alga, *Chlamydomonas nivalis*, from East Greenland and Spitzbergen (Hoham and Duval 1999). It does not appear, however, that heavy metal accumulation has any effect on snow algal life cycles.

5. References

Arthur, M.A. and Fahey, T.J. (1993) Soil Sci. Soc. Amer. 57, 1122-1130.
Bartuma, L.A., Cooper, S.D., Hamilton, S.K., Kratz, K.W. and Melack, J.M. (1990) Freshwat. Biol. 23, 571-586.
Bidigare, R.R., Ondrusek, M.E., Kennicutt II, M.C., Iturriaga, R., Harvey, H.R., Hoham, R.W. and Macko, S.A. (1993) J. Phycol. 29, 427-434.
Blackburn, S.I. and Tyler, P.A. (1987) Brit. Phycol. J. 22, 277-298.
Broady, P.A. (1996) Biodivers. Conserv. 5, 1307-1335.

144

Brooks, P.D., Schmidt, S.K., Sommerfeld, R. and Musselman, R. (1993) In: M. Ferrick and T. Pangburn (eds.) Proceedings of the Fiftieth Annual Eastern Snow Conference. Québec City, Québec, Canada, pp. 301-306.

Curl, H., Jr., Hardy, J.T. and Ellermeier, R. (1972) Ecology 53, 1189-1194.

Czygan, F.-C. (1970) Arch. Mikrobiol. 74, 69-76.

Davies, T.D., Delmas, R., Jones, H.G. and Tranter, M. (1989) New Sci. 122, 45-49.

Ettl, H. (1983) In: H. Ettl, J. Gerloff, H. Heynig and D. Mollenhauer (eds.) Süsswasserflora von Mitteleuropa 9. G. Fischer, Stuttgart.

Fjerdingstad, Einer, Vanggaard, L., Kemp, K. and Fjerdingstad, Erik (1978) Arch. Hydrobiol. 84, 120-134.

Fjerdingstad, Erik. (1973) Hydrologie 35, 247-251.

Fjerdingstad, Erik, Kemp, K., Fjerdingstad, Einer and Vanggaard, L. (1974) Arch. Hydrobiol. 73, 70-83.

Fogg, G.E. (1967) Phil. Trans. Roy. Soc. London, Ser. B, 40, 293-338.

Franzén, L.G., Hjelmroos, M., Kållberg, P., Brorström-Lundén, E., Juntto, S. and Savolainen, A.-L. (1994a) Atmospher. Environ. 28, 3587-3604.

Franzén, L.G., Mattsson, J.O., Mårtensson, U., Nihlén, T. and Rapp, A. (1994b) Ambio 23, 233-235.

Fukushima, H. (1963) J. Yokohama Munic. Univ., Ser. C, Nat. Sci. 43, 1-146.

Gallet, C. and Pellissier, F. (1997) J. Chem. Ecol. 23, 2401-2412.

Gamache, S. (1991) M.Sc. thesis, Institut national de la recherche scientifique, Ste Foy, Québec, Canada.

Germain, L. (1991) M.Sc. thesis, Institut national de la recherche scientifique, Ste Foy, Québec, Canada.

Gerrath, J.F. and Nicholls, K.H. (1974) Can. J. Bot. 52, 683-685.

Goodison, B.E., Louie, P.Y.T. and Metcalfe, J.R. (1986) In: E.M. Morris (ed.) Modelling Snowmelt-Induced Processes. IAHS Publ. No. 155, Internat. Assoc. Hydrolog. Sci., Budapest, Hungary, pp. 297-309.

Grinde, B. (1983) Polar Biol. 2, 159-162

Hoham, R.W. (1971) Ph.D. thesis, University of Washington, Seattle.

Hoham, R.W. (1975a) Arct. Alp. Res. 7, 13-24.

Hoham, R.W. (1975b) Phycologia 14, 213-226.

Hoham, R.W. (1976) Arct. Alp. Res. 8, 377-386.

Hoham, R.W. (1980) In: E.R. Cox, (ed.) Phytoflagellates. Elsevier North Holland, Inc., New York, pp. 61-84.

Hoham, R.W. (1987) In: J. Lewis (ed.) Proceedings of the Forty-fourth Annual Eastern Snow Conference. Fredericton, New Brunswick, Canada, pp. 73-80.

Hoham, R.W. (1992) In: B. Shafer (ed.) Proceedings of the Sixtieth Annual Western Snow Conference. Jackson Hole, Wyoming. Pp. 78-83.

Hoham, R.W. and Duval, B. (1999) In: H.G. Jones, J.W. Pomeroy, D.A. Walker and R.W. Hoham (eds.) Snow Ecology. Cambridge University Press, Cambridge, U.K., (in press).

Hoham, R.W., Kang, J.Y., Haselwander, A.J., Behrstock, A.F., Blackburn, I.R., Johnston. R. C. and Schlag, E.M. (1997) In: Proceedings of the 65th Annual Western Snow Conference. Banff Alberta, Canada. pp: 80-81.

Hoham, R.W., Kemp, K., Fjerdingstad, Erik and Fjerdingstad, Einer (1977) J. Phycol. 13 (Suppl.), 30 (Abstr. 167).

Hoham, R.W., Laursen, A.E., Clive, S.O. and Duval, B. (1993) In: M. Ferrick and T. Pangburn (eds.) Proceedings of the Fiftieth Annual Eastern Snow Conference. Québec City, Québec, Canada, pp. 165-173.

Hoham, R.W. and Mohn, W.W. (1985) J. Phycol. 21, 603-609.

Hoham, R.W. and Mullet, J.E. (1977) Phycologia 16, 53-68.

Hoham, R.W., Mullet, J.E. and Roemer, S.C. (1983) Can. J. Bot. 61, 2416-2429.

Hoham, R.W., Roemer, S.C. and Mullet, J.E. (1979) Phycologia 18, 55-70.

Hoham, R.W., Schlag, E.M., Kang, J.Y., Hasselwander, A.J., Behrstock, A.F., Blackburn, I.R., Johnson, R.C. and Roemer, S.C. (1998) Hydrolog. Process. 12, 1627-1639.

Hoham, R.W., Yatsko, C.P., Germain, L. and Jones, H.G. (1989) In: J. Lewis (ed.) Proceedings of the Forty-sixth Annual Eastern Snow Conference. Québec City, Québec, Canada, pp. 196-200.

Horner, R. A. (1985) Sea Ice Biota. CRC Press, Boca Raton, Florida.

Horner, R.A. (1996) Proc. NIPR Sympos. Polar Biol. 9, 1-12.

Hoshaw, R.W. (1961) Amer. Biol. Teach. 23, 489-499.

Jones, H.G. (1991) In: T.D. Davies, M. Tranter and H.G. Jones (eds.) NATO ASI Series G: Ecol. Sci., Vol 28, Seasonal Snowpacks, Processes of Compositional Changes. Springer-Verlag, Berlin, pp. 173-228.

Kawecka, B. (1983/1984) Acta Hydrobiol. 25/26, 281-285.

Kawecka, B. (1986) Polish Polar Res. 7, 407-415.

Kawecka, B. and Drake, B.G. (1978) Acta Hydrobiol. 20, 111-116.

Kessler, J.O., Hill, N.A., and Häder, D.-P. (1992) J. Phycol. 28, 816-822.

Kirst, G.O. and Wiencke, C. (1995) J. Phycol. 31, 181-199.

Kol, E. (1968) In: H.-J. Elster and W. Ohle (eds.) Die Binnengewässer. Vol. 24, E. Schweizerbart'sche Verlagsbuchhandlung, Stuttgart, 1-216.

Komárek, J., Hindák, F. and Javornický, P. (1973) Arch. Hydrobiol., Suppl. 41. Algol. Stud. 9:427-449.

Ling, H.U. (1996) Hydrobiologia 336, 99-106.

Ling, H.U. and Seppelt, R.D. (1990) Antarct. Sci. 2, 143-148.

Ling, H.U. and Seppelt, R.D. (1993) Eur. J. Phycol. 28, 77-84.

Ling, H.U. and Seppelt, R.D. (1998) Polar Biol. 20, 320-324.

Ling, H.U. and Seppelt, R.D. (1999) In: W. Davison (ed.) Antarctic Ecosystems: Models for Wider Ecological Understanding. Caxton Press, Christchurch. (in press).

Lukavský, J. (1993) Arch. Hydrobiol., Algol. Stud. Suppl. 69, 83-89.

Morita, R.Y. (1975) Bact. Rev. 39, 144-167.

Mosser, J L., Mosser, A.G. and Brock, T.D. (1977) J. Phycol. 13, 22-27.

Müller, T., Bleiss, W., Martin, C.-D., Rogaschewski, S. and Fuhr, G. (1998) Polar Biol. 20, 14-32.

Newton, A.P.W. (1982) Polar Biol. 1, 167-172.

Palmisano, A.C. and Garrison, D.L. (1993) In: E.I. Friedman (ed.) Antarctic Microbiology. Wiley-Liss, New York, pp. 167-218.

Pollock, R. (1970) Sierra Club Bull. 55, 18-20.

Rascher, C.M., Driscoll, C.T. and Peters, N.E. (1987) Biogeochemistry 3, 209-224.

Richardson, S.G. and Salisbury, F.B. (1977) Ecology 58, 1152-1158.

Roser, D.J., Seppelt, R.D. and Ashbolt, N. (1993) Soil Biol. Biochem. 25, 165-175.

Schofield, C.L. and Trojnar, J.R. (1980) In: T.Y. Toribara, M.W. Miller and P.E. Morrow (eds.) Polluted Rain. Environ. Sci. Res. Ser., Vol. 17, Plenum Press, New York, pp. 341-366.

Sze, P. (1993) A Biology of the Algae, 2nd ed., W. C. Brown Publ., Dubuque, Iowa.

Tazaki, K., Fyfe, W.S., Iizumi, S., Sampei, Y., Watanabe. H., Goto, M., Miyake, Y. and Noda, S. (1994a) Nat. Geogr. Res. and Explorat. 10, 116-117.

Tazaki, K., Fyfe, W.S., Iizumi, S., Sampei, Y., Watanabe. H., Goto, M., Miyake, Y. and Noda, S. (1994b) Clays and Clay Mineral. 42:402-408.

Thomas, W.H. (1972) J. Phycol. 8, 1-9.

Thomas, W.H. (1994) In: C. Troendle (ed.) Proceedings of the Sixty-second Annual Western Snow Conference. Santa Fe, New Mexico, pp. 56-62.

Thomas, W.H. and Duval, B. (1995) Arct. Alp. Res. 27, 389-399.

Tranter, M., Davies, T.D., Abrahams, P.W., Blackwood, I., Brimblecombe. P. and Vincent, C. E. (1987) Atmos. Environ. 21, 853-862.

Warren, S.G. (1982) Rev. Geophys. Space Phys. 20, 67-89.

Warren, G.J. (1994) Cryo-letters 15, 323-331.

Williams, M.W. (1993) In: M. Ferrick and T. Pangburn (eds.) Proceedings of the Fiftieth Annual Eastern Snow Conference. Québec City, Québec, Canada, pp. 239-245.

Yoshimura, Y., Kohshima, S. and Ohtani, S. (1997) Arct. Alp. Res. 29, 126-137.

Addendum: The Hoham and Duval reference listed as "in press" for 1999 will instead be released in 2000.

Biodata of **Jody W. Deming** author (with co-author A.L. Huston) of "*An Oceanographic Perspective on Microbial Life at Low Temperatures*".

Dr. **Jody W. Deming** is a Professor in the School of Oceanography, University of Washington, Seattle, WA, and former director of the UW Marine Bioremediation Program. She received her Doctor of Philosophy degree (awarded as the Best Ph.D. Thesis in Life Science) at the University of Maryland, College Park, MD, in 1981. She has received many honors and awards. Her research interests include microorganisms in cold environments, especially in the Arctic Ocean and deep sea; the molecular enzymatic basis for psychrophily in marine bacteria and its relevance to bioremediation, biotechnology and astrobiology; hydrostatic pressure in the evolution and ecology of marine bacteria; and the deep subsurface biosphere. Dr. Deming has published several dozens of Journal articles and book reviews. She has been involved in numerous Professional activities, including service on National and International Committees and as a reviewer and editor.

E-mail: **jdeming@u.washington.edu**

Biodata of **Adrienne L. Huston**, coauthor (with author Jody W. Deming) of the chapter on "*An Oceanographic Perspective on Microbial Life at Low temperatures.*"

Adrienne Huston is a doctoral candidate in the School of Oceanography, University of Washington, Seattle, WA, where she serves as a Graduate Research and Teaching Assistant. She received her MS degree at the University of Washington in 1999.

J. Seckbach (ed.), Journey to Diverse Microbial Worlds, 149-160.
© 2000 *Kluwer Academic Publishers. Printed in the Netherlands.*

AN OCEANOGRAPHIC PERSPECTIVE ON MICROBIAL LIFE AT LOW TEMPERATURES

With Implications for Polar Ecology, Biotechnology, and Astrobiology

J.W. DEMING and A.L. HUSTON

School of Oceanography, University of Washington, Seattle, WA 98195

1. Introduction

At the typical subzero temperatures of polar oceans, neither the growth nor survival strategies of heterotrophic microorganisms, in particular bacteria, are well understood. Protistan and viral predation on bacteria in high-latitude seas is a subject of current interest, but the potential inhibition of bacterial acquisition of food by very low temperatures remains the most debated issue in polar marine microbiology. If significant limitations are imposed on bacterial heterotrophy in seawater by near-freezing temperatures, then the flow of organic matter through the microbial loop and associated respiratory loss of carbon from the system (Yager, 1996) will also be limited. If the temperature limitations are differential (if phytoplankton growth can outcompete bacterial growth at subzero temperatures), as documented by Pomeroy and Deibel (1986), then the highly efficient food webs, often weighted towards higher trophic levels, that characterize polar marine ecosystems (Petersen and Curtis, 1980) are mechanistically explained.

Evidence exists, however, for bacterial strategies to overcome low temperature constraints on their ability to acquire the food they require in the form of dissolved organic matter (DOM), including the expression of membrane uptake systems with high affinity for the target substrate at subzero temperature (Yager, 1996; Yager and Deming, 1999) and the use of extracellular enzymes in cold porous matrices like sea ice, sinking detrital aggregates and seafloor sediments to increase the available supply of DOM (Vetter and Deming, 1994; Huston and Deming, 1999; Huston et al., 1999). A better understanding of these strategies may lead us not only to better predictions of polar ecosystem structure and vitality but also to new products and processes in biotechnology and bioremediation (Deming, 1998). A conceptual, if not quantitative, basis for predicting when and where microbial life may be found in very cold environments elsewhere in the Universe may also emerge.

2. Psychrophily in the Continuum of Thermal Classes

Prokaryotes (Bacteria and Archaea), unlike any class of higher organism (Eukarya), exist in forms that can grow not only at moderate temperatures, such as characterize the

150

thin surface of the Earth at low to middle latitudes, but also at both thermal extremes --
subzero temperatures (bacterial growth to about -5°C; Helmke and Weyland, 1995) and
temperatures well in excess of the boiling point of water at atmospheric pressure
(Deming and Baross, 1993; Baross and Deming, 1995). Both of these thermal extremes
on Earth are found in the ocean -- within the matrix of the (wintertime) sea ice that
covers high-latitude oceans and within the structures and fluids of hydrothermal vents
and related subsurface environments of the deep (pressurized) sea. Across the
continuum of thermal growth optima expressed by prokaryotic microorganisms, which
has been broken into discrete thermal classes for convenience (Fig. 1), representatives
from all classes, except the psychrophiles, have evolved means to grow equally rapidly
(generation times of approximately 30 min), whether their thermal optimum is 37 or
95°C. Under the best of conditions, maximal growth rates of psychrophilic bacteria, by
definition restricted to growth at temperatures below 20°C (and growth optima ≤15°C;
Morita, 1975), are measured in terms of hours (generation times of approximately 6 h or
longer).

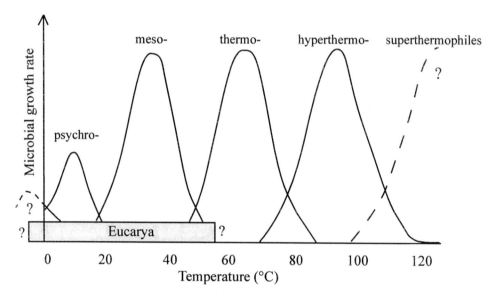

Figure 1. A generalized depiction of the thermal classes of Prokarya -- psychrophiles. mesophiles,
thermophiles, hyperthermophiles and the hypothesized superthermophiles (Deming and Baross, 1993) --
with a comparative scale bar for Eucarya. Classes have been designated historically on the basis of
minimal, optimal, and maximal temperatures for microbial growth in pure culture, but a thermal continuum
is expected to better represent the microbial world. Microorganisms are known to tolerate more extreme
temperatures than those permissive of growth; with more research, the range of growth permissive
temperatures is expected to expand (question marks).

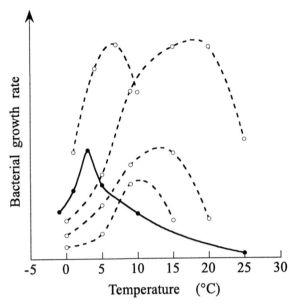

Figure 2. Comparative optimal temperatures and maximal rates of growth for psychrophilic bacterial strains in organic-rich marine media, based on published data (dashed lines) from Christian and Wiebe (1974) and Inoue (1977) and unpublished data by S. Carpenter and J. Deming (solid line).

Although strictly comparable data are difficult to find in the literature, higher maximal growth rates even among the psychrophiles tend to be associated with higher optimal growth temperatures (Fig. 2). The thermodynamic constraints imposed by increasingly cold temperature at the fundamental levels of enzyme reactivity and membrane fluidity (Gounot, 1991) have not been overcome by evolution at the organismal level. Instead, natural microbial populations confined to life in perennially cold regions of the ocean may have achieved system-level rates of activity on a par with warmer waters (Rich et al., 1997) by amassing more total biomass (or higher total enzyme concentration) to accomplish the same work load.

In assessing the activities of natural bacterial assemblages in the cold, the question of community structure and diversity invariably arises. With regard to thermal classes of Prokarya, psychrophiles might be expected to dominate at the lowest temperatures, just as only hyperthermophiles can thrive at the highest temperatures. Complicating polar-ocean scenarios, however, is the fact many non-psychrophilic marine bacteria with growth optima well above 20°C can also grow, albeit slowly, at near-freezing temperatures. Competitive tests between such psychrotolerant bacteria and true psychrophiles in the acquisition of food under environmentally relevant conditions are rare in the literature, but chemostat experiments (Harder and Veldkamp, 1971) have indicated that psychrophiles will outcompete psychrotolerant forms at subzero temperatures. The issue of thermal diversity in environmental samples is more difficult

152

to resolve, but recent tests in organic-rich subzero waters of an Arctic polynya indicated the dominance of psychrophilic bacterial populations (growth optima of 5-10°C; Yager, 1996), in keeping with evolutionary arguments for the dominance of psychrophiles when DOM is plentiful in a very cold environment (Baross and Morita, 1978). To accurately predict the role of bacteria in polar ecosystems -- in cycling the seasonal pulses of organic matter that characterize subzero waters -- obtaining more insight on the foraging and survival strategies of psychrophiles is in order.

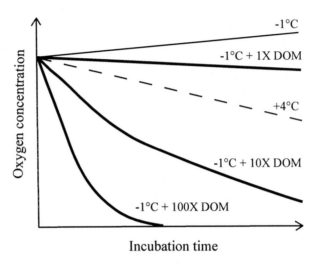

Figure 3. Generalized depiction of the substrate threshold phenomenon experienced by bacteria in near-freezing seawater (adapted from Pomeroy and Deibel, 1986; Pomeroy and Wiebe, 1993): aerobic heterotrophy (oxygen removal over time as DOM is consumed) is not detectable or competitive with photosynthesis (oxygen production, thin line) at -1°C without DOM enrichment (bold lines) or warming (+4°C, dotted line).

3. Acquiring Food at Low Temperature

Work by Pomeroy and colleagues (e.g., Pomeroy and Deibel, 1986; Pomeroy and Wiebe, 1993) has focused attention on the specific hypothesis that in near-freezing seawater natural bacterial assemblages have a higher threshold for substrate uptake than they do under warmer conditions. Supportive evidence has come from measuring oxygen concentrations over time in samples of Arctic seawater held at different temperatures and levels of DOM, as well as from related studies using pure cultures of bacteria from most thermal classes (Wiebe et al., 1992, 1993). The compelling conversion of a pelagic microbial system at -1°C from one dominated by photosynthesis to one dominated by aerobic heterotrophy, either by substrate amendment or slight warming (Fig. 3), has motivated further research on a substrate threshold for bacteria in near-freezing waters and organism (or population)-controlled strategies for overcoming it.

Bacteria faced with a substrate threshold in subzero marine environments have several options, intracellular energy stores permitting, for altering their circumstances to enable continued growth. They can express higher affinity membrane permeases to obtain food at concentrations below the threshold (Yager, 1996; Yager and Deming, 1999); they can move to an equally cold environment that has more food, either by chemotaxis or attachment to larger moving particles or organisms; they can move to a warmer environment (e.g., influenced by solar radiation); or they can mediate an increase in the amount of food available in the immediate vicinity. The latter refers primarily to the work of extracellular hydrolytic enzymes released by the individual bacterium and its neighbors and has the best chance of being effective over short (micrometer) distances in porous matrices like detrital aggregates and seafloor sediments (Vetter et al., 1998) or sea ice. Laboratory studies have shown that bacteria can make a living exclusively via the foraging action of released extracellular enzymes on otherwise inaccessible (due to size and other physical barriers) food particles (Vetter and Deming, 1999). Return of the dissolved food products from the hydrolysis sites to the bacteria is dependent on diffusion, a process sensitive to temperature (Jumars et al., 1993) but not sensitive enough to preclude this organism-controlled strategy for relief of food scarcity even at subzero temperatures. Because so little is known about high affinity permeases or chemotaxis in psychrophiles and because we and others have detected strong signals of extracellular enzyme activity (EEA) in a variety of subzero polar habitats, we focus on the use of extracellular enzymes by polar marine bacteria as a strategy for overcoming a substrate uptake threshold at cold temperatures.

4. The Intracellular-Extracellular Conundrum

To increase the supply of DOM for their reproductive benefit, psychrophilic bacteria might be expected to release extracellular enzymes with activities also optimized to low temperature. The literature indicates the contrary -- that the activities of exported enzymes are optimized to higher temperatures, sometimes much higher (by >30°C), than the intracellular processes of microorganisms (Fig. 4). This discrepancy between thermal optima for intracellular and extracellular processes characterizes natural assemblages of polar marine bacteria and their extracellular enzymes, as well as related laboratory cultures and enzyme preparations (Fig. 4); the conundrum remains unexplained at a mechanistic level. It has led, however, to the notion that for bacteria to benefit from EEA they must export copious amounts of enzymes to compensate for suboptimal activity at near-freezing temperatures (Reichardt, 1987), especially since evidence for export of fewer but cold-optimized enzymes has been lacking. Even if the compensation is incomplete, sufficient hydrolysate may be produced by enzymes operating at suboptimal rates to support some level of bacterial growth (Brenchley, 1996).

We have tried to resolve the intracellular-extracellular conundrum in two ways: 1) by searching polar marine habitats for the existence of extracellular enzymes that are truly psychrophilic (activity optima ≤15°C) (perhaps they had been missed by previous sampling or analytical approaches); and 2) by re-evaluating available information from a modeling perspective (perhaps the thermal optimum of EEA, typically based on

154

instantaneous maximum reaction rates, is not the most relevant parameter to the enzyme-releasing organism).

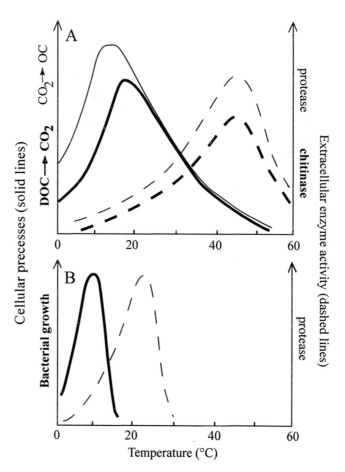

Figure 4. *A) Generalized examples of thermal optima for the cellular processes (solid lines) of heterotrophic respiration (DOC to CO_2, bold line) and photosynthesis (CO_2 to OC, thin line) and the extracellular activities (dashed lines) of the enzymes chitinase (bold dashes) and protease (thin dashes) in Antarctic sediments (adapted from Reichardt, 1987); B) thermal optima for growth and extracellular enzyme activity (partially purified, cell-free protease extract) for psychrophilic* Colwellia *strain 34H (Huston and Deming, 1999; Huston et al., 1999).*

In the search for EEA in environmental samples adapted to temperatures lower than previously reported in the literature, we have been successful in documenting the first and numerous examples (28% of 72 cases) of truly psychrophilic activity with thermal optima to unprecedented lows of 10°C, especially in samples of sinking detrital aggregates (Fig. 5) and sea ice (Huston and Deming, 1999). We have also isolated from Arctic sediments a psychrophilic *Colwellia* strain that releases a protease which, in crude extract, shows optimal activity at 20°C (Huston et al., 1999). Although this thermal optimum is lower than that for any other reported extracellular protease, it is still offset from the bacterial growth optimum by 12°C (Fig. 4B). The marked discrepancies between intracellular and extracellular thermal optima have been reduced by these search efforts (compare Fig. 4A and 4B), but the conundrum, as defined by enzyme activity optima, still exists.

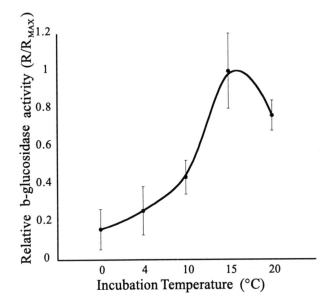

Figure 5. Example of psychrophilic EEA (optimum <20°C), measured via fluorescent substrate analog and scaled to maximum reaction rate (R), in a sample of sinking particle aggregates recovered in an unpoisoned 12-h floating sediment trap at -1°C and 75 m in the Northwater polynya, northern Baffin Bay (Huston and Deming, 1999); error bars indicate 95% confidence intervals

In our modelling effort (Huston and Deming, 1999), we have considered a critical parameter in bacterial foraging by extracellular enzymes to be the total amount of hydrolysate made available to the organism over time and thus the stability or lifetime of the enzyme. We have also relied upon prediction from current theory (Davail et al., 1994; Feller et al., 1996) that in order to retain sufficient activity at lower temperatures, a cold-active enzyme (e.g., a psychrophilic enzyme) must possess a lower activation energy (E_a) and deactivation energy (E_d) due to a more flexible and thus less stable structure, compared to more thermo-stable counterparts (e.g., less psychrophilic or mesophilic enzymes). Assuming that a constant fraction of hydrolysis product is available to the bacterium over the lifetime of the extracellular enzyme, then the amount of product made available to the organism will be a function of both the activity and lifetime of the enzyme.

A model based on an equation representing change in the concentration of hydrolysis product with time and accounting for enzyme activation and denaturation (adapted from Shuler and Kargi, 1992) allows predictions about the temporal and

156

spatial advantages of releasing less stable (more psychrophilic) versus more thermo-stable extracellular enzymes into the cold:

$$\frac{d[P]}{dt} = V = k_a[E]$$

where:

[P] = product concentration;

V = velocity of hydrolysis;

k_a = activation constant $= Ae^{-E_a RT}$;

[E] = enzyme concentration $= [E_0]e^{-k_d t}$; and

k_d = denaturation constant $= A_d e^{-E_d / RT}$.

Model runs (Fig. 6) for a psychrophilic enzyme, characterized by higher activation and denaturation constants, and for a less cold-adapted enzyme (with lower relative constants) reveal that the psychrophilic enzyme provides a greater benefit to the

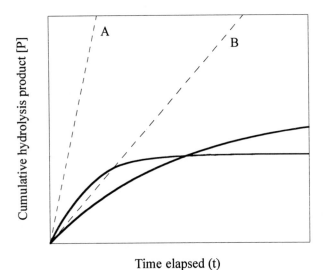

Figure 6. A model analysis of accumulated hydrolysis product [P] over time for enzymes A and B, where activation and denaturation constants for A are 4X the respective constants for B (making A more cold-adapted). In the absence of enzyme denaturation ([E] = [E₀]), product accumulates linearly (dashed lines); as the enzyme denatures over time, product accumulation slows (solid lines). Which enzyme results in maximal product accumulation and thus benefit to the bacterium depends upon time elapsed. Saturating substrate concentration was assumed in the model.

organism (higher cumulative hydrolysis product) over a shorter time period. Over longer time periods (e.g., greater diffusional distance for product return, less hydrolyzable target compounds requiring longer-lived enzymatic action), a less cold-adapted enzyme provides the greater return (Fig. 6). Different environments may thus select for different thermal classes of enzymes, depending on specific physical-chemical conditions. For example, in porous matrices characterized by short distances between bacteria and readily hydrolyzable food, fixed in space due to attachment or sorption to surfaces, psychrophilic extracellular enzymes are predicted to occur more frequently than in the water column. The presence of other organic polymers in these matrices may further stabilize and extend the lifetime of cell-free enzymes (Anchordoquy and Carpenter, 1996). Our detection of significantly more cases of psychrophilic EEA in environmental samples of organic-rich sea ice and sinking particle aggregates than in samples of seawater (Huston and Deming, 1999; unpublished) tends to verify the model prediction.

5. Some Implications of Pushing the Lower Temperature Limit

The coldest temperatures in the marine environment (Fig. 7), and on Earth (Fig. 8) that are still permissive of liquid habitats for life, can be found in wintertime sea ice. As the

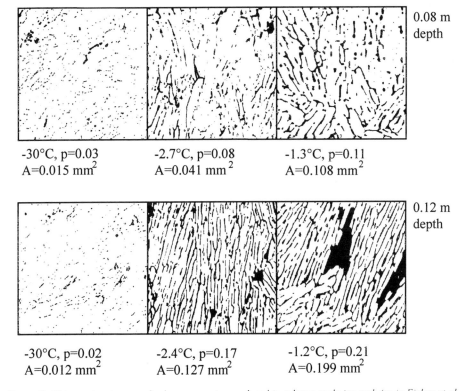

		0.08 m depth
-30°C, p=0.03 A=0.015 mm^2	-2.7°C, p=0.08 A=0.041 mm^2	-1.3°C, p=0.11 A=0.108 mm^2

		0.12 m depth
-30°C, p=0.02 A=0.012 mm^2	-2.4°C, p=0.17 A=0.127 mm^2	-1.2°C, p=0.21 A=0.199 mm^2

Figure 7. Thin-section images of columnar sea ice produced in a large-scale ice tank (as in Eicken et al., 1998; images courtesy of H. Eicken) Each 20 mm-wide panel of three sections shows the increase in porosity (p; pores shown in black) and pore-space connectivity with increase in temperature (-30 to -1.3°C). Also evident are changes in pore structure (A = mean pore cross-sectional area) with changes in temperature and depth in the ice sheet (upper panel from 0.08 m. lower from 0.12 m).

158

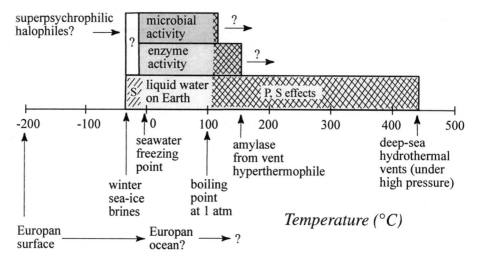

Figure 8. Generalized depiction of the temperature range of fluid regimes on Earth (and on Europa), enabled at the extremes by effects (hatched areas) of salinity (S) and pressure (P), and the narrower ranges of temperatures known to be permissive to enzymatic and microbial activity.

atmosphere above cools significantly in the absence of solar radiation, a temperature gradient establishes in the sea-ice matrix, from below -30°C near the air-ice interface to the near-freezing temperature (-1.7°C) of the seawater below (Fig. 7). As ice forms from seawater, salt is expelled into the increasingly small pore spaces within the ice matrix (Fig. 7), such that the remaining liquid is a concentrated brine solution. The intensity of the salt concentration (and parallel concentration of DOM) determines the freezing point of this remaining inhabitable liquid. Few studies of bacteria or other organisms in wintertime sea ice have been reported (none address the very coldest temperatures), but available information makes clear the potential for bacterial survival, growth and EEA in these remarkable habitats (Helmke and Weyland, 1995; K. Junge, C. Krembs, H. Eicken, and J. Deming, unpublished).

There is reason to speculate that microorganisms entrapped within sea-ice pockets may produce not only extracellular enzymes but also other polymeric substances (Krembs and Engel, 1999) which may serve as cryoprotectants both for the organisms and their enzymes (Hargens and Shabica, 1973; Anchordoquy and Carpenter, 1996). The environment is thus ripe for discovery of novel cold-active compounds (Gounot, 1991; Brenchley, 1996; Feller et al., 1996) produced microbially for ecological reasons, either as foraging tools (extracellular enzymes) or stabilizing agents to assist in the foraging process. Competitive survival in DOM-rich but microscale spatial zones (Fig. 7) may also involve production of unusual secondary metabolites (anti-microbial agents), a virtually unexplored possibility in sea ice with potentially significant biomedical application. The discovery of hydrolytic proteases, expressing the lowest activity optima yet recorded in cell-free extract (Huston et al., 1999), holds promise for the low-temperature cleaning industry and for discovery of other degradative enzymes well-suited to the in situ or ex situ bioremediation of contaminated sites at high latitudes or seasonally cold environments. The component of sediment grains and sorbed organic (and heavy-metal) contaminants often present in Arctic sea ice (Eicken et al., 1998) may have provided sufficient environmental pressure for the evolution of compound-specific degradative enzymes also of value in the realm of low-temperature bioremediation. In sharp contrast to all other marine environments, Arctic sea ice has yielded a remarkably high success rate in obtaining numerically important bacteria in laboratory culture (Junge et al., 1999); expectations for acquiring in culture those organisms responsible for producing any novel or useful compounds can be higher than usual.

A thermal class of microorganisms predicted from the nature of wintertime sea-ice habitats but not yet known to exist would be the superpsychrophilic halophiles (Fig. 8). Indeed, the presence of any type of cold-adapted Archaea in sea ice remains to be documented; for some types, the search has not been initiated. Finding novel microorganisms and compounds they may produce for cryoprotectant or other growth and survival means in sea ice will have implications for the pursuit of life elsewhere, particularly on the Jovian moon Europa speculated to harbor an ocean beneath its continually renewed ice cover (Squires et al., 1983; Chapman, 1999). Intensified study of the coldest environments on Earth can only facilitate the success of life-probing missions being planned for the Europan surface.

6. Acknowledgments

Preparation of this chapter was supported by grants from the WA State Sea Grant Program and the US National Science Foundation (OPP-LExEn and OPP-ANS for the International Northwater Polynya Project) and by the University of Washington through its Astrobiology Program. We are grateful to John Baross, Hajo Eicken, Karen Junge, Christopher Krembs, Barbara Krieger-Brockett, and Yves-Alain Vetter for insightful discussion and to Shelly Carpenter for bacterial growth data and assistance with graphics.

7. References

Anchordoquy, T.J., and Carpenter, J.F. (1996) *Arch. Biochem. Biophys.* **332**, 231-238.
Baross, J.A., and Deming, J.W. (1995) In: D.M. Karl (ed.), *The Microbiology of Deep-Sea Hydrothermal Vents*, CRC Press, Boca Raton, pp. 169-217.
Baross, J.A. and Morita, R.Y. (1978) In: D.J. Kushner (ed.). *Microbial Life in Extreme Environments*, Academic Press, London, pp. 9-71.
Brenchley, J.E. (1996) *J. Ind. Microbiol.* **17**, 432-437.
Chapman, C.R. (1999) *Science* **283**, 338-339.
Christian, R.R., and Wiebe, W.W. (1974) *Can. J. Microbiol.* **20**, 1341-1345.
Davail, S., Feller, G., Narinx, E., and Gerday, C. (1994) *J. Biol. Chem.* **269**, 17,448-17,453.
Deming, J.W. (1998) *Curr. Opinion Biotech.* **9**, 283-287.
Deming, J.W., and Baross, J.A. (1993) *Cosmochem. Geochem. Acta* **57**, 3219-3230.

160

Eicken, H., Weissenberger, J., Bussmann, I., Freitag, J., Schuster, W., Valero Delgado, F., Evers, K.-U., Jochmann, P., Krembs, C., Gradinger, R., Kindemann, F., Cottier, F., Hall, R., Wadhams, P., Reisemann, M., Kousa, H., Ikavalko, J., Leonard, G.H., Shen, H., Ackley, S.F., and Smedsrud, L.H. (1998) In: H.T. Shen (ed.), *Ice in Surface Waters*, Balkema, Rotterdam, pp. 363-370.

Feller, G., Narinx, E., Arpigny, J.L., Aittaleb, M., Baise, E., Genicot, S., and Gerday, C. (1996) *FEMS Microbiol. Rev.* **18**, 189-202.

Gounot, A.-M. (1991) *J. Appl. Bacteriol.* **71**, 386-397.

Harder, W., and Veldkamp, H. (1971) *Antonie von Leeuwenhoek J. Microbiol. Serol.* **37**, 51-63.

Hargens, A.R., and Shabica, S.V. (1973) *Cryobiol.* **10**, 331-337.

Helmke, E., and Weyland, H. (1995) *Mar. Ecol. Prog. Ser.* **117**, 269-287.

Huston, A.L., and Deming, J.W. (1999) Low-temperature activity optima for extracellular enzymes released by Arctic marine bacteria. *Abstract, American Society for Microbiology Annual Meeting*, May 1999, Chicago.

Huston, A.L., Krieger-Brockett, B.B., and Deming, J.W. (1999) Remarkably low temperature optima for extracellular enzyme activity from Arctic bacteria and sea ice. *Environ. Microbiol.* (submitted).

Inoue, K. (1977) *J. Gen. Appl. Microbiol.* **23**, 53-63.

Jumars, P.A., Deming, J.W., Hill, P.S., Karp-Boss, L., Yager, P.L., and Dade, W.B. (1993) *Mar. Microb. Food Webs* **7**, 121-159.

Junge, K., Staley, J.T., and Deming, J.W. (1999) Phylogenetic diversity of numerically important bacteria cultured at subzero temperature from Arctic sea ice. *Abstract, American Society for Microbiology Annual Meeting*, May 1999, Chicago (manuscript in preparation).

Krembs, C., and Engel, A. (1999) Abundance and variability of microorganisms and TEP across the ice-water interface of melting first-year sea ice in the Laptev Sea (Arctic). *Mar. Biol.* (submitted).

Morita, R.Y. (1975) *Bacteriol. Rev.* **39**, 144-167.

Petersen, G.H., and Curtis, M.A. (1980) *Dana* **1**, 53-64.

Pomeroy, L.R., and Deibel, D. (1986) *Science* **233**, 359-360.

Pomeroy, L.R., and Wiebe, W.J. (1993) *Mar. Microb. Food Webs* **7**, 101-118.

Reichardt, W. (1987) *Mar. Ecol. Prog. Ser.* **40**, 127-135.

Rich, J.H., Gosselin, M., Sherr, E., Sherr, B., and Kirchman, D.L. (1997) *Deep-Sea Res. II* **44**, 1645-1663.

Shuler, M.L., and Kargi, F. (1992) *Bioprocessing Engineering: Basic Concepts*, Prentice Hall, New Jersey.

Squires, S.W., Reynolds, R.T., Cassen, P.M., and Peale, S.J. (1983) *Nature* **301**, 225-226.

Vetter, Y.-A., Deming, J.W., Jumars, P.A., and Krieger-Brockett, B.B. (1998) *Microb. Ecol.* **36**, 75-92.

Vetter, Y.-A., and Deming, J.W. (1994) *Mar. Ecol. Prog. Ser.* **114**, 23-34.

Vetter, Y.-A., and Deming, J.W. (1999) *Microb. Ecol.* **37**, 86-94.

Wiebe, W.J., Sheldon, W.M., and Pomeroy, L.R. (1992) *Appl. Environ. Microbiol.* **58**, 359-364.

Wiebe, W.J., Sheldon, W.M., and Pomeroy, L.R. (1993) *Microb. Ecol.* **25**, 151-159.

Yager, P.L. (1996) *The Microbial Fate of Carbon in High-Latitude Seas: Impact of the Microbial Loop on Oceanic Uptake of CO_2*, Ph.D. Thesis, University of Washington, Seattle, 174 pp. (also available at http://wwwlib.umi.com/dissertations/main).

Yager, P.L., and Deming, J.W. (1999) Pelagic microbial activity in the Northeast Water Polynya: Testing for temperature and substrate interactions using a kinetic approach. *Limnol. Oceanogr.* (in press).

Biodata of **A. Aristides Yayanos** contributor of the chapter on *"Deep Sea Bacteria"*.

Dr. Art Yayanos is presently a Professor of Biophysics at Scripps Institution of Oceanography (SIO), University of California, San Diego, La Jolla, CA. He obtained his Ph.D. at the Pennsylvania State University in 1967 and is since in SIO. Dr. Yayanos research interests include deep-sea biology, high-pressure physical chemistry, radiation biology and the study of mass extinction. He is also associated with Life in Extreme Environments Program (LExEn) of the U.S. National Science Foundation. Dr. Yayanos is truly grateful for the support of grants from NASA, ONR, DOE, NIH and NSF that made his work possible over the pat thirty years.

E-Mail: ayayanos@ucsd.edu

J. Seckbach (ed.), Journey to Diverse Microbial Worlds, 161-174.

DEEP-SEA BACTERIA

A. ARISTIDES YAYANOS
Scripps Institution of Oceanography, University of California San Diego, La Jolla CA 92093-0202

1. Introduction

Hydrostatic pressure emerged as a significant environmental factor in ecology and evolution during the second half of the twentieth century. Pressure is not explicitly considered, for example, in Ekman's classic 1953 book on the biogeography of marine animals (Ekman 1953). The book, a revision of a much earlier edition, was probably completed just as the *Galathea* Expedition ended in 1952 and was uninfluenced by the discoveries of its scientists (Bruun 1957). ZoBell and Morita, as participants on the *Galathea* Expedition, showed that pressure must be considered as a significant environmental parameter affecting the distribution of bacteria (Yayanos 2000, ZoBell 1952; ZoBell and Morita 1959). Other members of the expedition found animals in the greatest ocean depths.

Since bacteria, archaea and metazoa occur in all of the deep-sea (Bruun 1957; Kato et al. 1997), the pressure ultimately limiting life processes is greater than 110 MPa, approximately the pressure found at the maximum ocean depth. Another observation of interest is that the inhabitants of one depth of the sea are easily discernible from those of another if the depth separation is a few thousand meters or more. The differences appear clearly in the physiological, biochemical and regulatory responses of marine organisms to changes in pressure relative to that of their habitat (Gage and Tyler 1991; Siebenaller et al. 1982; Siebenaller and Somero 1989; Somero 1992; Yayanos 1995). In terms of the greatest possible depth separation in the sea, no known organism lives at both 110 MPa and atmospheric pressure along the 2°C isotherm. Thus, a second inference is that the responses of organisms to pressure change relative to that of their habitat are strong indications that pressure is an agent of natural selection in the oceans. Finally, physiological studies of organisms from the Sulu and Mediterranean Seas (Yayanos 1995), which have warm bathyal and abyssal depths, and from submarine hydrothermal vents (Marteinsson et al. 1999) show that the temperature of a high pressure habitat modifies greatly the response of an organism to pressure change. Thus, the third and final point is that pressure and temperature are coordinate variables in ecology and evolution just as they are, for example, in thermodynamics.

Organisms brought from the deep-sea to the sea surface can experience decompression, warming, and illumination that potentially affect their survival. The continuous maintenance of a community of bacteria at the high pressure of its habitat is unnecessary for the purpose of ultimately cultivating at least some deep-sea bacteria. There are high pressure methods for the collection of bacteria from the deep-sea (Jannasch and Taylor 1984; Yayanos 1995). However, bacteria with an absolute pressure requirement in the sense that they instantly die if decompressed to atmospheric pressure, have not yet been isolated. I believe that the existence of such bacteria is highly plausible and that someday they will be found (Yayanos 1995). The bacteria isolated to date that grow only at high pressure either survive decompression or die slowly at atmospheric pressure (Chastain and Yayanos 1991; Yayanos and Dietz 1982; Yayanos and Dietz 1983).

Three requirements lead to the successful cultivation of a significant subset of the true deep-sea bacteria (Yayanos 1995). The first is the isothermal execution of sampling procedures to avoid the thermal inactivation of the deep-sea psychrophilic bacteria. The second is to avoid illuminating samples and media with UV light from sunlight and fluorescent lights. And, the third prerequisite is the application of the ecological method (enrichment culture technique) (Kluyver and van Niel 1956) at the temperature, pressure and darkness of the sampled environment. As a result of such efforts, representatives of the autochthonous bacteria of the cold deep-sea and of submarine hydrothermal vents are in pure culture (Horikoshi 1998; Marteinsson et al. 1999; Yayanos 1986;Yayanos 1995).

Because of the current and detailed character of recent reviews (Abe et al. 1999; Abe and Kato 1999; Bartlett 1992; Horikoshi 1998; Kato 1999; Kato and Bartlett 1997; Li and Kato 1999; Prieur 1992; Yayanos 1995; Yayanos 1998; Yayanos 1999), it seems prudent to avoid as much as possible undue repetition in this brief chapter. Therefore, I seek to develop a viewpoint that has been given insufficient attention. The underlying theme of this chapter is that many problems of microbial ecology in general and of adaptation to pressure in particular have a combinatorial essential quality.

2.   Piezophily and Barophily

ZoBell and Johnson define bacteria as barophilic if they grow or function at a high pressure better than they do at atmospheric pressure. A key problem with this definition is that the temperature of the test of a barophilic response is not defined. The concept of piezophily replaces that of barophily (Yayanos 1995; Yayanos 1998) and arises from an examination of *PTk* diagrams. The exponential growth rate constant, k, of a microorganism is a function of T, P, and the chemical composition of its environment and is given the following equation,

$$k = k(P, T, n_1, n_2 L \; n_N) \qquad (2.1)$$

where n_i represents the concentration of the ith chemical constituent. The form of this function can be determined for a given organism from experiments where nutrient levels are non-limiting and can be treated as constants. That is, a plot of k versus T and P is called a *PTk* diagram. An example is in Figure 1 and

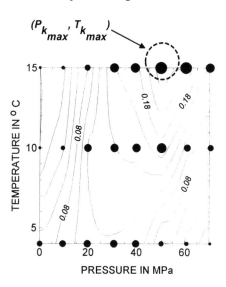

Figure 1. A *PTk* diagram for bacterial isolate DB6906 drawn from data in the paper by Kato, Inoue and Horikoshi (1996). Each contour represents a constant growth rate. Lines at 0.0693 and 0.1386 would represent growth rates of 10 and 5 h, respectively. The *PTk* diagram allows for easy approximate identification of the pressure--temperature combination, (P_{kmax}, T_{kmax}), where the growth rate is maximum, k_{max}, and allows for comparison of organisms from different *PT* habitats

shows there is a single k_{max}, defined as the maximum growth rate over all possible values of T and P. The value of P_{kmax} is also obtained from a *PTk* diagram. A microorganism is called a piezophile if

$$0.1 \text{ MPa} < P_{k\max} < 50 \text{ MPa} \qquad (2.2)$$

and a hyperpiezophile if

$$P_{k\max} > 50 \text{ MPa} \qquad (2.3)$$

The concept of piezophily blends nicely with that describing temperature adaptation so as to allow the classification of an organism with respect to its behavior as a function of both temperature and pressure. Thus, there are examples of hyperpiezospychrophiles, piezothermophiles and piezomesophiles (Yayanos 1998). Each of these is easily identified from its *PTk* diagram. The parameters, such as the values of P_{kmax} and T_{kmax}, derived from this diagram allow correlations with the pressure and temperature of the habitat of a microorganism that may help illuminate the nature of adaptation along gradients of temperature and pressure (Yayanos 1995).

3. A Melange of Adaptation Possibilities

3.1. MICROBIAL COMMUNITIES

The study of microbial communities is assisted immensely by isolation, sequencing, and sequence analysis of the genes that code for 16S rRNA. Analysis of bacterial genome sequences published so far suggests that the phylogenies based on 16S rRNA might not be faultless (Doolittle 1999). These phylogenies, nevertheless, allow for the subdivision of a community of cells into operationally defined sub-types and, thereby, for an evaluation of the number of possible communities having the same phylogenetic components. If a bacterial community of N cells in a sample from the sea contains Y different types as determined by 16S rRNA sequences or some other agreed upon method such as ability to use a given reductant, then

$$N = s_1 + s_2 + \cdots + s_Y \qquad (3.1.1)$$

where s_i is the number of cells of subtype i. The total number of ways, M_c, such a community of cells can be constructed is

$$M_c = C\binom{N+Y-1}{Y-1} = \frac{(N+Y-1)!}{(Y-1)!N!} \qquad (3.1.2)$$

M_c is the number of possible sets of communities under the conditions of equation (3.1.1) and $0 \le s_i \le N$. A few typical values for $N = 10^3$ are as follows: $Y = 2$, $M_c = 10^3+1$; $Y = 4$, 1.7×10^8; and, $Y = 5$, $M_c = 4.2 \times 10^{10}$. These large numbers of possible communities become astronomical as N and Y increase. Thus, for $N=10^6$ cells and $Y=5$, there are 4.2×10^{22} possible microbial communities. In a microbial community comprised of seven different classes of microorganisms (as defined by 16S rRNA phylogenetic analysis. e.g.) and a total of 10^8 cells, there are 1.14×10^{45} hypothetically distinguishable communities.

The number of microbial communities actually realized is less than the calculated value for reasons both known and unknown. For example, the presence of a cell type at very low numbers may be inconsequential to the function of the community of cells. Thus, those possible communities having some cell types at low cell concentrations are probably indistinguishable from communities where these cell types are absent. Even so, the calculated number of possible communities remains large enough to generate formidable difficulties for ecological analysis. That is, we need more knowledge to be able to answer the question of why a particular community is observed in nature. An improved understanding of how microorganisms interact with both each other and their environment may someday provide the means for dynamical models that show how a particular community of cells arises and persists. It is not clear at this time how the pressure dependence of processes both internal and external to cells leads to differences in communities along the pressure gradient in the sea. The description of microbial community structure along depth gradients with 16S rRNA phylogentic analysis has begun and will surely provide important clues (DeLong 1997; Kato et al. 1997; Li and Kato 1999).

3.2. ALLOCHTHONOUS—AUTOCHTHONOUS

Microbial communities suffer pressure change when they move vertically by natural processes and when they are moved to the laboratory for study (Alongi 1992; Deming 1986; Rowe and Deming 1985). When a community of Y types of cell (a natural population) is disturbed with a pressure change, an observed physiological response may be difficult to interpret for at least two reasons. First, the particular community is one of many possible composed of the same Y types of cell. Each of the Y types may have a unique pressure response. Furthermore, the community at one particular depth acquires new cells not only by growth of its autochthonous bacteria but also by transport of cells from other depths of the ocean. A sediment sample collected at a depth of 5,800m, for example, will contain not only indigenous sediment bacteria but also bacteria from shallower depths such as 3,000m. The organisms from different depths will perform differently in response to a change in pressure.

In general, microbial communities contain invader (allochthonous) organisms and those that are true inhabitants (autochthonous). At any depth of the sea, the allochthonous organisms can come from different ocean depths and from terrestrial sources. A microorganism native to a warm surface water habitat will function marginally or not at all as an allochthonous member of a cold microbial community at a depth of 5,000m (Trent and Yayanos 1985; Turley 1993; Turley and Carstens 1991). On the other hand, admixture of bacteria from different but more closely spaced isothermal depths, e.g., 1000m and 5,000m, could result, at least for some period of time, in a bacterial community with comparably functioning autochthonous and allochthonous bacteria.

A comparison of the possible contributions of autochthonous and allochthonous bacteria to the uptake of a carbon compound by a community of bacteria illustrates the problem. Let us suppose that we can describe the rate of incorporation of a substrate, S, by a particular bacterial species, i, with a single rate constant, $r_i=r_i(T,P)$, a function of temperature and pressure. Let us assume we have a sample from the deep sea that contains two classes (simplified for purposes of discussion) of bacteria: the class $i=DS$ of true deep-sea bacteria; and, the class $i=ML$ comprising the bacteria from the mixed layer of the sea. We assume that the DS class and the ML class do not interact (syntrophy, e.g.) in any way. The deep-sea bacteria have an uptake constant r_{DS} and the upper ocean bacteria have an uptake constant r_{ML}. These constants are pressure and temperature dependent. The rate of uptake of S by a community of these two classes of cells would be

$$\frac{dS}{dt} = r_{DS} N_{DS} + r_{ML} N_{ML} \qquad (3.2.1)$$

where N_{DS} is the number of deep-sea bacterial cells and N_{ML} is the number of upper ocean bacterial cells. Field experiments usually measure r, the uptake by the community of cells which is given by

$$\frac{dS}{dt} = rN \qquad (3.2.2)$$

where N is the total number of cells ($N=N_{DS}+N_{ML}$). Therefore,

$$r = r_{DS} \frac{N_{DS}}{N} + r_{ML} \frac{N_{ML}}{N} \qquad (3.2.3)$$

Laboratory and *in situ* studies of uptake of radiolabelled substrates by microbial communities of the cold deep sea have been done often at a single substrate concentration (Tabor et al. 1981; Wirsen and Jannasch 1986). In those studies, the observed uptake rate, r in equation (3.2.2), is usually found to decrease with increasing pressure. Several hypotheses have been proposed to explain this lack of or capricious occurrence of pressure stimulation in older experiments. One hypothesis is that warming of samples during their retrieval kills or severely attenuates the indigenous deep-sea bacteria (Deming and Colwell 1985; Yayanos and Dietz 1982) whereas some of the allochthonous ones recover their ability to grow. That is, N_{DS} decreases while N_{ML} increases and N first decreases and then increases. Another hypothesis (Jannasch and Wirsen 1984) is that "the presence of barotolerant bacteria in deep water...causes the non-barophilic response of natural populations." That is, all N_{DS} bacteria are not piezophilic. A similar but distinct hypothesis is that the deep-ocean harbors autochthonous bacteria that are piezophilic under high nutrient conditions but that possess a spectrum of pressure-dependent growth responses manifested at particular nutrient conditions. For example, bacterial isolate PE36, which is a piezopsychrophile, loses its pressure dependent growth response when grown on glycerol in a salts medium

(Yayanos and Chastain 2000). The pressure dependence of uptake kinetics has not been determined with a sufficient variety of deep-sea bacteria, however, to test fully this hypothesis.

As a hypothetical example, the uptake rate constants of allochthonous and autochthonous components in a microbial community from 5,000m and 2°C might have the following relationship at 25°C and 0.1 MPa,

$$r_{DS} = r_{ML} \qquad (3.2.4)$$

whereas the opposite would be true at 2°C and 50 MPa,

$$r_{DS} \; ? \; r_{ML} \qquad (3.2.5)$$

Equation (3.2.4) describes the rates of processes in a decompressed and warmed sample whereas equation (3.2.5) describes the rates in a microbial community on the sea floor (Lampitt 1985). More data are needed, however, to judge both the degree of the above inequalities, the pressure and temperature dependence of uptake rate constants (r_{DS} and r_{ML}), and the variability of vertical cell transport by buoyant particle flux and sedimentation. Because the sea is an open and interconnected system, its microbial communities are comprised of true inhabitants and visitors. The interpretation of the activities of such microbial communities to include the influence of pressure changes and a variable nutrient regime is the purpose of a recent paper by Patching and Eardly (1997). Unlike much of the early work in the field showing an absence of pressure-facilitated uptake rates by deep-sea microbial communities (Jannasch and Taylor, 1984), their work shows greater rates of incorporation of thymidine and leucine at high than at low pressures.

3.3. SPECIES INTERACTIONS

The parameterization of species interactions with the Lotka-Volterra equations results in a combinatorial problem, as pointed out by Renshaw (Renshaw 1991). He notes that the six parameters in the Lotka-Volterra equations describing a two species interaction can assume postive, negative or zero values and that, therefore, there are 3^6 different models describing the two species community. Studies are needed to assess how interspecies interactions are influenced by the temperatures, pressures, and nutrient conditions in the sea. I am unaware of studies on this problem.

3.4. CELLULAR BIOCHEMISTRY

Many biochemical reactions occur in a coupled fashion with the utilisation or generation of ATP, reductants, and ion gradients. The coupled reactions, for example, in the phosphotransferase system (PTS) that transports glucose across the bacterial cell membrane can be described with approximately 20 rate constants (Kholodenko et al.

1998). From this standpoint alone, there may be many sets of rate constant values that would enable the adaptation allowing a deep-sea cell to tranport glucose at high pressures. There have not been enough studies of deep-sea bacteria to to reveal the spectrum of adaptive strategies that have evolved. But it is clear that different strategies are used by different species (DeLong and Yayanos 1987). Furthermore, there is a proposal that PTS proteins form complexes attributable to macromolecular crowding within cells (Rohwer et al. 1998). Since pressure affects the volumetric properties of proteins and protein-protein interactions (Heremans 1993; Heremans 1999), it will likely affect macromolecular crowding. Unravelling the properties of the PTS in bacteria from different depths will be both tricky and fascinating.

3.5. MEMBRANE LIPIDS

The fatty acid composition of many microbial membranes varies as a function of growth temperature and pressure. For example, *Staphylococcus aureus* adapts to different temperatures not only by changing its membrane fatty acid composition (Joyce et al. 1970) but also in the case of a temperature decrease by synthesising membrane cartoneoids. A striking feature of many deep-sea bacterial membranes is the presence of polyunsaturated fatty acids (PUFAs) (Allen et al. 1999; DeLong and Yayanos 1986; Hamamoto et al. 1995; Wirsen et al. 1987; Yano et al. 1997; Yazawa et al. 1988). Recent studies present evidence that PUFA-deficient mutants of strain *SS9* can grow at different pressures as well as the PUFA-producing wild type. Since some deep-sea bacteria do not have membrane PUFAs (DeLong and Yayanos 1985), it is clear that their presence is not an obligatory facet of deep-sea bacterial adaptation to high pressure. Nevertheless, the level of membrane unsaturated fatty acids, PUFAs included, varies with pressure and temperature.

The most common way of describing the lipids of bacterial membranes is to list their fatty acid composition. The incompleteness of such a description is evident since the work of Thompson (1989) who showed that phospholipid isomerization (switching of fatty acid chains between the *sn-1* and *sn-2* positions of a phospholipid) can occur in the membrane of *Tetrahymena* immediately following a temperature shift. Isomerization occurs without new lipid synthesis, leads to immediate changes in membrane fluidity, and is likely a phylogenetically ubiquitous process not yet demonstrated in bacteria.

Thus, there are several levels of membrane lipid structure that need to be examined to understand how a pressure– or temperature–induced alteration in fatty acid composition leads to an adapted membrane. The problem is again a combinatorial one. There are PL types of each phospholipid class in a bacterial membrane where

$$PL = n^2 \qquad (3.5.1)$$

and n is the number of different types of fatty acid. If there are 7 different principal fatty acids, then there are 49 possible phospholipids for each phospholipid class. One estimate places the number of phospholipid molecules in the membrane of *E. coli* at 1.6 x 10^7. Equation (3.1.2) applies also to this problem so that for N=1.6 x 10^7 phospholipid molecules and Y=49 types of phospholipid, there are 5.06 x 10^{284} possible membranes differing from each other in phospholipid composition. This amazingly large number is even larger if the variety and variability of membrane proteins is included. Notwithstanding the limitations imposed by any selection rules to exclude certain phopholipid molecular species, there remains a huge set of possible membranes. Cells use the ability to alter their membrane structure not only in response to pressure and temperature but also under the influence of a changing nutrient and chemical environment. Starvation, for example, has been shown to cause a change in membrane fatty acid composition (Rice and Oliver 1992) and growth on different carbon sources alters the membrane proteins that respond to pressure change (Yayanos and Chastain 2000). Elucidation of these processes may require studies on phospholipid molecular species and their arrangement in a membrane.

3.6. MEMBRANE PROTEINS

The levels of several types of membrane protein respond to pressure change. Outer membrane proteins of the bacterium SS9, ompH (Bartlett et al. 1989) and ompL, are being studied by Bartlett and his colleagues (Chi and Bartlett 1993; Chi and Bartlett 1995). TW12 is a mutant of SS9 created by transposon mutagenesis that does not synthesize OmpL and produces high levels of OmpH at all pressures. Welch and Bartlett (1998) describe this as a pressure-sensing mutant. The mutation is in the *toxRS* operon. TW30 is a constructed mutant of SS9 from which 55% of the *toxR* coding sequence was removed. Pressure affects the synthesis of OmpL and of OmpH by TW30 in exactly the same way as in TW12. A plasmid containing the *toxRS* operon and inserted into TW30 restores the pressure-dependent synthesis of OmpL and OmpH. Welch and Bartlett thus conclude that the *toxRS* operon participates in OmpL induction and in OmpH repression by pressure. Other membrane protein studies include the description of a pressure-regulated synthesis of cytochromes in deep-sea bacteria (Qureshi et al. 1998). An interesting finding is that cytochrome c-552 occurs only in cells grown at atmospheric pressure whereas cytochrome c-551 is in both high and low pressure cultures.

4.   **Summary**

An organism that lives at a given depth of the sea acquires properties reflecting the ambient pressure at that depth (Childress and Thuesen 1993; George 1985; George and Strömberg 1985; Somero 1992; Young et al. 1996; Young and Tyler 1993). Microbial community structure, bacterial cell structures, biochemical functions, and gene expression all respond to pressure change in a way reflecting adaptation to the conditions of temperature, pressure, and nutritional state of the native habitat. Analyses

172

with thermodynamics, molecular dynamics, and population dynamics will all likely help to find the rules selecting one biological state over another as an adaptive strategy or as part of an evolutional sequence of organisms. Work over the past fifty years shows that pressure is a significant selection factor affecting the biogeography of marine organisms. Recent work establishes temperature and pressure as coordinate environmental parameters.

5. References

Abe, F. and Kato, C. (1999) *in:* K. Horikoshi. and K. Tsujii (eds.) Extremophiles in Deep-Sea Environments, Springer-Verlag, Tokyo. pp. 227-248.

Abe, F., Kato, C., and Horikoshi, K. (1999) *Trends in Microbiology* 7, 447-453.

Allen, E.E., Facciotti, D., and Bartlett, D.H. (1999) *Appl.Environ.Microbiol.* 65, 1710-1720.

Alongi, D.M. (1992) *Deep-Sea Res.* 39, 549-565.

Bartlett, D., Wright, M., Yayanos, A.A. and Siverman, M. (1989) *N* 342, 572-574.

Bartlett, D.H. (1992) *Sci.Prog.,Oxf.* 76, 479-496.

Bruun, A.F. (1957) *Geol.Soc.Am.,Mem.* 67, 641-672.

Chastain, R.A. and Yayanos, A.A. (1991) *Appl.Environ.Microbiol.* 57, 1489-1497.

Chi, E. and Bartlett, D. H. (1993) *J.Bacteriol.* 175, 7533-7540.

Chi, E. and Bartlett, D.H. (1995) *Molecular Microbiology* 17, 713-726.

Childress, J.J. and Thuesen, E.V. (1993) *Limn.Ocean.* 38, 665-670.

DeLong, E.F. (1997) *Trends Biotech.* 15, 203-207.

DeLong, E.F. and Yayanos,A.A. (1985) *Science* 228, 1101-1103.

DeLong, E.F. and Yayanos,A.A. (1986) *Appl.Environ.Microbiol.* 51, 730-737.

DeLong, E.F. and Yayanos,A.A. (1987) *Appl.Environ.Microbiol.* 53, 527-532.

Deming, J.W. (1986) *Microbiol.Sci.* 3, 205-211.

Deming, J.W. and Colwell,R.R. (1985) *Appl.Environ.Microbiol.* 50, 1002-1006.

Doolittle, W.F. (1999) *Science* 284, 2124-2128.

Ekman, S. (1953) Zoogeography of the Sea, Sidgwick and Jackson, Ltd., London.

Gage, J.D. and Tyler,P.A. (1991) Deep-Sea Biology: A Natural History of Organisms at the Deep-Sea Floor, Cambridge University Press, Cambridge, England.

George, R.Y. (1985) *in:* P.E. Gibbs (ed.) Proceedings of the Nineteenth European Marine Biology Symposium, Plymouth, 1984, Cambridge University Press, Cambridge, UK. pp. 173-182.

George, R.Y. and Strömberg, J.-O. (1985) *Polar Biol.* 4, 125-133.

Hamamoto, T., Takata, N., Kudo, T. and Horikoshi, K. (1995) *FEMS Microbiology Letters* 129, 51-56.

Heremans, K. (1993) *in:* R. Winter and J. Jonas (eds.) High Pressure Chemistry, Biochemistry and Materials Science, Kluwer Academic Publishers, Dordrecht, The Netherlands. pp. 443-469.

Heremans, K. (1999) *in:* R. Winter and J. Jonas (eds.) High Pressure Molecular Science, The Netherlands, Dordrecht. pp. 437-472.

Horikoshi, K. (1998) *Current Opinion In Microbiology* 1, 291-295.

Jannasch, H.W. and Taylor, C.D. (1984) *Ann.Rev.Microbiol.* 38, 487-514.

Jannasch, H.W. and Wirsen, C.O. (1984) *Arch.Microbiol.* 139, 281-288.

Joyce, G.H., Hammond, R.K., and White, D.C. (1970) *J.Bacteriol.* 104, 323-330.

Kato, C. (1999) *in:* K. Horikoshi and K. Tsujii (eds.) Extremophiles in Deep-Sea Environments, Springer-Verlag, Tokyo. pp. 91-111.

Kato, C. and Bartlett, D.H. (1997) *Extremophiles* 1, 111-116.

Kato, C., Li, L., Tamaoka, J., and Horikoshi, K. (1997) *Extremophiles* **1**, 117-123.

Kholodenko, B.N., Rohwer, J.M., Cascante, M., and Westerhoff, H.V. (1998) *Mol.Cell Biochem* **184**, 311-320.

Kluyver, A.J. and van Niel, C.B. (1956) The microbe's contribution to biology, Harvard University Press, Cambridge.

Lampitt, R.S. (1985) *Deep-Sea Res.* **32**, 885-897.

Li, L. and Kato, C. (1999) *in:* K. Horikoshi and K. Tsujii (eds.), Extremophiles in Deep-Sea Environments. Springer-Verlag, Tokyo. pp. 55-88.

Marteinsson, V.T., Birrien, J.L., Reysenbach, A.L., Vernet, M., Marie, D., Gambacorta, A., Messner, P., Sleytr, U.B., and Prieur, D. (1999) *Int.J.Syst.Bact.* **49**, 351-359.

Patching, J.W. and Eardly, D. (1997) *Deep-Sea Research Part I Oceanographic Research Papers* **44**, 1655-1670.

Prieur, D. (1992) *in:* R.A. Herbert and R.J. Sharp (eds.), Molecular Biology and Biotechnology of Extremophiles Chapman and Hall, New York. pp. 163-202.

Qureshi, M.H., Kato, C., and Horikoshi, K. (1998) *FEMS Microbiology Letters* **161**, 301-309.

Renshaw, E. (1991) Modelling Biological Populations in Space and Time, Cambridge University Press, Cambridge.

Rice, S.A. and Oliver, J.D. (1992) *Appl.Environ.Microbiol.* **58**, 2432-2437.

Rohwer, J.M., Postma, P.W., Kholodenko, B.N., and Westerhoff, H.V. (1998) *Proc.Nat.Acad.Sci Usa* **95**, 10547-10552.

Rowe, G.T. and Deming, J.W. (1985) *J.Mar.Res.* **43**, 925-950.

Siebenaller, J.F. and Somero, G.N. (1989) *Rev.Aquat.Sci.* **1**, 1-25.

Siebenaller, J.F., Somero, G.N., and Haedrich, R.L. (1982) *Biol.Bull.* **163**, 240-249.

Somero, G.N. (1992) *Annu.Rev.Physiol.* **54**, 557-577.

Tabor, P.S., Deming, J.D., Ohwada, K., Davis, H., Waxman, M., and Colwell, R.R. (1981) *Microb.Ecol.* **7**, 51-65.

Thompson, G.A., Jr. (1989) *J.Bioenerg.Biomembr.* **21**, 43-60.

Trent, J.D. and Yayanos, A.A. (1985) *Mar.Biol.* **89**, 165-172.

Turley, C.M. (1993) *Deep-Sea Res.* **40**, 2193-2206.

Turley, C.M. and Carstens ,M. (1991) *Deep-Sea Res.* **38**, 403-413.

Welch, T.J. and Bartlett, D.H. (1998) *Molecular Microbiology* **27**, 977-985.

Wirsen, C.O. and Jannasch, H.W. (1986) *Mar.Biol.* **91**, 277-284.

Wirsen, C. O., Jannasch, H. W., Wakeham, S. G., and Canuel, E. A, (1987) *Current Microbiol.* **14**, 319-322.

Yano, Y., Nakayama, A., and Yoshida, K. (1997) *Appl.Environ.Microbiol.* **63**, 2572-2577.

Yayanos, A.A. (1986) *Proc.Natl.Acad.Sci.USA* **83**, 9542-9546.

Yayanos, A.A. (1995) *Ann.Rev.Microbiol.* **49**, 777-805.

Yayanos, A.A. (1998) *in:* K. Horikoshi and W.D. Grant (eds.) Extremeophiles, John Wiley & Sons, New York. pp. 47-92.

Yayanos, A.A. (1999) *in:H.* Ludwig (ed.) Advances in High Pressure Bioscience and Biotechnology: Proceedings of the International Conference on High Pressure Bioscience and Biotechnology, Heidelberg, August 30-September 3, 1998. Springer-Verlag, Berlin. pp. 3-9.

Yayanos, A.A. (2000) *in:* C.R. Bell, M. Brylinsky and P. Johnson-Green (eds.) Microbial Biosystems: New Fronteirs. Proceedings of the 8th Eighth International Symposium on Microbial Ecology, Atlantic Canada Society for Microbial Ecology, Halifax, Canada (in press).

Yayanos, A.A. and Chastain, R.A. (2000) *in:* C.R. Bell, M. Brylinsky and P. Johnson-Green (eds.) Microbial Biosystems: New Fronteirs. Proceedings of the 8th Eighth International Symposium on Microbial Ecology (Atlantic Canada Society for Microbial Ecology, Halifax, Canada (in press).

Yayanos, A.A. and Dietz, A.S. (1982) *Appl.Environ.Microbiol.* **43**, 1481-1489.

Yayanos, A.A. and Dietz,A.S. (1983) *Science* **220**, 497-498.

Yazawa, K., Araki, K., Watanabe, K., Ishikawa, C., Inoue, A., Kondo, K., Watabe, S. and Hashimoto, K. (1988) *Nippon Suisan Gakkaishi-Bulletin of the Japanese Society of Scientific Fisheries* **54**, 1835-1838.

Young, C.M., Tyler, P.A. and Gage, J.D. (1996) *J.Mar.Biol.Assoc.U.K.* **76**, 749-757.

Young, C.M. and Tyler, P.A. (1993) *Limnol.Oceanog.* **38**, 178-181.

ZoBell, C.E. (1952) *Science* **115**, 507-508.

ZoBell, C.E. and Morita, R.Y. (1959) *Galathea Report* **1**, 139-154.

V

ALKALIPHILES

Biodata of **W.D. Grant** co-author (with B.E. Jones) of the chapter on "*Microbial Diversity and Ecology of Alkaline Environments*".

Dr. **Bill Grant** is Professor of Environmental Microbiology at the University of Leicester (UK). He obtained his Ph.D. from the University of Edinburgh (UK) in 1968. He has long-standing interest in the ecophysiology of microorganisms that live in hypersaline and/or alkaline environments, notably the soda lakes of the East Africa rift valley. Dr. Grant also has an interest in exobiology and has been examining microbial populations in ancient evaporite deposits as terrestrial analogues of presumed similar extra-terrestrial sites on Mars.

E-Mail: WDG1@leicester.ac.uk

Bill Grant

Brian Jones

Biodata of **Brian E. Jones**, author (with Bill Grant as color-author) of "*Microbial Diversity and Ecology of Alkaline Environments*".

Dr. **Brian Jones** is a staff scientist at Genencor International, Leiden, The Netherlands. He is responsible for global screening issues and currently is a Visiting Professor with the Environmental Microbiology Group at the University of Leicester, UK. Dr. Jones is a graduate of Loughborough University of Technology with a Ph.D. in microbial chemistry from the University of Manchester, UK (1976). He has spent most of his professional life in industrial research—mainly screening for new products and processes. A major research interest centers on the microbiology of extreme environments especially thermal and alkaline habitats as a source of novel genes and microorganisms. Dr. Jones joined Genencor International, an industrial biotechnology company, in 1995, where he heads a team involved in microbial and molecular screening.

E-Mail: Bjones@genencor.com

J. Seckbach (ed.), Journey to Diverse Microbial Worlds, 177-190.
© 2000 *Kluwer Academic Publishers. Printed in the Netherlands.*

MICROBIAL DIVERSITY AND ECOLOGY OF ALKALINE ENVIRONMENTS

B.E. JONES[1] and W.D. GRANT[2]
[1]*Genencor International B.V. Archimedesweg 30, 2333 CN Leiden.*
The Netherlands, and [2]Department of Microbiology, University of
Leicester Leicester LE1 9HN United Kingdom

1. Introduction

1.1. ALKALIPHILES AND ALKALIPHILY

Alkaliphiles are usually defined as organisms that grow at alkaline pH, with pH optima for growth around two pH units above neutrality (Kroll, 1990) although growth conditions may influence the pH range for growth (Horikoshi 1998). In this chapter, the term alkaliphile is used only for Bacteria that grow optimally at pH values above 9, and which do not grow at neutral pH. These are sometimes referred to as obligate alkaliphiles (Grant *et al.*, 1990). Organisms that grow at alkaline pH, but grow more rapidly at around neutrality or less are described as alkalitolerant. Investigations of microbial populations flourishing at pH values above 9 have been largely restricted to surveys of prokaryotes (both Bacteria and Archaea), although it is certain (largely from microscopic observation) that there are eukaryotic microbes capable of growth and activity at high pH values - to date these have not been examined in detail.

1.2. TYPES OF ALKALINE ENVIRONMENTS

Alkalinity arising through biological activity such as ammonification or sulfate reduction must be a wide spread feature of all heterogeneous environments, which probably explains the occurrence of alkaliphiles in many soils (Horikoshi, 1991), sewage treatment systems (Wiegel, 1998) and marine sediments (Takami *et al.*, 1997), environments that would not be considered to be particularly alkaline (at least in terms of "bulk" pH measurements). Commercial processes such as cement manufacture and those that involve alkali-treatment steps such as electroplating and some vegetable preparation also generate alkaline wastes, although little microbiological information is available for most of these environments. There is, however, a particular bacterium that colonizes the waste generated by the alkaline removal of potato skins (Collins *et al.*, 1983).

Alkaline conditions are also generated by the low temperature weathering of the minerals olivine and pyroxene, where CO_2-charged surface water decompose these, releasing Ca^{2+} and OH^- into solution under reducing conditions generated by the concomitant production of hydrogen. Here the alkalinity is produced by sparingly soluble Ca(OH)$_2$ in equilibrium with the solid phase. Whether or not these kinds of environments exemplified by the Oman Springs examined by Bath *et al.* (1987) actually have an indigenous population of microbes is not clear since the culture conditions employed have, in general, not matched the environments sampled.

Soda lakes and soda deserts represent the major type of naturally-occurring highly alkaline environment. These environments are characterized by the presence of large amounts of sodium carbonate (or complexes of the salt) formed by evaporative concentration. These environments are the most alkaline on earth with pH values >11.5 in some cases. Despite the apparent harshness of the conditions, soda lakes harbor a much more diverse population of alkaliphilic microbial types. This review is largely concerned with the microbial ecology of soda lakes, notably those of the East African Rift Valley. For details of alkaliphiles isolated from other sites, in the main sewage systems and soils, Wiegel (1998) and Horikoshi (1991), respectively, should be consulted.

2. Soda Lakes

2.1. GENESIS OF SODA LAKES

Soda lakes are highly alkaline aquatic environments. Often ephemeral in nature, their terrestrial equivalents û soda deserts û represent their desiccated remains. Although soda lakes have a worldwide distribution, they are mainly confined to tropical and sub-tropical latitudes in continental interiors or rain-shadow zones. Owing to their hostile nature they are often remote from the main centers of human activity and perhaps for this reason they have been little studied. The best studied regions are the lakes and solonchaks of the Central Asian (Siberia) steppes and the Rift Valley of Eastern Africa.

The Great Rift Valley running through East Africa is an arid tropical zone where tectonic activity has created a series of shallow depressions. These are often closed basins with no obvious outflow where ground water and (ephemeral) streams flowing from the surrounding highlands on the margins of the Rift Valley collect to form (semi) - permanent standing bodies of water. Surface evaporation rates exceed the rate of inflow of water allowing the dissolved minerals to concentrate into a caustic alkaline brine with CO_3^{2-} and Cl^- as major anions (Grant *et al.*, 1990; Jones *et al.*, 1994). The precise mineral composition of individual lakes may depend on local geology. In the Kenya-Tanzania section of the Rift Valley the bedrock is composed of alkaline trachyte lavas laid down during the Pleistocene period. These lavas contain carbonate and silicates and have a very low calcium and magnesium content. In groundwater containing an excess of $CO_2/CO_3^{2-}/HCO_3^-$, the alkaline earth cations are removed from solution as insoluble minerals, allowing the dissolved salts to concentrate by evaporation creating a pH of 8.5 to >12. This is quite unlike the situation in most other aquatic environments where the concentration of Ca^{2+} and Mg^{2+} usually considerably exceeds that of CO_3^{2-}/HCO_3^-, buffering these environments to close to neutrality through the removal of CO_3^{2-} (like the Dead Sea or the Great Salt Lake) (Figure 1).

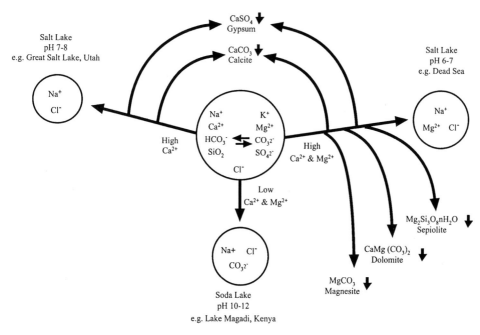

Figure 1 Schematic representation of the genesis of saline and alkaline lakes (modified from Grant *et al.* (1998)

As evaporative concentration progresses, Cl^- ions also dominate in solution. The salinity of these lakes varies from around 5% (w/v) total salts in the northern lakes (Bogoria, Nakuru, Elmenteita, Sonachi) to saturation in parts of the Magadi-Natron basin in the south. These extreme pH environments, in spite of being so apparently hostile, are, in fact, extremely productive because of high ambient temperatures, high light intensities and an unlimited supply of CO_2. Primary production rates of >10 g C $m^{-2} day^{-1}$ (Melack and Kilham, 1974) have been recorded, making soda lakes among the most productive aquatic environments, in terms of biomass, anywhere in the world. This remarkable photoautotrophic primary production stands at the top of the food chain and is presumably the driving force behind all biological processes in what is essentially a closed environment.

A systematic examination of chosen examples of the many strains brought into laboratory culture reveals a remarkable diversity of prokaryotes with alkaliphiles represented in most of the major taxonomic groups. This has permitted the identification of many of the major trophic groups responsible for the recycling of carbon, sulfur and nitrogen in the lakes, using the obvious parallels with better characterized aquatic ecosystems, especially thalassohaline salt lakes. Many of the microorganisms so far characterized from soda lakes have relatives in salt lakes but with the added difference that they are all alkaliphilic or at the very least highly alkalitolerant (Jones *et al.*, 1994). However, salinity is also an important defining factor in the alkaline lakes. There is a distinct difference in composition of the microbial communities between the hypersaline, alkaline lakes such as parts of Lake Magadi with salinity approaching saturation or higher, compared with the more dilute lakes like

Nakuru. Several hundreds of strains of non-phototrophic aerobic organotrophs have been isolated from the environs of Rift Valley soda lakes on a variety of media. About 100 of these have been examined phenotypically and chemotaxonomically in some considerable detail for the purpose of numerical taxonomy (Jones *et al.*, 1994). Phylogenetic analysis has been performed on around 30 isolates (Duckworth *et al.*, 1996). Many of the strains isolated in pure culture can be assigned to existing taxa as new species or novel genera, but some isolates have no close phylogenetic ties with known microbes and appear to be separate lines of evolutionary descent perhaps peculiar to the soda lake environment.

2.2. PRIMARY PRODUCTION IN SODA LAKES

In the moderately saline lakes (Nakuru, Elmenteita, etc.) cyanobacteria are the main contributors to primary production, notably *Spirulina* spp., although a considerable number of other species of alkaliphilic cyanobacteria have now been described (Dubinin *et al.*, 1995; Gerasimenko *et al.*, 1996) Not only do *Spirulina* spp. support the large numbers of organotrophic Bacteria (10^6 cfu ml^{-1}) but it is the major food source for the vast flocks (>10^6 individuals) of the Lesser Flamingo (*Phoeniconaias minor*) that graze on these lakes. An essential role of *Spirulina* is in the fixation of N_2 and the production of O_2. The significance of the other species of cyanobacteria that have been recorded, e.g. *Cyanospira*, *Chroococcus*, *Synechococcus*, is probably minor in comparison with *Spirulina* spp. There is also a contribution to primary production by anoxygenic phototrophic Bacteria of the genus *Ectothiorhodospira* that is likely to be significant but is as yet unquantified (Grant and Tindall, 1986).

In hypersaline lakes like Magadi it is rather uncertain what organisms are responsible for primary production, especially in the trona beds that are usually dominated by organotrophic haloalkaliphilic Archaea. This is a challenging environment for cyanobacteria because O_2 has a low solubility in concentrated salt solution. Cyanobacterial blooms do occur occasionally in the lagoon waters but only after an unusually wet rainy season has caused substantial dilution of the brine. Filamentous cyanobacteria like *Spirulina* clearly have a preference for lower salinities and although unicellular forms of cyanobacteria have been demonstrated at high salinities, they show alterations in morphology with changing conditions that complicates their taxonomic identification (Dubinin *et al.*, 1995). Halophilic *Ectothiorhodospira* spp., recently reclassified as *Halorhodospira* spp. (Imhoff and Süling, 1996) have been isolated from Magadi brines, but they are unlikely to be the only contributors to primary production.

2.3. CARBON CYCLE IN SODA LAKES

2.3.1. *Aerobes*
Even though these data are incomplete it is possible to make some predictions as to the roles played by the different organisms in the recycling of nutrients in these highly alkaline environments, especially since there are many parallels with the better studied thalassohaline salt lake systems.

Especially abundant in Kenyan soda lakes are Gram-negative Bacteria related to (>96% 16S rDNA sequence similarity), but clearly different from members of the *Halomonadaceae* family of moderately halophilic Bacteria found in a range of terrestrial and marine pH neutral saline environments.

Isolates have been characterised by phenotypic, chemotaxonomic and phylogenetic analysis, indicating that they fall into three groups along with known halomonad species, probably representing several novel species (Duckworth *et al.*, 1996)(Figure 2).

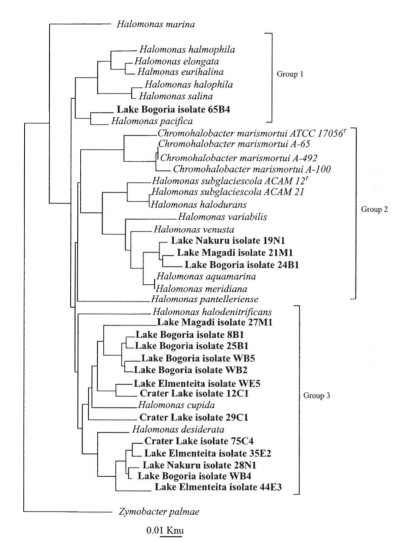

Figure 2. Unrooted phylogenetic tree of soda lake halomonads and related isolates (modified from Fish *et al.*, 1999).

It is clear that the type species of halomonads (isolated from neutral sites) are somewhat different from the soda lake types. Some of these soda lake isolates have novel plasmids (Fish *et al.*, 1999). While it has been clear that halomonads often constitute a major group in moderately saline environments, it is now obvious that they are equally prominent in the alkaline, saline environment. They are also likely to be widespread elsewhere, since similar strains have been isolated from an alkaline, saline lake in the Pacific Northwest of the USA (M. Mormile, personal communication), sewage works (Berendes *et al.*, 1996) and alkaline soils (Romano *et al.*, 1996).

As might be expected considering the nutrient rich habitat that they inhabit, the soda lake isolates are biochemically reactive. Besides hydrolyzing proteins and polymeric carbohydrates such as starch, they utilize a very wide range of sugars, organic and amino acids, and are nutritionally less demanding than the neutrophilic halomonads. One of the features of the soda lake isolates is their capacity for anaerobic growth in the presence of nitrate and nitrite (Duckworth *et al.*, in preparation). Since these organisms form a major part of the soda lake microbial community they presumably play an important and substantial role in the nitrogen cycle.

Other groups of Bacteria with an uncertain affiliation within the enteric/*Vibrio/Aeromonas* part of the Gamma 3 subdivision of the Proteobacteria appear strongly proteolytic, whereas other isolates that can be firmly assigned to the genus *Pseudomonas sensu strictu* (rRNA group I) have a strong preference for lipids (Duckworth *et al.*, 1996)

There is wide diversity among the Gram-positive isolates that are found in both the high G+C and low G+C divisions of the Gram-positive lineage. Especially abundant are Bacteria associated with the diverse *Bacillus* spectrum. Some of these can be included in a group which also includes *Bacillus alcalophilus*, an alkaliphile originally isolated from soil.(Figure 3). However, other strains are associated with *Bacillus agaradhaerens* and *Bacillus clarkii* that are probably also derived from a soda lake, forming a distinct but related group of alkaliphilic bacilli (Duckworth *et al.*, 1996).

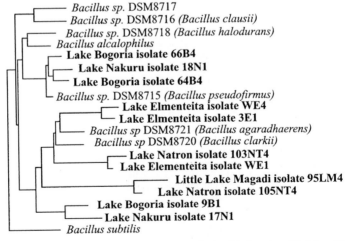

Figure 3. Unrooted phylogenetic tree of soda lake bacilli and related isolates (modified from Duckworth *et al.*, 1996)

These alkaliphilic *Bacillus* spp. probably play an important role in the breakdown of biopolymers since they have an arsenal of extracellular hydrolytic enzymes -

proteinases, cellulases, xylanases and other enzymes capable of degrading complex carbohydrates to their component sugars. There is some evidence to suggest that the two groups of bacilli may be segregated within the soda lake environment. The strains associated with *Bacillus alcalophilus* seem to predominate in shore line muds and dried foreshore soils that are subject to fluctuating conditions of alkalinity and salinity as water levels change with the seasons. The other group of strains which also have a high absolute Na^+ requirement for growth, are seemingly more abundant in lake waters and sediments. These alkaliphilic spore-forming organisms probably enjoy a worldwide distribution and are likely to be dispersed widely by wind blown dust and migratory birds.

Also among the littoral muds and dry soda soils surrounding the lakes are novel *Streptomyces* spp. growing at pH 10 (Jones *et al.*, 1998). These too hydrolyze proteins and carbohydrates. There is further diversity of high G+C Gram-positive Bacteria present in the soda lake environment including a novel genus *Bogoriella* (Groth *et al.*, 1997), a new species of *Dietzia*, *Dietzia natronolimnaios* (Duckworth *et al.*, 1998) and uncharacterized relatives of *Nesterenkonia* (formerly *Micrococcus halobius*) (Duckworth *et al.*, 1996; Ventosa *et al.*, 1998).

A quite different population of prokaryotes is present in the trona (sodium sesquicarbonate) beds and concentrated alkaline brines of hypersaline lakes such as those found in the Magadi-Natron basin (Kenya/Tanzania). Lake Magadi, where salt concentrations achieve saturation levels, is the center of a commercial enterprise. The vast trona deposits are excavated by dredger and are kilned producing soda ash (anhydrous sodium carbonate) for glass manufacture. Common salt is also produced in a series of solar evaporation ponds. The salt-making ponds at Magadi provide the most extreme alkaline environment at pH >12 and are dominated by bright red blooms of organotrophic haloalkaliphilic Archaea. Since the inorganic components of the environment have accumulated by evaporative concentration it seems reasonable to suppose that organic material also accumulates and this alone may be sufficient to support a substantial community of organotrophs where photosynthetic primary production plays a lesser role. The organisms most commonly cultured from the hypersaline, alkaline lakes have been assigned to a distinct physiological group of the archaeal halophile lineage. These organisms not only require high concentrations of NaCl but also high pH (8.5 - 11) and low Mg^{2+} (<10 mM). Originally assigned to only two genera, *Natronobacterium* and *Natronococcus*, they have recently been recognized as representing considerable diversity and have been reclassified on the basis of 16S rRNA sequence comparisons, with the creation of two new additional genera, *Natrialba* and *Natronomonas* and the assignment of one of the original *Natronobacterium* species to the genus *Halorubrum* which had been considered to harbour only non-alkaliphilic types (Kamekura *et al.*, 1997). Recently, a further genus, *Natronorubrum*, has been reported as being characteristic of Tibetan soda lakes (Xu *et al.*, 1999). A superficial examination of the phenotypes of randomly selected examples of the cultivated diversity from sites at Lake Magadi suggests that rod-shaped *Natronobacterium* / *Natrialba* / *Natronomonas* spp. are the most numerous types. However, 16S rDNA sequencing of selected isolates presents an altogether different picture indicating an even greater phylogenetic diversity (Duckworth *et al.*, 1996). The only other organisms that have been cultivated from this hypersaline environment are haloalkaliphilic

Bacillus spp. Although these isolates are related phylogenetically to the Na⁺- requiring species, they are phenotypically quite distinct, being able to grow well in 25-30% (w/v) NaCl with a minimum requirement for at least 15% (w/v) NaCl. The distinctness of this group of bacilli is also evident from signature sequences in the 16S rRNA gene (Jones *et al.*, 1998).

2.3.2. *Anaerobes*

The monomers produced by the hydrolysis of complex polymers presumably form the substrates for not only alkaliphilic halomonads, but also for anaerobic fermentative Bacteria. These Bacteria are likely to have a worldwide distribution, for example *Spirochaeta* spp. found at Lake Magadi (Kenya) and Lake Khatyn (Central Asia) (Zhilina *et al.*, 1996b). These organisms utilize a wide variety of pentoses, hexoses and disaccharides producing acetate, lactate, ethanol and hydrogen. Our own work shows that anaerobic enrichment cultures performed on complex polymeric substrates yields almost exclusively facultative anaerobes while obligate anaerobes could be isolated on glucose supplemented with appropriate amounts of NaCl at the correct pH. Viable counts on the black, anoxic soda lake sediments indicate they contain more than 10^6 cfu ml^{-1} (at 37^0C) chemoorganotrophic alkaliphilic anaerobes. In the soda lake environment, obligate anaerobes are probably 'secondary organotrophs' utilizing the products of primary hydrolysis provided by aerobes and facultative anaerobes. Among the obligate anaerobes partially characterized are isolates that were assigned to the *Clostridium* spectrum by phylogenetic analysis (Jones *et al.*, 1998). Isolates from the moderately saline lakes were associated with *Clostridium* group XI. They are phenotypically quite diverse, fermenting a variety of simple sugars or amino acids to acetate and propionate or butyrate and with a maximum salt tolerance ranging from 4 to 12% (w/v) A totally different selection of isolates was obtained from the hypersaline habitats at Lake Magadi. These strains have a high tolerance for NaCl (about 25% w/v) and a minimum requirement of at least 12-16% (w/v) for growth that is consistent with the conditions of the Lake Magadi habitat. These too, ferment a range of sugars producing mainly iso-valeric acid with smaller amounts of iso-butyric and acetic acid as end products. Detailed phylogenetic analysis also placed these isolates within the *Clostridium* spectrum but as a separate, well-defined group (Jones *et al.*, 1998). It is likely that these represent a new genus of obligately anaerobic haloalkaliphiles. It is probable that fatty acids are consumed by organisms similar to the homoacetogen *Natroniella acetogena* that was also isolated from a soda lake (Zhilina *et al.*, 1996a). The organism *Tindallia magadii,* isolated from Lake Magadi, is a fermentative haloanaerobe that ammonifies amino acids and thus additionally has a role in the nitrogen cycle (Kevbrin *et al.*, 1998).

There is plenty of evidence for biogenic methane production in alkaline lakes, although so far in the East African lakes it is mainly anecdotal. Methanogenic Bacteria isolated to date from other soda lakes are mainly obligately methylotrophic, utilizing a variety of C1 compounds and not H_2/CO_2 as energy yielding substrates. Compounds such as methanol, methylamine and dimethyl sulfide are probably abundant in the alkaline environment being derived from the anaerobic digestion of algal mats, and from compatible solutes such as betaine and ectoine which have been detected in some alkaliphilic organotrophs (unpublished). This suggests that, in general, methanogens are not functioning as a H_2 sink and are not competing with sulfate-reducing Bacteria for resources in soda lakes, although there may be exceptions. The methanogens characteristic of soda lakes have all been assigned to *Methanohalophilus* (*Methanosalsus*) as a separate genus within the *Methanomicrobiales* (Kevbrin *et al.*, 1997).

The fate of methane in the soda lake system has recently also become clear. Aerobic methane-oxidizing Bacteria have been isolated from Central Asian soda lakes (Khmelenina *et al.*, 1997), and more recently from the Rift Valley soda lakes (Sorokin *et al.* unpublished). Methane oxidation has also been detected in Mono Lake, a less alkaline and saline United States soda lake (Joye *et al.*, 1999).

2.4. SULPHUR CYCLE IN SODA LAKES

The active role of *Ectothiorhodospira* and *Halorhodospira* spp. in anaerobic primary production has already been noted. These Bacteria require a reduced sulfur species as photosynthetic electron acceptor which is oxidized to sulfur and then sulfate. These observations point to an active sulfur cycle that is confirmed by recent data. Growth may be photoautotrophic by the fixation of CO_2 via the ribulose-bisphosphate pathway or photoorganootrophically on volatile fatty acids. These Bacteria are also a major source of the nitrogenous compounds, glycine betaine and ectoine which function as intracellular compatible solutes in these extreme halophiles (Galinski, 1995). Although geochemists have postulated sulfate-reducing Bacteria activity to explain the relative depletion of sulfate in soda lake brines, it is only recently that their activity at alkaline pH has been confirmed. *Desulfonatronovibrio hydrogenovorans* is a H_2-utilising sulfate-reducing Bacteria isolated from the trona beds of Lake Magadi which functions optimally at pH 9.5 and 3% (w/v) NaCl (Zhilina *et al.*, 1997). *Desulfononatrun lacustre* is another recent isolate (Pikuta *et al.*, 1998). As a H_2 sink, sulfate-reducing Bacteria form an important link between the carbon and sulfur cycles, providing sulfide for anaerobic photoautotrophic primary production.

The sulfur cycle also has an aerobic component. In most aquatic systems H_2S is poorly soluble and is spontaneously oxidized in the oxic zone, so that aerobic sulfur oxidizing Bacteria occupy a narrow zone where H_2S and O_2 mix. In highly alkaline soda lakes sulfide is present as the soluble HS^- ion. These lakes are very shallow but they are not stratified because there is considerable mixing from adiabatic wind action. This combination of circumstances may be particularly advantageous to the alkaliphilic sulfur-oxidizing Bacteria that have recently been isolated from soda lakes (Sorokin *et al.*, 1998a). Some strains appear to be related to typical chemolithotrophic sulfur oxidizers such as *Thiomicrospira*, whereas others appear as a new lineage in the Gamma Proteobacteria.

2.5. NITROGEN CYCLE IN SODA LAKES

The nitrogen fixing and denitrification activities of *Spirulina* spp. and the alkaliphilic halomonads, respectively, have already been noted. Recent data point to further novel alkaliphiles that are involved in the cycling of nitrogen compounds. Several strains of

188

an autotrophic, CO_3^{2-} - dependent nitrite-oxidizing bacterium have been isolated from Siberian and Kenyan soda lakes. Although isolated from diverse locations, the strains form a compact species group related to *Nitrobacter* but distinct from known species (Sorokin *et al.*, 1998b). At alkaline pH ammonia is volatile and leads to loss of nitrogen from the environment. The methane-oxidizing Bacteria from the Kenyan soda lakes mentioned above are also able to oxidize ammonia to nitrite optimally at pH 10-10.5. Ammonia oxidation was also noted in Mono Lake (Joye *et al.*, 1999). These data suggest that the nitrogen flux remains in balance and that losses due to an "unfavorable" pH are adequately compensated.

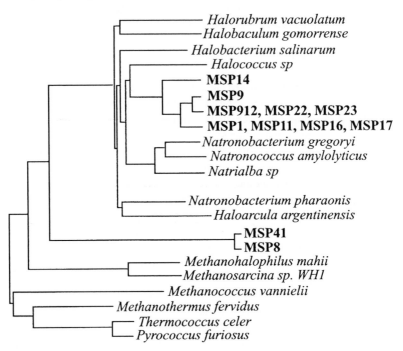

Figure 4. Unrooted phylogenetic tree showing the relationship between phylotypes (MSP) recovered from Lake Magadi alkaline salterns and archaeal types in culture (modified from Grant *et al.*, 1999).

2.6. UNCULTIVATED SODA LAKE PROKARYOTES?

Clearly the ability to culture organisms is a major hurdle to our complete understanding of soda lake microbial communities and this is further biased by our manner of sampling, culture conditions employed and the choice of isolates for study. An examination of the sequences of PCR amplified 16S rRNA genes from DNA extracted on site in East Africa revealed an unexpected diversity even in the salterns at Lake Magadi - the most extreme environment in terms of alkalinity and salinity and therefore expected to be restricted in terms of microbial diversity. Nevertheless, in common with all other environments examined in this way to date, including the related neutral hypersaline environments (Benlloch *et al.*, 1996), none of the sequences (phylotypes) recovered matched those of the cultivated haloalkaliphilic Archaea that appeared to dominate these sites (at least in terms of frequency of isolation). In particular, two of the phylotypes exhibited little relatedness to reported species, with only 76% identity with known Archaea, placing them as members of a deeply branching group of the Euryarchaeota (Figure 4). Signature sequence analysis revealed a number of features in common with the extremely halophilic Archaea and a relationship to the methanogens, but without any clear relationship with any of the three major methanogen families (Grant *et al.*, 1999).

It is evident from these preliminary studies that the full extent of the soda lake microbial diversity, community structure and the roles played by individual organisms has yet to be wholly revealed. It is an enterprise that is not aided by the remoteness of these extraordinary extreme environments.

3. References

Bath, A.H., Christofi, N., Neal, C., Philp, J.C., Cave, M.R., McKinley, I.G. and Berner, U. (1987) Rep. Fluid Processes Group Br. Geol Surv. FLPU 87-2.

Benlloch, S., Acinas, S.G., Martínez, A.J. and Rodríguez-Valera, F. (1996) Hydrobiologia **329**, 19-31.

Berendes, F., Gottschalk, G., Hemedobbernack, E., Moore, E.R.B and Tindall, B.J. (1996) Syst. Appl. Microbiol. 19, 158-167.

Collins, M.D., Lund, B.M., Farrow, J.A.E. and Schleifer, K-H. (1983) J. Gen. Microbiol. **129**, 2037-2042.

Dubinin, A.V., Gerasimenko, L.M. and Zavarzin, G.A. (1995) Microbiology **64**, 717-721.

Duckworth, A.W., Grant, W.D., Jones, B.E . and Van Steenbergen, R. (1996) FEMS Microbiol. Ecol **19**, 181-191.

Duckworth, A.W., Grant, S., Grant, W.D., Jones, B.E. and Meijer, D. (1998) Extremophiles **2**, 359-366.

Fish, S.A., Duckworth, A.W. and Grant, W.D. (1999) Plasmid **41**: in press

Galinski, E.A. (1995) Adv. Microbial Physiol. **37**, 273-328.

Gerasimenko, L.M., Dubinin, A.V. and Zavarzin, G.A. (1996) Microbiology **65**, 736-740.

Groth, I., Schumann, P., Rainey, F.A., Martin, K., Schuetze, B., and Augsten, K (1997) Int. J. Syst .Bacteriol. **47**, 788-794.

Grant, S., Grant, W.D., Jones, B.E., Kato, C and Li, L. (1999) Extremophiles, **3**, 139-146.

Grant, W.D. and Tindall, B.J. (1986) In: R.A. Herbert and G.A. Codd (eds.) *Microbes in extreme environments*, Academic Press, London, pp. 22-54.

Grant, W.D., Jones, B.E., and Mwatha, W.E. (1990) FEMS Microbiol.Rev. **75**, 255-270.

Grant, W.D., Gemmell, R.T. and McGenity, T.J. (1998) In: K. Horikoshi and W.D. Grant (eds.) *Extremophiles:microbial life under extreme conditions*, Wiley-Liss, New York, pp. 93-132.

Horikoshi, K. (1991) *Microorganisms in Alkaline environments*. Kodansha- VCH, Tokyo, 275 pp.

Horikoshi, K. (1998) In: K.Horikoshi and W.D. Grant (eds.) *Extremophiles: microbial life under extreme conditions*, Wiley-Liss, New York, pp. 155-180.

Imhoff, J.F. and Süling, J. (1996) Arch. Microbiol. **165**, 106-113.

Jones, B.E., Grant, W.D., Collins, N.C. and Mwatha, W.E. (1994) In: F.G. Priest, A. Ramos-Cormenzana and B.J. Tindall (eds.) *Bacterial diversity and systematics*, Plenum Press, New York, pp. 195-229.

Jones, B.E., Grant, W.D., Duckworth, A.W. and Owenson, G.G. (1998) Extremophiles **2**, 191-200.

Joye, S.B., Connell, T.L., Miller, L.G., Oremland, R.S. and Jellison, R.S. (1999) Limnol. Oceanogr. **44**, 178-188.

Kamekura, M., Dyall-Smith, M.L., Upsani, V., Ventosa ,A. and Kates, M. (1997) Int. J. Syst. Bacteriol. **47**, 853-857.

Kevbrin, V.V., Lysenko, A.M. and Zhilina, T.N. (1997) Microbiology **66**, 261-266.

Kevbrin, V.V., Zhilina, T.N., Rainey, F.A. and Zavarzin, G.A. (1998) Curr. Microbiol. **37**, 94-100.

Khmelenina, V.N., Kalyuzhnaya, M.G., Starostina, N.G., Suzina, N.E. and Trotsenko, Y.A. (1997) Curr. Microbiol. **35**, 257-261.

190

Kroll, R.G. (1990) In: C. Edwards (ed.), *Microbiology of Extreme Environments*, McGraw-Hill, New York, pp. 55-92.

Melack, J.M. and Kilham, P. (1974) Limnol. Oceanogr. **19**, 743-755.

Pikuta, E.V., Zhilina, T.N., Zavarzin, G.A., Kostrikina, N.A., Osipov, G.A. and Rainey, F.A. (1998) Microbiology **67**, 105-115.

Romano, I., Nicolaus, B., Lana, C., Manca, M.C. and Gambacorta, A. (1996) Syst. Appl. Microbiol. **19**, 326-333.

Sorokin, D.Y., Jones, B.E., Robertson, L.A., Jetten, M.S.M. and Kuenen, J.G. (1998a) Abstr. ISME 8 Congress Halifax, p307.

Sorokin, D.Y., Muyzer, G., Brinkhoff, T., Kuenen, J.G. and Jetten, M.S.M (1998b) Arch. Microbiol. **170**, 345-352.

Takami, H., Inoue, A., Fuji, F. and Horikoshi, K. (1997) FEMS Microbiol. Letts. **152**, 279-285.

Ventosa, A., Marquez, M.C., Garabito, M.J and Arahal, D.R. (1998) Extremophiles **2**, 297-304.

Wiegel, J. (1998) Extremophiles **2**, 257-268.

Xu, Y., Zhou., P.J. and Tian, X.Y. (1999) Int J. Syst. Bacteriol. **49**, 261-266.

Zhilina, T.N., Zavarzin, G.A., Detkova, E.N. and Rainey, F.A. (1996a) Curr. Microbiol. **32**, 320-326.

Zhilina, T.N., Zavarzin, G.A., Rainey, F.A., Kevbrin, V.V., Kostrikina, N.A. and Lysenko, A.M. (1996b) Int. J. Syst. Bacteriol. **46**, 305-312.

Zhilina, T.N., Zavarzin, G.A., Rainey, F.A., Pikuta, E.N., Osipov, G.A. and Kostrikina, N.A. (1997) Int. J. Syst. Bacteriol. **47**, 144-149.

Biodata of **Georgiy Alexandrovich Zavarzin** the author (with co-author Dr. **Tatjana Zhilina**) of the chapter entitled: "***Anaerobic Chemotrophic Alkaliphiles***:"

Dr A.G. Zavarzin is a head of the Department at the Institute of Microbiology Russian Academy of Sciences in Moscow. He obtained his Dr. Sc. in 1966 and is a full member of the Russian Academy of Sciences. Dr. Zavarzin published about 200 articles and his main interest is in studies of the functional microbial diversity related to geochemical processes. He described budding bacteria, lithotrophic microorganisms including nitrifieres, iron, manganese and hydrogen bacteria as well as methanogens. His present interest is in the trophic interactions within microbial communities in relation to the gaseous composition of the atmosphere.

E-mail: zavarzin@inmi.host.ru.

Biodata of **Tatjana Nikolaevna Zhilina** the co-author (with author **G.A. Zavarzin**) of the chapter entitled: "***Anaerobic Chemotrophic Alkaliphiles***:"

Dr. **T. N. Zhilina** is a Leading Scientist at the Institute of Microbiology RAS in Moscow. She obtained her Dr. Sci. in. 1992. from Russian Academy of Sciences. Her main interest is Biodiversity of the Microbial World, including Extremophiles, anaerobic decomposition of osmolytes in the halophilic community. She studies hydrogen bacteria, methanogens, and haloanaerobs. Dr. Zhilina is an author of 120 papers and studies currently the diversity of extremely alkaliphilic anaerobic chemotrophic bacteria.

E-Mail: zhilina@inmi.host.ru.

J. Seckbach (ed.), Journey to Diverse Microbial Worlds, 191-208.
© 2000 *Kluwer Academic Publishers. Printed in the Netherlands.*

ANAEROBIC CHEMOTROPHIC ALKALIPHILES

G.A. ZAVARZIN and T.N. ZHILINA
Institute of Microbiology RAS
Prospect 60-letja Octjabrja 7/2, 117811 Moscow, Russia

1. Introduction

In spite of the long history of research on alkaline habitats and the microbial communities inhabiting them, the most important anaerobic metabolic decomposition pathways operating at high pH have not yet been characterized. Only a few anaerobic alkaliphiles have been isolated as yet, most of them being photoautotrophs. We have attempted to describe an alkaliphilic microbial community as a trophic system by isolating representatives of the key functional groups. This work resulted in the discovery of a number of new genera belonging to different phylogenetic branches. It appears that the diversity of prokaryotes is extremely high, even at the highest alkalinity of water overlaying soda deposits, and they represent an entire world organized into an autonomous community. Insight into the properties of this community may provide us with a better understanding of microbial processes that may have occurred in inland waters of the Proterozoic.

2. Habitats

Soda lakes are model habitats for alkaliphiles. These lakes are typically found in the inner parts of the continents under a semi-arid climate. The main mechanism of soda lake formation includes CO_2- induced leaching of rocks by water, precipitation of calcium carbonate and, after the calcium reservoir is exhausted, accumulation of sodium carbonate as a result of evaporation. A pH of 10.2, due to the high concentration of sodium carbonates, is characteristic of this type of habitat. The salt concentration, increased by evaporation, may be as high as 20%, close to saturation, and depends on temperature. Due to the variations in the nature of the source rocks, the chemistry of water in these athalassic lakes varies considerably compared to that of oceanic hyperhaline lagoons, which are uniform in ionic composition all over the world. The main processes of the formation of alkaline waters are essentially similar in the inner parts of the continents due to the CO_2- induced leaching by meteoric water at ambient temperature. This process was applied by Kempe and Degens (1985) for the description of the composition of primary waters on Earth as well as on Mars (Kempe and Kazmierczak 1997). At salt concentrations above 7-10% development of eukaryotic protists and invertebrates is prevented, thus making the ecosystem essentially prokaryotic. At temperatures over 50 ^{0}C , which sometimes develop in the tropics due to heliothermic heating (the solar lake effect), conditions are close to those used in the laboratory for the hydrolysis of bacterial cells, and should be considered extreme even for microorganisms. However, in spite of the harsh conditions all microbial transformations known from conventional environments do occur.

The question may be asked whether soda lakes harboring prokaryotic communities can be considered as analogues of the terrestrial communities of the ancient continents of the early Proterozoic, which due to their high salt concentration remained unoccupied by eukaryotes, and whether soda lakes may thus represent a refuge for the relict communities (Zavarzin 1993). Could alkaliphilic communities represent the beginning of the line of terrestrial evolution from cyanobacteria to green algae and to plants? Is there a possibility to consider the reverse line of development from the terrestrial, or more precisely from the epicontinental, "water ponds" to the saline marine environment with stromatolitic belts and cyanobacterial mats expanding from the continents into the sea? To what extent does the alkaliphilic community include the main branches of the phylogenetic system of the prokaryotes? What bioenergetic mechanisms are employed by microorganisms developing in a high sodium - low proton medium? To what extent is the alkaliphilic community functionally sufficient to close the main cycles of the elements, and thus maintain itself for long periods of time? How could this community function in the anoxic atmosphere of the early Proterozoic?

3. Model Ecosystems

While alkaline lakes have attracted considerable attention among limnologists, there are only a few alkaline lakes which have served as models for microbiologists (see e.g. Javor 1989). The first systematic studies of microbial life in alkaline steppe lakes were performed in the 1920s by Issatchenko (1951) in the Kulunda steppe south-west of the Irtysh river. He documented the development of cyanobacterial and algal blooms, the presence of the main physiological groups of decomposers, dominated by sulfate reducers, and mass development of anoxygenic phototrophic bacteria. The next site investigated was Wadi-el-Natrun, west of the Nile delta (Egypt). Abd-el-Malek ascribed the origin of alkalinity in this environment partly to sulfate reduction (Abd-el-Malek and Rizk 1963). Trüper and Imhoff studied the purple alkaliphile *Halorhodospira* (formerly named *Ectothiorhodospira*) (Imhoff and Trüper 1981, Imhoff et al. 1981), which closes the sulfur cycle, but they also mentioned briefly that these alkaline lakes contain representatives of the main physiological groups, including nitrifiers (Imhoff et al. 1979). The biogeochemistry of the alkaline Big Soda Lake (Nevada) and Mono Lake (California) was studied by Oremland (Oremland et al. 1982, 1987; Kiene et al. 1986), who noticed the minute contribution made to decomposition by methanogens. From Wadi-el-Natrun and alkaline lakes in Oregon the extremely alkaliphilic methylotrophic methanogen *Methanosalsus* (*Methanohalophilus*) *zhilinae* (Boone et al. 1986; Mathrani et al. 1988; Boone et al. 1993a) and *Methanolobus* (*Methanohalophilus*) *oregonensis* (Liu et al. 1990; Boone et al 1993a) were isolated. The estuarine methylotrophic methanogen *Methanolobus taylorii* (Oremland and Boone 1994) is not an obligate alkaliphile, since it can also grow at pH 7.

We have studied microbes from different alkaline lakes in the cryoarid zone of Central Asia. However, the main site of the studies reported here was equatorial Lake

Magadi, Kenya, which is believed to represent the ongoing process of soda precipitation. Lake Magadi is situated in the East African Rift Valley, an area with strong volcanic activity. It represents a model for soda accumulation processes (Eugster 1980; Garret 1992). Lake Magadi is unusual due to the considerable contribution of volcanic activity in the rift zone and the unique geothermal gradient, leaching of fresh volcanic products, sometimes rich in carbonatites (igneous alkaline rocks rich in carbonates). The bottom of the lagoons contains a 20-30 m thick layer of trona ($NaHCO_3.Na_2CO_3.12H_2O$), which covers the dry parts as a white precipitate. The biological properties of the system strongly depend on the season. During the dry season it is partly covered, as by ice, with a few centimeters thick white trona crust, with a cherry-red *Halorhodospira* bloom below. During the rainy period mass development of various planktonic cyanobacteria occurs, together with floating sheets of benthic forms. At the end of cyanobacterial bloom the water is dirty-green, and planktonic forms of *Rhabdoderma* dominate (Dubinin et al. 1995). In spite of the saturation with soda, the alkaliphilic community is most variable in time. The lake is eutrophic due to the extremely high content of phosphate (about 40 mg/l), which is not precipitated due to the absence of calcium and iron. Development of cyanobacteria results in a very high content of nitrogenous compounds: as high as 120 mg/l of nitrate was measured in September 1998. Cyanobacteria are most efficient primary producers, supplying the system with oxygen and dead biomass. This prokaryotic paradise without grazers is disturbed by birds, especially by flamingoes, which represent important consumers and litter the water with excreta.

Stromatolitic structures have been reported from Lake Magadi as well as from its Late Archaean analogue (Karpeta 1989). Observations made by Guerrero et al. (1992) on cyanobacterial mats in the Iberian peninsula provide an obvious analogy with the stromatolite-forming communities in lacustrine environments. Phototrophic benthic communities are well known for a long time. For instance Gmelin, during his expedition in Asia in 18th century, described two types of lacustrine macrostructures: "skin" (cyanobacterial) and "felt" (algal), as quoted by Issatchenko (1951). It may be concluded that macrostructures of alkaliphilic cyanobacterial communities, when lithified, could have produced stromatolites in the ancient epicontinental water bodies, and such communities still exist in the inner parts of continents. Thus, stromatolitic structures are not necessarily relevant to the oceanic hydrochemistry, but could have developed in shallow epicontinental water bodies (Krylov and Zavarzin 1988).

4. Biodiversity of Alkaliphiles

The microbiological aspects of the community in Lake Magadi were comprehensively reviewed by Tindall (1988). He described *Natronobacterium* and *Natronococcus* (Tindall et al. 1984), representing a new group of alkaliphilic Archaea. More recent studies were reported by Grant and Jones, who investigated aerobic organotrophic Proteobacteria and their phylogeny (Grant et al. 1990; Jones et al. 1994; Duckworth et al. 1996). These studies, together with the classic studies of alkaliphilic bacilli from alkaline soils by Horikoshi (1996), form the basis for our present knowledge of the aerobic alkaliphiles. The bioenergetics of aerobic microorganisms in high-sodium environments was thoroughly studied by Skulachev (1989), using *Vibrio alginolyticus* as a model, and by Krulwich et al. (1990), using alkaliphilic bacilli. The sodium pump and the Na^+/H^+ -antiporter are the main components of the bioenergetic system (Padan and Schuldiner 1994).

Our studies of anaerobic alkaliphilic communities from soda lakes are an extension of our earlier work on the halophilic cyanobacterial community in the hypersaline lagoons of Sivash, Crimea (Zavarzin et al. 1993). As one can judge from direct microscopic observation, the biodiversity of the alkaliphilic microbial community in soda lakes is extremely high. In addition to various alkaliphilic cyanobacteria adapted to high pH and high salinity (Dubinin et al. 1995; Gerasimenko et al. 1996), one can observe a variety of morphologically different bacteria, including spore-forming and asporogenous rods of various shape and size, vibrios, methanogens, spirochetes, prosthecate bacteria, and various phototrophs (Fig. 1).

The morphological diversity of these extremely alkaline habitats is by no means less than that observed in neutral freshwater ponds, the only significant difference being the virtual absence of eukaryotes. There is one conspicuous exception: a small green unicellular eukaryotic alga, tentatively identified as "*Chlorella minutissima*", has now been isolated from many alkaline lakes as a minor component in cyanobacterial communities (Dubinin et al. 1995; Gerasimenko et al., manucript in preparation). This observation indicates that there is nothing incompatible with the existence of eukaryotes here; however, eukaryotes did not massively invade this type of habitat, the limiting factor being the salt concentration. Cyanobacteria do not develop under the trona, where they are substituted by an extremely dense population of purple bacteria, mainly *Ectothiorhodospiraceae*. These oxidize hydrogen sulfide, and thus represent secondary producers. The microbial community in eutrophic Lake Magadi and in the other Rift Valley lakes is very rich. Similar communities were found in the less saline lakes in Central Asia. Microscopical examination is sufficient to confirm the great biodiversity among alkaliphilic prokaryotes; most of the known morphotypes of bacteria can be found in even the most extremely alkaline environments.

It may be concluded that in alkaline lakes, including those with extremely high salt concentrations, a most diverse prokaryotic community proliferates, similar in complexity to the communities found in thalassic habitats.

5. Trophic Relationships within the Community

The alkaliphilic microbial community acts as a cooperative trophic entity. We do not intend to present here a full biogeochemical descriptions of its activity, realizing that without seasonal monitoring and description of its components such work is of limited value. Our aim was to understand the metabolism of the anaerobic alkaliphilic community and to characterize microorganisms catalyzing certain specific steps, rather than to isolate pure cultures at random for subsequent studies (Zhilina and Zavarzin 1994). Exploiting the diversity of the lakes, we used these often quite small water bodies as a kind of "natural enrichments" (Zavarzin et al. 1996). Organic deposits, which at times may be quite luxurious, usually do not accumulate on the bottom of these lakes, and it is thus obvious that an effective microbial decomposition is operative.

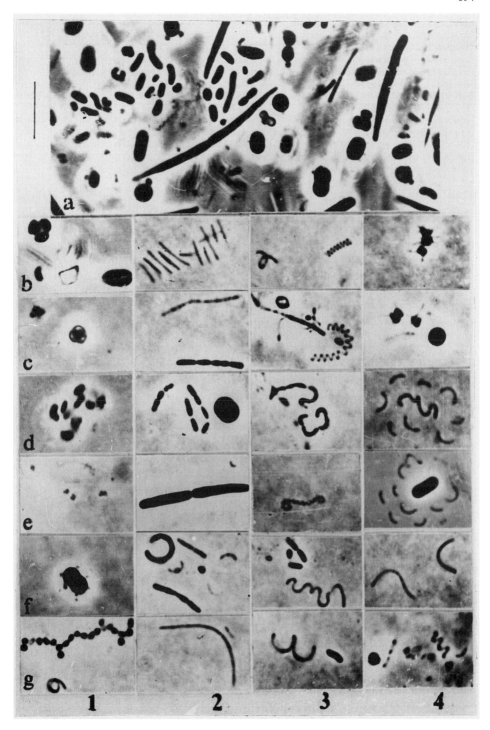

Fig. 1. Morphological diversity within the alkaliphilic prokaryotic community from Lake Magadi, Kenya. Phase contrast. Bar, 10 μm for all photos. Sample from western lagoon were taken in September 1998 (total salt concentration 26%; 35 °C; pH 10.2).
a. Overview of the planktonic community. The large long cells belong to the cyanobacterium *Rhabdoderma*, the round cells are *Aphanocapsa*, and the curved large rods are anoxygenic photosynthetic bacteria, mostly representing *Halorhodospira*.
b-g. Morphological diversity of alkaliphiles: **1b** - flat rectangular cells similar to Archaea; **1c** - "*Chlorella minutissima*"-like cell; **1d** - methanogens; **2b** - flat cells in palisade; **3b, 3c** - *Heliospira*-like cells; **4b** - prosthecate bacteria with large prosthecae; **4c** - prosthecate bacteria and *Synechocystis*; **4e** - *Aphanothece* surrounded by curved rods; **2c, 2d** - different types of rods; **2e** - *Rhabdoderma lineare*; **3d, 4d, 4c** - curved rods of *Tetrarcus* and *Brachyarcus* types; **1e** - *Microspira*-type cells; **3e** - spirochetes; **1f** - minute *Bdellovibrio*-type cells attached to *Aphanocapsa*; **1d** - streptococcal arangement of cells; **2f, 3f, 4f, 3g, 4g** - curved rods and spirilla of various types.

Decomposition is performed in part by various functional groups of alkaliphilic copiotrophic organotrophic aerobes capable of decomposing the main components of the dead biomass. They belong to the Proteobacteria, and include new genera closely related to *Halomonas* (Duckworth et al. 1996). However, the anaerobic food web leading to sulfidogenesis is quantitatively more important.

Four main trophic groups are involved in the decomposition process: hydrolytic microorganisms that decompose particulate organic matter, copiotrophic fermenters utilizing soluble compounds, dissipotrophic fermenters that utilize oligomeric compounds dissipating from the sites of their origin, and secondary anaerobes utilizing fermentation products. The food web is most easily studied by the use of elective enrichment media. The main substrates for the secondary anaerobes are H_2 and acetate, driving hydrogenotrophic and acetotrophic pathways of metabolism within the community; other compounds such as volatile fatty acids are produced in small quantities. The hydrogen sink provided by anaerobic hydrogenotrophs is the most important parameter determining the nature of the metabolism of the entire anaerobic community.

Curiously, alkaliphilic sulfate reducers have not previously been isolated in pure culture, in spite of their obvious role in soda lakes. We used liquid media designed to approximately reflect the composition of the main mineral components in the lakes under study. We were quite cautious in single colony isolation of extremophiles on agar medium because use of agar often leads to the isolation of undesired contaminants or mixed colonies. We usually isolated single colonies from pure or highly purified cultures. Enrichment cultures inoculated with black mud from Lake Magadi drainage channels showed rapid and strong sulfidogenesis with hydrogen as electron donor in selective high strength sodium carbonate medium, but not with lactate or acetate. The hydrogenotrophic sulfate reducer was isolated in pure culture, and was found to be a small motile vibrio which utilized only H_2 or formate as electron donors and sulfate, and sulfite, but not elemental sulfur as electron acceptors. Thiosulfate was fermented to sulfide and sulfate. This isolate required yeast extract or acetate for anabolic processes, and thus was analogous to the conventional type of obligatory lithoheterotrophic sulfate reducers. However, its pH range was from 8 to 10.2, with an optimum at pH 9.7, being much above the values reported for any of the numerous neutrophilic sulfate reducers described, and it is the first truly alkaliphilic sulfate-reducing bacterium. Its salinity range extended up to 12% with an optimum at 3%. The strain needed both Na^+ and carbonate ions for growth. Its phylogenetic position among the sulfate reducers represent a new lineage in the delta-subclass of the *Proteobacteria*. The alkaliphilic strain clusters only with the recently isolated extremely halophilic *Desulfohalobium* (Ollivier et al. 1991) and strain TD3, a yet unnamed hydrocarbon-utilizing marine sulfate reducer (Rueter et al. 1994). The alkaliphilic hydrogenotrophic sulfate-reducing vibrio has now been described as a representative of a new genus *Desulfonatronovibrio hydrogenovorans* (Zhilina et al. 1997) (Fig. 2a). Strains with more than 90% DNA-DNA homology with the type strain were isolated from different lakes in Central Asia (Pikuta et al. 1997), demonstrating that alkaliphilic *Desulfonatronovibrio* is ubiquitous from the equatorial tropics to the cold arid steppe in alkaliphilic communities of lakes with different salt concentrations.

In an attempt to obtain organisms performing the C-2 metabolic pathway of sulfidogenesis, a larger oligotrophic alkaliphilic sulfate-reducing vibrio was isolated from lake Khadyn (Tuva), growing optimally at pH 10 and about 1.5% salt. In addition to hydrogen and formate the isolate could oxidize ethanol with the production of acetate. No growth was found below pH 8. It possessed the same lithoheterotrophic type of metabolism as *Desulfonatronovibrio*, but it was phylogenetically distinct, representing a separate line among the sulfate reducers within the delta-branch of the Proteobacteria. The new alkaliphilic sulfate reducer was named *Desulfonatronum lacustre* (Pikuta et al. 1998) (Fig. 2b).

Cooperative development of the hydrogen-scavenging sulfidogens with a dissipotrophic fermentative spirochete was demonstrated in a co-culture with an alkaliphilic spirochete that produced hydrogen and acetate from hexoses, with in addition some ethanol. In co-culture with *Desulfonatronum* neither hydrogen nor ethanol were produced and the amount of acetate increased, demonstrating the characteristic acetogenic shift (Pikuta et al. 1998).

Acetate accumulates in the anaerobic alkaline environment. The pathway of its consumption remains unclear (Zavarzin 1993). Acetate was found to be very slowly utilized for sulfidogenesis by alkaliphiles from lake Khadyn. Sulfur reduction, which is not a specific process, and is catalyzed by various bacteria (Bonch-Osmolovskaya 1994), leads to the formation of yellow polysulfide in alkaline medium. We observed this process with H_2 or other electron donors in enrichments from different soda lakes.

In anaerobic environments the sulfur cycle is closed by the photosynthetic production of sulfate by purple sulfur bacteria most characteristic of the habitat. In Lake Magadi *Halorhodospira* spp. are known to produce heavy blooms even at soda saturation. In lakes of Central Asia a diversity of purple bacteria was found, *Ectothiorhodospira* or *Halorhodospira* developing quite commonly in enrichments with H_2 or Na_2S. These were not the sole purple bacteria found, and depending on the locality and conditions many other an oxygenic bacteria such as *Amoebobacter* developed mass blooms in addition to the minute purple "*Heliospira daurica*" from lake Khylgatyn (Gorlenko et al. 1997). It is currently under description (Bryantseva et al., manuscript in preparation). Sulfate can also be formed aerobically from sulfide. New alkaliphilic aerobic sulfur-oxidizing bacteria have been isolated from lake Khadyn and other soda lakes (Sorokin et al. 1996a, 1996b). We did not observe the colorless sulfur bacteria described by Issatchenko in our extremely alkaline environments.

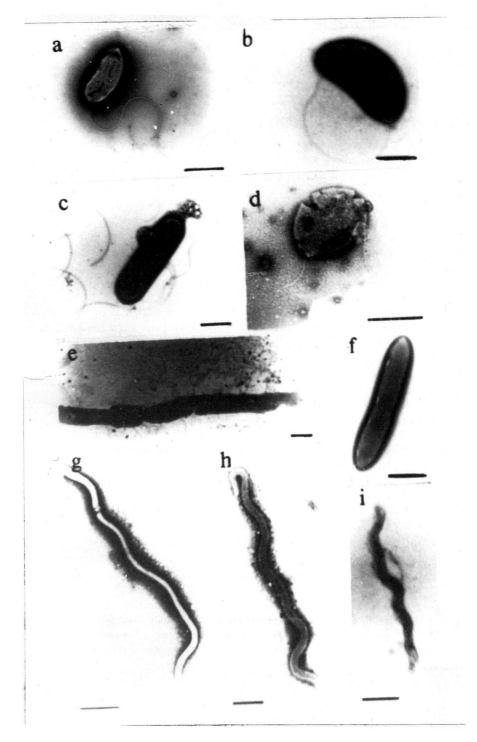

Fig. 2. Obligately anaerobic prokaryotes from Lake Magadi and continental soda lakes of Central Asia. Electron microscopy. Bars, 1 μm.
a. *Desulfonatronovibrio hydrogenovorans* Z-7952; **b.** *Desulfonatronum lacustre* Z-7951^T; **c.** *Natronoincola histidinovorans* Z7940^T; **d.** *Methanosalsus zhilinae* Z-7936; **e.** *Natroniella acetigena* Z-7937^T; **f.** *Tindallia magadii* Z-7934^T; **g.** *Spirochaeta africana* Z-7962^T; **h.** *Spirochaeta alcalica* Z7491^T; **i.** *Spirochaeta asiatica* Z-7591^T.

6. Acetogenesis

Many organotrophic anaerobes produce acetate as the main fermentation product. A specific way of acetate production is realized by homoacetogenic bacteria that use the Ljungdahl-Wood pathway of metabolism with CO-dehydrogenase as the key enzyme (Drake 1994).

From Lake Magadi (pH 10) we isolated the first haloalkaliphilic acetogen, *Natroniella acetigena* (Zhilina et al. 1996a) (Fig. 2c). This bacterium belongs to the order *Haloanaerobiales,* representing together with the extremely halophilic homoacetogen *Acetohalobium* a separate branch in the family *Halobacteroidaceae* (Zhilina and Zavarzin 1990; Rainey et al. 1995). It was isolated from an acetogenic enrichment with hydrogen, but hydrogen is not utilized. *Natroniella* is a large motile spore-forming rod with a gram-negative cell wall structure. Its pH range is from 8.1 to 10.7 with an optimum at pH 9.7-10. The organism obligately depends on Na^+ and carbonate ions. Its salinity optimum is at 12-15%, with slow growth after a prolonged lag phase at 20-26%. No growth occurred below 10% NaCl. The only compounds used were lactate, ethanol, glutamate, and propanol. Acetate was the sole product, with a yield of 0.46 g biomass per mol of acetate formed, accordingly to the equations:

$$2\ CH_3CH(OH)COO^- \rightarrow 3\ CH_3COO^- + H^+$$
$$2\ CH_3CH_2OH + 2\ HCO_3^- \rightarrow 3\ CH_3COO^- + H^+ + 2\ H_2O$$

When the substrate is exhausted the cells rapidly lyse. For biosynthetic processes the organism relies on the CO-dehydrogenase pathway (Pusheva et al. 1999a). Phosphorylation is of the oxidative type. Sodium and proton pumps, as well as Na^+/H^+ antiport, operate in concert. *Natroniella* invariably reacted by immediate lysis to various inhibitors and ionophores, except for valinomycin (Pusheva et al. 1999b). It may be speculated that this haloanaerobe adapted itself to extreme alkalinity by a mechanism which includes energization of the membrane, clearly demonstrated by light scattering experiments (Mitchell and Moyle 1969; Kreke and Cypionka 1994).

In the lakes of Tuva, providing a spectrum of habitats with different pH and salinity, we found that acetogenesis from hydrogen correlates with close to neutral pH- and low salt concentrations. Acetogenesis outcompeted sulfidogenesis when lactate was used as substrate (Zavarzin et al. 1996). A methylotrophic alkaliphilic homoacetogen was isolated by us from Lake Khadyn, and is currently under study. As already mentioned, alkaliphilic hydrogenotrophic homoacetogens do exist, but they are under strong competitive pressure exerted by sulfate reducers. There are various acetogens, including those possessing the CO-dehydrogenase pathway, which utilize different substrates; some of these are discussed below.

7. The Methane Cycle

In the high-salinity environment, the methylotrophic pathway of methanogenesis dominates (Oremland et al. 1982; Zeikus 1983). From hypersaline lagoons a series of methylotrophic methanogens have been isolated, covering the whole salinity range, from the moderately halophilic *Methanohalophilus* species to the extremely halophilic *Methanohalobium* (Zhilina 1986; Zhilina et Zavarzin 1987; Boone et al. 1993b). It was demonstrated that their metabolism is linked to the breakdown of osmolytes such as glycine betaine, following degradation to methylamines by haloanaerobes (Zhilina and Zavarzin 1991). From the Wadi-el-Natrun soda lakes the first methylotrophic alkaliphilic methanogen, *Methanosalsus zhilinae*, was isolated (Mathrani et al. 1988), a strain phylogenetically distant from the other halophilic genera. A new motile strain Z-7936 of the same species, represented by irregular cocci, was found in Lake Magadi (Fig. 2d) , and its physiology was described in some detail (Kevbrin et al. 1997). It uses methanol, dimethylsulfide and methylamines, which supported growth only at low concentrations due to the toxicity of ammonium for this organism at high pH. Methanogenesis was completely inhibited by H_2. Hydrogenotrophic methanogenic rods were found in Wadi-el-Natrun (Boone et al. 1986), but no further description is available. In our experiments, hydrogenotrophic sulfidogenesis outcompeted methanogenesis in alkaline environments. The enrichment culture technique was not successful in our hands for the isolation of aceticlastic or hydrogenotrophic alkaliphilic methanogens. However, it may be premature to conclude that they are absent. When degradation of cellulose was studied at high pH, methane was formed, and careful examination of the cultures under the fluorescence microscope revealed both rods and sarcinas with green fluorescence. It may be speculated that these methanogens could develop in a well-balanced community, but failed to grow in selective media.

Until recently, alkaliphilic methanotrophs were not known. By now, two such organisms have been isolated. The first, *Methylobacter alcaliphilus,* was isolated from Lake Khadyn (Khmelenina et al. 1997), and is a moderate haloalkaliphile. The second organism, the obligately methylotrophic *Methylomicrobium alcaliphilum*, isolated from the sediments of Kenyan soda lakes (Sorokin et al., manuscript in preparation), is quite unusual in its ability to oxidize both methane and ammonium, based on the functional similarity of methane and ammonium monooxygenases. Here we have a branching point between the cycles of methane and nitrogen. The biological methane ("Soehngen") cycle, not so prominent as in fresh water basins or in the ocean, is closed for the alkaliphilic community, especially in the low salinity habitats.

8. Ammonification

Ammonium appears as the end product of the proteolytic process. Under aerobic conditions it can be rapidly volatilized into the air from alkaline medium, as documented for urealytic bacteria present in pastures. For anaerobic ammonifiers it is less easy to get rid of ammonium, toxic at high pH. When studying proteolytic processes in soda lakes we anticipated the predominance of the bacteriolytic loop and used "Gaprin" - a trade mark for steam-treated biomass of methanotrophs - as a substrate (Zhilina et Zavarzin, 1994). This substrate was not very convenient for microscopic work because it contains a large amount of non-lysed cells and undecomposed refractory cell walls. Extractable protein was degraded anaerobically by diverse copiotrophic rods, which were isolated using a yeast extract-peptone medium (pH 10) that approximately reflected the mineral composition of Lake Magadi brine. Using this procedure, two new genera of anaerobic ammonifiers were found, fermenting a number of amino acids. These bacteria grew rather poorly on peptone, but proliferated in media containing certain amino acids and a low concentration of yeast extract for anabolic needs. The requirement for monomeric substrates indicates that they belong to the group of dissipotrophs rather than participating in the initial steps of the proteolytic pathway.

Natronoincola histidinovorans (Zhilina et al. 1998), a moderately haloalkaliphilic obligately anaerobic fermentative bacterium (Fig. 2e), used only histidine or glutamate with the production of acetate and ammonia:

$$2 \text{ Histidine} + 12 \text{ H}_2\text{O} \rightarrow 5 \text{ CH}_3\text{COO}^- + 2 \text{ HCO}_3^- + \text{H}^+ + 6 \text{ NH}_4^+$$

It should be recalled that the halophilic acetogen *Acetohalobium,* among a few substrates utilized, also ferments these two amino acids. *Natronoincola* is a true alkaliphile with a pH range from 8.0 to 10.5 and an optimum at pH 9.4. Its optimal salinity is at 10% NaCl, with a range from 4% to 16%. Morphologically it is a gram-positive rod with peritrichous flagellation. The type strain (Z-7940), isolated from the mud of Lake Magadi during a dry period, was asporogenous, the second strain (Z-7939), isolated from a desiccated floating cyanobacterial mass during the rainy period, was oligosporic. Spores were resistant to the desiccation but not to heating. *Natronoincola* belongs to cluster XI of the low G+C gram-positive bacteria. It is a true acetogen, forming acetate via the CO-dehydrogenase pathway. In spite of the fermentative type of metabolism, the organism obligately depends on ionic pumps (Pusheva et al. 1999a).

The second acetogenic ammonifier is *Tindallia magadii* (Kevbrin et al. 1998) (Fig. 2f), named in the honor of B.J. Tindall, who first investigated the Lake Magadi microbes. It belongs to the same phylogenetic cluster as *Natronincola.* It ferments mainly amino acids of the ornithine cycle (arginine, ornithine, citrulline) and a few oxyacids, producing acetate, propionate, H_2, and NH_3. Arginine may be formed during degradation of cyanophycin - a nitrogenous reserve compound in cyanobacteria - but other substrates rather originate from the excreta of birds, primarily flamingoes. *Tindallia* is a true alkaliphile and a moderate halophile.

These two new genera of anaerobic alkaliphiles represent examples of acetogenic ammonifiers able to ferment amino acids as the sole substrate for energy generation.

9. The Saccharolytic Pathway

A large variety of carbohydrates is fermented by alkaliphilic anaerobes, even at extreme pH and saturation with sodium carbonates characteristic of Lake Magadi and in less saline Central Asia lakes. Cellulose is decomposed by minute rods which cover its filaments as a slimy monolayer. They are accompanied by a number of other microorganisms, including sulfate reducers, methanogens and various rods. Quite remarkable are also the spirochetes.

Spirochetes are typical anaerobic dissipotrophs fermenting oligomeric carbohydrates. Neutrophilic spirochetes are numerous in anaerobic methane tanks, thermophilic spirochetes are associated with decomposing kelps in marine hydrothermal areas. However, they have not earlier been isolated from alkaline environments in spite of being easily observed by direct microscopy. We have now isolated alkaliphilic spirochetes from three sources: from springs feeding Lake Magadi from under the lava flows in which patches of cyanobacterial mat were found, from purple blooms underneath the trona layer on the surface of shallow lagoons in Lake Magadi, and from low-salinity lake Khadyn. All three isolates were found to be separate species within the genus *Spirochaeta* showing both phylogenetic and phenotypic differences: the halophilic *S. africana* from the spring (Fig. 2g), the alkaliphilic *S. alcalica* from the water of Lake Magadi (Fig. 2h), and the haloalkaliphilic *S. asiatica* (Fig. 2i) from lake Khadyn (Zhilina et al. 1996b). These spirochetes fermented various oligosaccharides with sucrose as the most easily degraded substrate. The obligately anaerobic *S. asiatica* significantly differs phenotypically and does not produce hydrogen. Fermentation products varied, with acetate, lactate, ethanol, and H_2 being the main products (Kevbrin et al. 1997). In co-cultures with hydrogenotrophic sulfate reducers spirochetes demonstrated an acetogenic shift with acetate as the single product. They were found to be extremely tolerant to sulfide.

Among other representatives of the saccharolytic pathway we isolated alkaliphilic organisms utilizing soluble polymeric carbohydrates such as pectin, xylane, and starch. These are mainly motile rods of various shape and size. Activity of lipolytic anaerobes, which have some advantage in highly alkaline medium, leads to luxurious sulfidogenesis and, if light is sufficient, to blooms of various purple bacteria.

All processes mentioned above lead to the formation of acetate as the end product. The question remains what the fate of acetate in the alkaliphilic anaerobic community is, and why it accumulates in enrichment cultures on different substrates but not in microcosms.

10. Aceticlastic Alkaliphilic Anaerobes

At present there are a number of possible mechanisms for acetate consumption within the alkaliphilic community, but no information is available on the extent to which they occur. Acetate is easily consumed by many aerobic alkaliphiles, and rod-shaped acetate oxidizing aerobes growing at pH 10 can easily be isolated. In the anaerobic community acetate is utilized as substrate of anabolism both by photolithotrophs such as purple bacteria and by chemolithoheterotrophs such as sulfate reducers. However, mechanisms for acetate catabolism by the alkaliphilic anaerobic community, particularly at high salt concentrations, remain unknown. There are many sources contributing to the pool of acetate, but no obvious sinks. This makes the question of the C-2 pathway the main unsolved problem in the description of the metabolism of the anaerobic alkaliphilic community. Aceticlastic methanogens are present only in low salt systems, and acetotrophic sulfate reducers are slow-growing creatures, even in the less saline alkaline lakes. Photoorganotrophs may be important in the system.

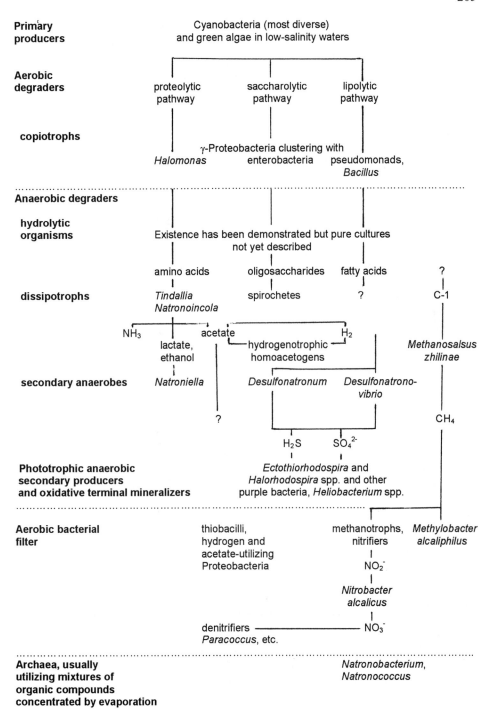

Fig. 3. The trophic network in the alkaliphilic community

Acetate may easily be oxidized in the aerobic zone together with other reduced so-"bacterial filter" system comprised by the microorganisms utilizing gases. Hydrogen or hydrogen sulfide are utilized by the lithotrophic microorganisms such as *Natronohydrogenobacter thiooxidans*, that can grow lithoheterotrophically on acetate + H_2 or on acetate + HS⁻ at pH 10 were isolated by Sorokin (Sorokin and Kuenen, 1999).

11. Conclusion

Recent intensive studies allow us to present a preliminary picture of the trophic structure of the alkaliphilic community. Alkaline habitats are most variable, and even under extreme conditions of saturation with soda the microbial community is quite diverse. The community represents an autonomous entity with a limited dependence on inputs from outside. The trophic structure of the alkaliphilic community with the positions of the described representatives of the key functional groups is given in Fig. 3. The community includes: (I) oxygenic cyanobacterial primary producers with accompanying aerobes; (II) an anaerobic degradative community in which sulfidogenesis dominates; (III) a oxidative bacterial filter system, that oxidizes reduced compounds and includes aerobic bacteria and/or anaerobic anoxygenic phototrophs.

Alkaline athalassic lakes may be considered as refuges of ancient terrestrial microbial communities. However, it is not known which type of such habitats can best be regarded as analogous to the ancient ones. Lakes in the rift valleys with volcanic activity seem to be one of the most promising sites. Alkaliphilic cyanobacteria are quite diverse, and terrestrial soda lakes with a high phosphate content appear to be more suitable for them than the ocean. However, the reverse line of development from the continental into marine environment cannot be excluded for cyanobacterial community. (Zavarzin et al., 1999). The diversity of alkaliphilic microorganisms includes the main branches of the phylogenetic tree. Most probably part of the microorganisms in this community, such as the haloanaerobe *Natroniella*, are not true alkaliphiles, and only tolerate alkalinity. Even with the present limited knowledge, our picture of the trophic system of the alkaliphilic community is sufficiently complete to close the main biogeochemical cycles. This indicates that this type of community is sufficient to support itself in a geologically reasonable time scale. However, it should also be realized that the community studied here is an extreme one, which means that it survived in a specialized protected environment, preventing its competition with the representatives of the main line of evolution.

12. Acknowledgments

The work was supported by a grant from the Russian State Program "Biodiversity", project "Biodiversity of extremophilic microorganisms and their communities" and RFBR 99-04-48056. The authors thank Dr. A. Lebedinsky and Dr. A. Oren for editorial improvements of the English text and Drs. V.M. Gorlenko and D. Yu. Sorokin for making unpublished information available.

13. References

Abd-el-Malek, Y. and Rizk, S.G. (1963) J. Appl. Bacteriol. 26: 20-26.

Bonch-Osmolovskaya, E.A. (1994) FEMS Microbiol. Rev. 15: 65-77.

Boone, D.R., Worakit, S., Mathrani, I.M. and Mah, R.A. (1986) Syst. Appl. Microbiol. 7: 230-234.

Boone, D.R., Whitman, W.B. and Rouvière, P. (1993a) In: J.G. Ferry (ed.) *Methanogenesis*. Chapman and Hall, London. pp. 35-80.

Boone, D.R., Mathrani, L.M., Liu, Y., Menaia, J.A.G.F., Mah, R.A. and Boone, J.E. (1993b) Int. J. Syst. Bacteriol. 43: 430-437.

Drake, H.L. (1994) In: H.L. Drake (ed.) *Acetogenesis*. Chapman & Hall, New York. pp. 3-60.

Dubinin, A.V., Gerasimenko, L.M. and Zavarzin, G.A. (1995) Microbiologiya 64: 717-721. (Engl. transl).

Duckworth, A.W., Grant, W.D., Jones, B.E. and van Steenbergen, R. (1996) FEMS Microbiol. Ecol. 19: 181-191.

Eugster, H.P. (1980) In: A. Nissenbaum (ed.) *Hypersaline Brines and Evaporitic Environments*. Elsevier, Amsterdam. pp. 195-232.

Garret, D.E. (1992) *Natural soda ash: occurrences, processing, and use*. Van Nostrand Reinhold, N.Y.

Gerasimenko, L.M., Dubinin, A.V. and Zavarzin, G.A. (1996) Microbiologiya 65: 736-740. (Engl. transl).

Gorlenko, V.M., Bryatseva, I.A. and Kompantseva (1997) Abstr. Workshop, «Green and Heliobacteria». Urbino, Italy. p. 6.

Grant, W.D., Jones, B.E. and Mwatha, W.E. (1990) FEMS Microbiol. Rev. 75: 255-270.

Guerrero, M.-C. and de Wit, R. (1992) Limnetica 8: 197-204.

Horikoshi, K. (1996) FEMS Microbiol. Rev. 18: 259-270.

Imhoff J.F., Sahl, H.G., Soliman, G.S.H. and Trüper, H.G. (1979) Geomicrobiol. J. 1: 213-234.

Imhoff, J.F. and Trüper, H.G. (1981) Zbl. Bakt. I Abt., Orig. 2: 228-234.

Imhoff, J.F., Tindall, B.J., Grant, W.D. and Trüper, H.G. (1981b) Arch. Microbiol. 130: 238-242.

Issatchenko, B.L. (1951) Selected works, v.2. Academia Nauk, Moscow, pp. 143-162. (Russ).

Javor, B. (1989) *Hypersaline Environments: Microbiology and Biogeochemistry*. Springer-Verlag, Berlin

Jones, B.E., Grant, W.D., Collins, N.G. and Mwatha, W.E. (1994) In: F.G. Priest, A. Ramos-Cormenzana and B.J. Tindall. (eds.) Bacterial diversity and Systematics. Plenum Press, N.Y. pp. 195-229.

Karpeta, W.P. (1989) South Afr. J. Geol. 92: 29-36.

Kempe, S.and Degens, E.T. (1985) Chem. Geol. 53: 95-108.

Kempe, S. and Kazmierczak, J. (1997) Planet. Space Sci. 45: 1493-1499.

Kevbrin, V.V., Zhilina, T.N. and Zavarzin, G.A. (1997) Microbiologiya 66: 39-42. (Engl. transl).

Kevbrin, V.V., Zhilina, T.N., Rainey, F.A. and Zavarzin, G.A. (1998) Curr. Microbiol. 37: 94-100.

Kevbrin, V.V., Lysenko, A.M. and Zhilina, T.N. (1997) Microbiologiya 66: 261-266. (Engl. transl).

Khmelenina, V.N., Kalyuzhnay, M.G., Starostina, N.G., Suzina, N.E. and Trotsenko, Yu.A. (1997) Curr. Microbiol. 35: 257-261.

Kiene, R.P., Oremland, R.S., Catena, A., Miller, L.G. and Capone, D.G. (1986) Appl. Environ. Microbiol. 52: 1037-1045.

Kreke, B. and Cypionka H. (1994) Arch. Microbiol. 161: 55-61.

Krylov, I.N. and Zavarzin, G.A. (1988) Dokl. Akad. Nauk SSSR 300: 1123-1125. (Russ.).

Krulwich, T.A., Guffanti, A.A. and Seto-Young, D. (1990) FEMS Microbiol. Rev. 75: 271-278.

Liu Y., Boone, D.R. and Choy, C. (1990) Int. J. Syst. Bacteriol. 40: 111-116.

Mathrani, I.M., Boone, D.R., Mah, R.A., Fox, G.E. and Lau, P.P. (1988) Int. J. Syst. Bacteriol. 38: 139-142.

Mitchell, P. and Moyle, J. (1969) Eur. J. Biochem. 9: 149-155.

Ollivier, B., Hatchikian, C.E., Prensier, G., Guezennec, J. and Garcia, J.L. (1991) Int. J. Syst. Bacteriol. 41: 74-81.

Oremland, R.S., Marsh, L.M. and Desmarais, D.J. (1982) Appl. Environ. Microbiol. 43: 462-468.

Oremland, R.S., Miller, L.G. and Whiticar, M.J. (1987) Geochim. Cosmochim. Acta 51: 2915-2929.

Oremland, R.S. and Boone, D.R. (1994) Int. J. Syst. Bacteriol. 44: 573-575.

Padan, E. and Schuldiner, S. (1994) Biochim. Biophys. Acta 1185: 129-151.

Pikuta, E.V., Lysenko, A.M. and Zhilina, T.N. (1997) Microbiologiya 66: 216-221. (Engl. transl).

Pikuta, E.V., Zhilina, T.N., Zavarzin, G.A., Kostrikina, N.A., Osipov, G.A. and Rainey, F.A. (1998) Microbiologiya 67: 105-113. (Engl. transl).

Pusheva, M.A., Pitryuk, A.V., Detkova, E.N. and Zavarzin, G.A. (1999a) Microbiologiya (in press).

Pusheva, M.A., Pitryuk, A.V. and Netrusov, A.A. (1999b) Microbiologiya (In press).

Rainey, F.A., Zhilina, T.N., Boulygina, E.S., Stackebrandt, E., Tourova, T.P. and Zavarzin, G.A. (1995) Anaerobe 1: 185-199.

Rueter, P., Rabus, R., Wilkes, H., Aeckersberg, J., Rainey, F.A., Jannasch, H.W. and Widdel, F. (1994)

208

Nature 372: 455-458.

Skulachev, V.P. (1989) FEBS Lett. 250: 106-114.

Sorokin, D.Yu., van Steenbergen, R., Robertson, L.A., Jones, B. and Kuenen, J.G. (1996a) Abstr. International Congress on Extremophiles. Estoril, Portugal, p. 204.

Sorokin, D.Yu., Lysenko, A.M. and Mitiushina, L.L. (1996b) Microbiologiya 65: 370-383.

Sorokin, D.Yu. and Kuenen, J.G. (1999) Extremophiles. (In press).

Tindall, B.J., Ross, H.N.M. and Grant, W.D. (1984) Syst. Appl. Microbiol. 5: 41-57.

Tindall, B.J. (1988) In: F.Rodriguez-Valera (ed.) *Halophilic bacteria*. CRC Press, Inc, Boca Raton, pp. 31-67.

Zavarzin, G.A. (1993) Microbiologiya 62: 473-479. (Engl. transl).

Zavarzin, G.A., Gerasimenko, L.M. and Zhilina, T.N. (1993) Microbiologiya 62: 645-652. (Engl. transl).

Zavarzin, G.A., Zhilina, T.N. and Pikuta, E.V. (1996) Microbiologiya 65: 480-486. (Engl. transl).

Zavarzin, G.A., Zhilina, T.N. and Kevbrin, V.V. (1999) Microbiologiya 68: 579-599. (Russ)

Zeikus, J.G. (1983) Symp. Soc. Gen. Microbiol. 34: 423-462.

Zhilina, T.N. (1986) Syst. Appl. Microbiol. 7: 216-222.

Zhilina, T.N. and Zavarzin, G.A. (1987) Dokl. Akad. Nauk SSSR 293: 464-468. (Russ.)

Zhilina, T.N. and Zavarzin, G.A. (1990) Dokl. Akad. Nauk SSSR 311: 745-747. (Russ.)

Zhilina, T.N. and Zavarzin, G.A. (1991) Zh. Obshch. Biol. 52: 302-318. (Russ.)

Zhilina T.N. and Zavarzin G.A. (1994) Curr. Microbiol. 29: 109-112.

Zhilina, T.N., Zavarzin, G.A., Detkova, E.N. and Rainey, F.A. (1996a) Curr. Microbiol. 32: 320-326.

Zhilina, T.N., Zavarzin, G.A., Rainey, F.A., Kevbrin, V.V., Kostrikina, N.A. and Lysenko A.M. (1996b) Int. J. Syst. Bacteriol. 46: 305-312.

Zhilina, T.N., Zavarzin, G.A., Rainey, F.A., Pikuta, E.V., Osipov, G.A. and Kostrikina, N.A. (1997) Int. J. Syst. Bacteriol. 47: 144-149.

Zhilina, T.N., Detkova E.N., Rainey, F.A., Osipov, G.A., Lysenko, A.M., Kostrikina, N.A. and Zavarzin, G.A. (1998) Curr. Microbiol. 37: 177-185.

Biodata of **Sammy Boussiba** author of "*Alkaliphilic Cyanobacteria*" (together with X. Wu and Z. Zarka).

Dr. **Sammy Boussiba** is an associate Professor at the Ben-Gurion University of the Negev and the head of the Albert Katz Department of Dryland Biotechnologies at the Jacob Blaustein Institute (Israel). He received his PhD. from the Ben-Gurion University in the Negev in 1981. Dr. Boussiba's main scientific interests are in cyanobacteria in extreme alkaline environments; N_2-fixating cyanobacteria; and model organisms to study biotechnological aspects of the carotenogenesis process and production of astaxanthin rich algae.
E-mail: **sammy@bgumail.bgu.ac.il**

J. Seckbach (ed.), Journey to Diverse Microbial Worlds, 209-224.
© 2000 *Kluwer Academic Publishers. Printed in the Netherlands.*

ALKALIPHILIC CYANOBACTERIA

S. BOUSSIBA, X. WU and A. ZARKA
Microalgal Biotechnology Laboratory
The Jacob Blaustein Institute for Desert Research
Ben-Gurion University of the Negev
Sede-Boker Campus 84990, Israel

1. Introduction

In almost every corner of the biosphere on the earth, life thrives in a vast number of diverse forms. Various microorganisms can be found in extreme environmental conditions at temperatures from below 0°C to above 100°C, and at pH from below 2 to above pH 11. Microorganisms also exist in soils from mountain heights to the bottom of the ocean, and from tropic to polar regions. Generally, an environment may be regarded as "extreme" if it imposes physiological limitations and stress to most biota. A few specialized groups of microbes, however, may consider such an environment as optimal for growth and survival. The ability to adapt to those extreme environmental conditions relies on the high genetic plasticity of most microorganisms.

Natural alkaline environments include desert soils and springs (Grant and Tindall, 1986). The most alkaline natural environments are soda lakes and deserts where pH values higher than 10 have been recorded. Two such areas are well known: Wadi Natrum in Egypt and Lake Magadi in Kenya (Grant and Tindall, 1986). The alkaline and saline environment is a rich source of a phylogenetically diverse range of organisms. Microorganisms inhabiting alkaline environments may be classified as alkaliphiles, including a wide spectrum of both photosynthetic and non-photosynthetic bacteria, fungi, yeasts and cyanobacteria. Alkaliphiles may be further divided into obligate alkaliphiles, facultative alkaliphiles and alkalitolerants. Alkalitolerant organisms can withstand alkaline environments, growing optimally at a lower pH between 8.5 and 9, but cannot normally grow at above pH 9.5. Such strains can also grow under neutral conditions. In contrast, obligate alkaliphiles are organisms with an obligate requirement for alkaline growth conditions (i.e. above pH 9), they do not grow at neutral pH and often grow optimally at pH values in excess of 10. Facultative alkaliphiles grow well at neutral pH, but optimally at pH 10 or above (Krulwich and Guffanti, 1989a).

In order to colonize alkaline environments, alkaliphilic microorganisms have evolved highly specialized physiological processes. The ecological and molecular physiology of alkaliphilic microorganisms were reviewed by Horikoshi and Akiba (1982), Krulwich and Guffanti (1983), Grant and Tindall (1986), Sharp and Munster (1986), and Krulwich and Guffanti, (1989a) in the eighties. Several elegant reviews on various aspects of alkaliphilic microorganisms were also published in this decade (Krulwich, 1995;

Krulwich et al., 1994, 1996; Padan and Schuldiner, 1994). Marked progress on the molecular aspects of the adaptation of alkaliphilic microorganisms to high pH has been achieved, especially several relevant antiporter genes have been cloned and sequenced from them. These achievements are mainly based on the pioneering work using *Escherichia coli* as a model (Horikoshi, 1986; Ivey et al., 1991, 1992; Waser et al., 1992; Padan and Schuldiner, 1994; Shimamoto et al., 1994; Krulwich, 1995; Krulwich et al., 1994, 1996).

The groups of microbes in extreme environments are diverse. Among them, the cyanobacteria present a unique group (Tandeau de Marsac and Houmard, 1993). Little research has been done so far on the molecular aspects of alkaliphilic cyanobacteria. Alkaliphilic cyanobacteria not only represent a rich source of research material for future fundamental research into microbial adaptation, but also contain valuable genetic resources, the potential applications of which are now gradually being appreciated. Much effort in our group has been devoted trying to elucidate mechanisms that allow alkaliphilic cyanobacteria to grow in extreme alkaline environments. In this chapter, recent progress in the study on alkaliphilic cyanobacteria has been summarized. Most of the contents will deal with physiological adaptation processes (non-genetic adaptation, acclimation) of cyanobacteria to the natural alkaline environments.

2. Ecology

2.1. ALKALINE SOIL AND WATER

The most stable and significant, naturally occurring alkaline environments are caused by a combination of geological, geographical and climatic conditions. A variety of alkaline lakes and deserts so produced are geographically widely distributed (Grant and Tindall, 1986). The Wadi Natrun containing a dozen small lakes in Egypt is one of the well-known and well-characterized locations (Jannasch, 1957). The mechanisms contributing to the formation of alkalinity have been reviewed by a number of researchers (Cole, 1968; Eugster, 1970; Hardie and Eugster, 1970; Eugster and Hardie, 1978). Environments such as these usually developed by concentration resulting from evaporation or by leaching of metal bicarbonates washed by CO_2-charged water from surrounding rocks; this is concentrated to produce a $NaHCO_3$-Na_2CO_3 brine (Hardie and Eugster, 1970). The environment may also contain high concentrations of other salts such as NaCl. In most cases the carbonate is derived from solution of atmospheric or respired CO_2. One of the most significant factors governing the evolution of saline brines is the relative concentration of divalent cations (Mg^{2+} and Ca^{2+}) to anions (CO_3^{2-} and SO_4^{2-}). Even in situations of low evaporative concentration, ground water becomes rapidly saturated with respect to $CaCO_3$, resulting in the deposition of calcite (often with coprecipitation of $MgCO_3$). It is this initial step, apparently universal to all drainage systems, which seems to determine whether alkalinity will be generated (Eugster and Hardie, 1978).

2.2. EVOLUTION AND DISTRIBUTION OF ALKALIPHILIC CYANOBACTERIA

2.2.1. *General Acclimation Ability*

Cyanobacteria have successfully colonized a wide range of ecological niches in the course of their evolution. Today they are found virtually in all aquatic ecosystems (from freshwater to hypersaline, and from hot springs to ice), on naked rock or in the soil of deserts and even in the air (Fogg et al., 1973). They owe their worldwide distribution to the acquired wide range of morphological and physiological properties and their remarkable capacity to adapt to changes in a variety of environmental factors (Stanier and Cohen-Bazire, 1977; Tandeau de Marsac and Houmard, 1993).

In their natural environment, cyanobacteria generally exist under growth limiting conditions frequently facing stresses of light, salinity, nutrient availability and temperature. In the past few years, many of the adaptation mechanisms used by cyanobacteria were described down to the molecular level. The detailed, in depth studies include chromatic adaptation to changes in response to varying light quality and nitrogen deficiency adaptation (heterocyst differentiation) (Tandeau de Marsac, 1991; Buikema and Haselkorn, 1993; Tandeau de Marsac and Houmard, 1993).

2.2.2. *Distribution*

Cyanobacteria can be found in widely diverse environmental conditions; the filamentous *Spirulina platensis*, which thrives at extreme alkaline habitats, is an example of this phenomenon (Ciferri, 1983; Vonshak, 1997). In the mid-1960s, one of the dominant species in an algal bloom in the evaporation ponds of a sodium bicarbonate facility in a lake near Mexico City was identified as one of the strains of *Spirulina*, *S. maxima* (Vonshak, 1997). Even among cyanobacteria, however, its adaptation to high pH values is unique: most cyanobacteria can tolerate alkaline pH, but grow optimally in neutral media (Stanier and Cohen-Bazire, 1977). This cyanobacterium fails to grow at pH 7, but grows optimally at pH 9 to 10. Even at pH 11.5, growth rates reach nearly 80% of the optimal values (Belkin and Boussiba, 1991; Schlesinger et al., 1996). It can, therefore, be clearly defined as an obligate alkaliphile (Krulwich and Guffanti, 1989a).

The criterion which assigns alkaliphiles to one or other group is very arbitrary and is often further complicated by differing pH profiles obtained using different substrates (Guffanti et al., 1978). Recent studies imply that alkaliphilic microorganisms are more ubiquitous than first considered with the isolation of alkaliphilic bacteria from both neutral and acidic (pH 4) soils (Horikoshi and Akiba, 1982). Other organisms present in visible accumulations in alkaline environments are the cyanobacteria-*Chloroflexus* mats, characteristically associated with alkaline hot springs throughout the world (Brock, 1978; Tindall, 1980). Cyanobacteria prefer alkaline environments (Fogg, 1956), although they are not generally believed to be obligate alkaliphiles. Physiological studies of *Microcystis aeruginosa* and *Synechococcus* suggest that they are alkalitolerant (McLachlan and Gorham, 1962; Kallas and Castenholz, 1982a,b).

As in other extreme environments, species diversity in alkaline environments decreases as the salinity and/or alkalinity increases. Thus, in the less saline soda lakes of the Kenyan Rift Valley, there is a large variety of invertebrates, algae and cyanobacteria (Grant and Tindall, 1986). The phytoplankton is dominated by cyanobacteria, although diatoms of several genera are often present in considerable numbers, particularly in less concentrated and less alkaline conditions (Hecky and Kilham, 1973). Soda lakes in Tuva are characterized by a wide diversity of alkaliphilic cyanobacteria; 16 genera and 34 species were found. In contrast to Lake Magadi, filamentous benthic forms predominate among cyanobacteria (Gerasimenko et al., 1996). The highly saline and alkaline regions of Lake Magadi of East Africa (a classic arid zone soda lake) and the lakes of Wadi Natrun in Egypt do not appear to contain organisms ranking higher than protozoa (Imhoff et al., 1979). At Lake Magadi the central part of the lake consists of a thick deposit of trona some hundreds of meters deep. Digging down through the upper layers reveals a clear stratification. The uppermost region is orange-pink owing to the presence of archaebacterial halophiles. Below this are several distinct green- or purple-red-colored horizons. Continuing downward is a green pigmented layer owing to the presence of cyanobacteria (Tindall, 1980; Grant and Tindall, 1986). The next horizon is purple-red and contains a number of different species of photosynthetic bacteria (Tindall, 1980). In the more saline lakes of Wadi Natrun, halophilic members of this genus are also apparent (Imhoff et al., 1979).

In the less saline lakes of the Kenyan Rift Valley, the predominant species are the filamentous *Anabaenopsis arnohiji*, *Spirulina platensis*, or a related species *Spirulina maxima*, and certain unicellular species, which may be *Chroococcus* sp., *Synechococcus* sp. or *Synechocystis* sp. The species composition is subject to seasonal changes, and probably dependent on the salinity of the lake (Brown, 1959, Tindall, 1980). Both *S. platensis* and *S. maxima* can grow optimally between pH 8 and 10, and are inhibited by high levels of divalent cations, such as Mg^{2+}. Therefore, they may be separated from other members of this genus, as they appear to be typical alkaliphiles.

One of the striking features of naturally formed alkaline lakes is the predominance of microorganisms, particularly phototrophs, as permanent or seasonal blooms (Melack, 1978; Imhoff et al., 1978, 1979; Tindall, 1980). This is reflected in the high primary productivity commonly encountered in these lakes, due to the presence of dense populations of cyanobacteria (Talling et al., 1973; Melack and Kilham, 1974). Cyanobacteria commonly participating in bloom formation include heterocystous filamentous *Spirulina* sp., *Anabaenopsis* sp., recently reclassified as *Cyanospira* sp. (Florenzano et al., 1985). Unicellular cyanobacteria, which may be *Synechococcus* or *Gloeocapsa*, are also common and may occasionally dominate particular soda lakes (Grant et al., 1990), while in the highly saline lakes of Wadi Natrun or Lake Magadi, anoxygenic phototrophic bacteria as well as cyanobacteria may be bloom forming (Jannasch, 1957; Imhoff et al., 1978, 1979; Tindall, 1980; Grant and Tindall, 1986).

2.2.3. Evolution

Cyanobacteria, found in the primitive terrestrial environment where they evolved, are among the world's oldest known living organisms. With fossil records dating back about 3×10^9 years, the Proterozoic (2.5×10^9 until 5.7×10^7 years ago) represents about two-thirds of the recorded history of life as 'the age of cyanobacteria' (Painter, 1993).

In central Poland, mats of benthic coccoid cyanobacteria in a sequence of Late Jurassic open marine sediments were calcified in situ and formed limestone. Such an intensive calcification of marine cyanobacteria could have proceeded only in environments more supersaturated than modern seawater with respect to calcium carbonate minerals. The excess alkalinity in the ancient marine world is proposed as the main factor enhancing colonization of extensive sea bottom areas by the alkaliphilic cyanobacteria and in promoting their *in vivo* calcification (Kazmierczak et al., 1996).

3. Physiology

3.1. PROPERTIES OF ALKALIPHILES

Several properties are shared by all the extreme alkaliphiles studied to date. These include: a) the unique composition of membrane lipids and the membrane lipid/protein ratio, often associated with very high levels of respiratory-chain components in the membrane. b) the capacity to maintain intracellular pH homeostasis. c) possession of exclusive Na^+/H^+ antiporters. d) presence of Na^+/solute symporters operated via a Na^+ cycle. e) a generally more acidic amino acid composition of proteins that are exposed to or excreted into the external milieu. The protein-synthesizing machinery exhibits no significant differences in alkaliphilic and neutrophilic bacteria.

A feature of most alkaliphiles is their absolute requirement for Na^+ (Horikoshi and Akiba, 1982; McLaggan et al., 1984). The dependence of alkaliphiles on sodium has been well-documented (Padan et al., 1981; Booth, 1985; Krulwich and Guffanti, 1989b). Studies on a number of organisms have indicated that Na^+ is important for nutrient uptake through Na^+/solute symporters in alkaliphiles (Koyama et al., 1976; Krulwich et al., 1979; Karazanov and Ivanovsky, 1980; Garcia et al., 1983; Krulwich and Guffanti, 1983), presumably because of the scarcity of protons and the difficulty of establishing a normal proton gradient at alkaline pH.

3.2. MAINTAINING INTRACELLULAR pH HOMEOSTASIS

The most important common feature of alkaliphiles living in extreme alkaline pH values seems to be their ability to control their internal pH (pH_i) via a sodium cycle that facilitates solute uptake and pH homeostasis (Booth, 1985). The intracellular pH of most alkaliphilic bacteria appears to be between 7 and 8.5, although the external pH of the medium may be in excess of pH 10 (Padan et al., 1981; Booth, 1985; Belkin and Boussiba, 1991). They maintain their cytoplasm at a more acidic level than the external medium by exchange of intracellular Na^+ for external H^+ (Krulwich et al., 1994). This exchange is mediated by membrane-associated Na^+/H^+ antiporter activity which, at alkaline pH, appears to operate in an electrogenic fashion that is driven by the membrane potential. A kinetic analysis showed that the V_{max} of the transport system was much increased in higher pH medium. Also, the Na^+/H^+ antiport activity greatly increased with an increase in pH of the assay medium (Kuroda et al., 1994).

So far, experimental data are accumulating in support of the presence of a Na^+/H^+ antiporter in alkaliphilic cyanobacteria (Miller et al., 1984; Ritchie, 1991, 1992; Belkin and Boussiba, 1991; Buck and Smith, 1995; Schlesinger et al., 1996), but there is still lack of solid evidence, and no such antiporter genes have been isolated from cyanobacteria. In eubacteria, however, several Na^+/H^+ antiporter genes have been cloned and sequenced (Horikoshi, 1986; Ivey et al., 1991, 1992; Waser et al., 1992; Padan and Schuldiner, 1994;

Shimamoto et al., 1994; Krulwich, 1995; Krulwich et al., 1994, 1996). Evidence supporting the role of a Na^+/H^+ antiporter in pH regulation was also derived from mutant studies. A mutant of *Bacillus alcalophilus* which grew only at neutral pH and could not grow above pH 9 showed no evidence of Na^+ efflux, suggesting that the antiporter has a role in pH_i regulation (Krulwich et al., 1979).

The pH_i may be measured by a number of methods (Mizushima et al., 1964; Harold et al., 1970; Hsung and Haug, 1975; Thomas et al., 1976; Ogawa et al., 1978). However, only the following methods are suitable for the measurement of pH_i in alkaliphilic microorganisms: a) the weak base method - the weak base methylamine is usually used for measuring pH_i of organisms growing at high pH (Rottenberg et al., 1972; Padan et al., 1981; Miller et al., 1984). Measurement of pH_i in the dark was recommended for avoiding the complexity in the light, because most of the charged methylamine accumulated in the intrathylakoidal space through ammonium carrier and not in the cytoplasm (Miller et al., 1984; Boussiba, 1989; Boussiba and Gibson, 1991). b) the electron spin resonance (ESR) method - both the cytoplasmic and thylakoid pH can be determined in the dark by ESR spin-probe spectroscopy. It is used with a combination of weak acid, weak base (amine) and neutral nitroxide spin probes (Belkin et al., 1987). c) the ^{31}P nuclear magnetic resonance (NMR) method (Kallas and Dahlquist, 1981; Buck and Smith, 1995). d) the use of fluorescent probes whose fluorescence changes according to pH_i after being incorporated into cells (Dwivedi et al., 1994).

For an external pH 8-8.5, the cytoplasmic pH of the cyanobacterium *Anacystis nidulans* is about 7.5, and the $\Delta\mu H^+$ near zero (Falkner et al., 1976; Peschek et al., 1985; Ritchie, 1991, 1992). With a pH-sensitive fluorescent probe 2,7'-bis-(2-carboxyethyl)-5-carboxyfluorescein acetomethyl ester (BCECF-AM), the pH_i of both the alkaliphilic cyanobacteria *Microcystis aeruginosa* and *Hapalosiphon welwitschii* was found to be 7.5 and 7.8 respectively at external pH (pH_{out}) of 10 (Dwivedi et al., 1994). It was also shown that a certain amount of Na^+ is required for such pH homeostasis. In illuminated *Synechocystis* cells grown in 2 mM Na^+ a lower pH_i of 7.7 was observed by ^{31}P-NMR method than that in the dark with pH_i of 7.9-8.0 (Buck and Smith, 1995). Based on the ESR method, the pH_i variation of *S. platensis* was measured in accordance with the variation of external pH (Belkin and Boussiba, 1991). At the range of pH_{out} tested, the average pH_i was relatively stable, changing from 6.8 (at pH_{out} of 7) to 7.9 (at pH_{out} of 11.2). At the same time, the cytoplasmic pH increased from 7.2 to 8.5, while thylakoid pH changed from 6 to 7 (Belkin and Boussiba, 1991). This is one of the most delicate measurements of pH_i that can distinguish between cytoplasmic and thylakoid pH.

Essentially, an organism that is apparently capable of remarkable pH homeostasis might differ from a conventional one, such as a neutrophile, in one or more of the three distinct elements: a) primary generation of $\Delta\mu H^+$ via a respiration chain, b) a Na^+/H^+ antiporter with a stoichiometry of less than 1, and c) Na^+-coupled symporters and other related channels for Na^+ reentry (Krulwich et al., 1994).

3.3. HIGH pH AND Na^+ EFFECTS

The absence of Na^+ had pleiotropic effects upon the growth and metabolism of cyanobacterial cells at high pH. In some strains capable of growing at both alkaline and neutral pH, a requirement for NaCl is observed at lower pH which is not apparent at high pH, presumably because of the abundance of Na_2CO_3. Since the activity of an antiporter entails the excretion of Na^+, operation of a multitude of Na^+-coupled solute porters may provide a means for Na^+ reentry to facilitate the sodium cycle in these cyanobacteria (Kaplan et al., 1989; Buck and Smith, 1995).

The most significant function of Na^+ involves maintaining cellular pH homeostasis which is exclusively required for facilitating solute uptake. However, there is uncertainty with regard to both H^+ and Na^+ chemiosmotic circuits, whether they are primarily based on protons, similar as in plant cells, or upon a sodium-based circuit, as found in animal cells (Kaplan et al., 1989; Buck and Smith, 1995).

The first proposal suggests that the primary generation of $\Delta\mu H^+$ is effected by a plasma membrane H^+-ATPase and/or H^+-translocating redox chains (Krulwich et al., 1994). This H^+ would then be converted into a $\Delta\mu Na^+$ by the H^+-importing/Na^+-extruding antiporter (Paschinger, 1977; Nitschmann and Peschek, 1982; Nitschmann et al., 1982; Blumwald et al., 1984; Scherer et al., 1984; Molitor and Peschek, 1986; Molitor et al., 1986; Padan and Vitterbo, 1988; Peschek et al., 1988; Nicholls et al., 1992; Nitschmann, and Packer, 1992; Gabbay-Azaria et al., 1992, 1994). The $\Delta\mu Na^+$ thus created could in turn drive secondary Na^+-dependent transport systems. This situation appears especially to apply for cyanobacteria exposed to external pH 7 or below, but does not apply to high external pH, because the $\Delta\mu H^+$ values are usually too small to play a significant role. For example, the $\Delta\mu H^+$ of *A. nidulans* is near zero at external pH 8-8.5, which according to this mode is insufficient to be used as the driving force for Na^+-dependent transport systems (Falkner et al., 1976; Peschek et al., 1985; Ritchie, 1991, 1992).

In contrast, the second proposal suggests the active extrusion of Na^+ by a primary pump driven by ATP hydrolysis or light (Batterton and van Baalen, 1971; Brown et al., 1990; Skulachev, 1994). In this case, the Na^+/H^+ antiporter would generate a $\Delta\mu H^+$ at the expense of the $\Delta\mu Na^+$ (Ritchie, 1992). This seems to be the prevalent situation for cyanobacteria at alkaline pH, which maintains a fairly constant cytoplasmic pH lower than that of the growth medium (Wolk, 1973; Miller et al., 1984; Belkin and Boussiba, 1991; Buck and Smith, 1995).

3.3.1. *The Essential Physiological Role of Na^+ Ions*

Although very high concentrations of inorganic ions are toxic to cyanobacteria, low Na^+ concentrations are essential for many cellular processes. a) Intracellular pH regulation. It is generally agreed that the Na^+/H^+ antiporter's involvement in regulating internal pH in cyanobacteria for maintaining pH homeostasis (Kaplan et al., 1984) is essential for life at high pH (Miller et al., 1984; Buck and Smith, 1995; Schlesinger et al., 1996) or for alkalitolerance. However, a concentration of 5-10 mM is sufficient to satisfy most of these needs (Buck and Smith, 1995; Thomas and Apte, 1984). b) Nutrient uptake. The driving force for nutrient uptake is the electrochemical potential of Na^+ (Kitada and Horikoshi, 1977, 1992). Concentrations of Na^+ required for solutes uptake of Cl^-, PO_4^{3-}, NO_3^-, CO_2, and HCO_3^-, etc. via Na^+/solute symporters varied from 10 μM to 40 mM (Krulwich, 1986; Maeso et al., 1987; Miller et al., 1988; Espie et al., 1988, 1991; Espie and Kandasamy, 1994; Fernández and Avendanño, 1993). c) Photosynthesis. Cleavage of water by photosystem II has been proven to be a Na^+-dependent process, as has carbon transport in cyanobacterial photosynthesis (Zhao and Brand, 1988). d) Nitrogen fixation. The essential role played by low Na^+ concentrations in maintaining a functioning nitrogen fixation system has been known for some time, Na$^+$ ions apparently being needed for the activation of nitrogenase (Thomas and Apte, 1984). e) Motility. It has been shown that the active motility of marine isolates of *Synechococcus* WH8113 is energized directly by a Na$^+$-motive force (Willey et al., 1987), similar to the case in bacteria (Sugiyama et al., 1985).

In alkaliphilic heterotrophic bacteria, it has been repeatedly shown that in the absence of sodium the internal pH equalled that of the medium, with a subsequent loss of viability (Kitada et al., 1982; Krulwich et al., 1984). Several physiological effects caused by incubation in sodium-deficient media have also been reported in cyanobacteria, which will be described in the following sections.

3.3.2. *Growth and viability*
The alkaliphilic cyanobacterium *S. platensis* could grow optimally at pH 9 to 10 in a complete medium containing 250 mM Na$^+$. It failed to grow however, at Na$^+$ less than 50 mM when tested at pH 8 to 11 (Boussiba 1989; Belkin and Boussiba 1991; Schlesinger et al., 1996). At pH 7, as expected, no growth occurred at any sodium concentration. At higher pH, growth rate and Na$^+$ requirements were pH-dependent. At pH 8, growth rates were similar at sodium concentrations of 75 mM and above, at pH 10 or 11 higher concentrations (150 or 250 mM Na$^+$ respectively) were required to maintain a maximal growth (Schlesinger et al., 1996). When sodium was completely removed from the growth medium (pH 10), the cultures failed to grow, turned yellow and the cells lysed in less than 60 minutes. This occurred also when lithium (250 mM) was substituted for sodium in the medium (Schlesinger et al., 1996).

Complete loss of viability in the absence of Na$^+$ was observed in alkalitolerant *Synechococcus leopoliensis* after about 1 h incubation at pH 9.8, while minimum 0.5 mM of sodium was sufficient to maintain optimal growth rates at pH 9.3, and 2 mM allowed maximal photosynthetic oxygen evolution rates at pH 9.6 (Miller et al., 1984). In the absence of Na$^+$, cell division did not occur at any pH, but metabolic activity such as chlorophyll synthesis and photosynthesis did continue. A large membrane potential ($\Delta\psi$) was maintained by cells at pH 9.6. Immediately upon shift-up of the external pH, a substantial increase of the magnitude of $\Delta\psi$ occurred in compensation for the decreased pH (Miller et al., 1984).

It seems that the sensitivity of cyanobacteria at alkaline pH to the presence of Na$^+$ is different in the light and in the dark. A marked difference in requirement of Na$^+$ for maintaining viability at pH 10 was observed in alkaliphilic *Synechocystis* sp. which was isolated from soil (Buck and Smith, 1995) and in *Spirulina* (Schlesinger et al., 1996). At pH 10, cell death of *Synechocystis* was not observed at 5 mM Na$^+$ in the light, but occurred in the dark, and greater than 50 mM Na$^+$ is necessary for preventing from such cell lysis. Illumination was found to decrease the pH$_i$ in these case (Buck and Smith, 1995; Falkner et al., 1976).

3.3.3. *Photosynthesis*
Differential effects on PSI and PSII. Synechococcus leopoliensis in Na$^+$-free medium could maintain its full photosynthetic activity at pH 8.5, while activity was reduced to 10% at pH 9.6 and no activity was found at pH 10. Addition of 2 mM NaCl allowed maximal photosynthetic oxygen evolution rates at pH 9.6 (Miller et al., 1984). In these as

well as in many other cases, the effects of sodium deficiency were attributed to the inhibition of bicarbonate or carbon dioxide transport (Kaplan et al., 1984, 1994). In contrast, Becker and Brand (1982, 1985) reported on a loss of PSII activity in *Anacystis nidulans*, reversible by the addition of submillimolar amounts of either Ca^{2+} or Na^+. Zhao and Brand (1988, 1989) reported a similar situation in *Synechocystis*, with loss of oxygen evolution within 1 h, restorable with sodium and calcium but unaffected by the presence of inorganic carbon.

In none of these cases was the effect of sodium deprivation as drastic as in *Spirulina* cultures. When sodium was completely omitted from the suspension medium at pH 10, a very rapid sequence of events occurred. Oxygen evolution rate was reduced drastically after 15 to 20 min, and completely lost within 1 h. This was accompanied by extensive bleaching of cellular phycocyanin and cellular lysis. From the kinetic analysis of loss of partial reactions it was concluded that PSII rather than PSI activity was inhibited. The Na^+-deficiency effect was more pronounced at higher light intensities. However, the Na^+-deficiency effects were also observed in the dark at pH 10, but not at pH 8.5. The addition of 200 μM DCMU to *Spirulina* cultures deprived of sodium and kept in the light indeed caused total quenching of O_2 evolution but prevented cell lysis (Schlesinger et al., 1996).

The Na^+ concentration required for optimal growth of *S. platensis* at pH 10 appeared to be between 150 to 250 mM. However, far lower concentrations were required for short-term oxygen evolution. In fact, 2.5 mM sodium was sufficient to sustain normal activity for at least 2 hours at pH 10. In the absence of sodium, the decay rate of oxygen evolution was pH dependent. The higher the pH, the faster the quenching occurred (Schlesinger et al., 1996). While it seems likely that the effect of Na^+ deprivation on photosynthetic electron transport was secondary, and caused by the cells' disintegration, it is also possible that the damage to the oxygen-evolving apparatus occurred independently. This may be supported by the observation that reversibility of the effects caused by the lack of sodium was limited. When sodium was added, at any concentration, later than 15 minutes after the initial suspension in the Na^+-free medium, the effect was irreversible. After shorter periods within 15 min from the time of sodium removal, the degree of recovery depended upon the sodium concentration added. While addition of 100 mM within this period could restore full activity, 5 mM prevented cell lysis without a regain of oxygen evolution, at least within 2 hours.

In most of the cases, Li^+ could not replace Na^+ for growth. The same fate of cell lysis of *S. platensis* grown at high pH was found in the presence of either 5 or 250 mM Li^+ as in the absence of Na^+ (Schlesinger et al., 1996). However, in *S. leopoliensis*, the Na^+ requirement could be partially met by Li^+ to reach 14% of the growth rate of those in NaCl, although Li^+ also exhibited an inhibitive effect by extending the lag period and reducing 20% of photosynthetic activity in the presence of 5 mM Na^+ (Miller et al., 1984).

Cell lysis and pigment decomposition. In *S. platensis*, deterioration in photosynthetic oxygen evolution caused by the lack of sodium was accompanied by two physical phenomena: a yellowing of the culture and a complete lysis of the cells. The change in pigmentation was caused by bleaching of phycocyanin. While the absorption of methanol extracts of the cells show no significant difference, in the water extracts the 620 nm absorption peak, characteristic of phycocyanin, disappears. The reduction in absorbance of the phycocyanin could be attributed to an exposure of the phycobiliprotein to high pH, when the cells lost their ability to maintain a stable pH gradient (Schlesinger et al., 1996). Phycocyanin is pH-sensitive, and loses its specific 620 nm absorbance at pH 10.

The reasons for the rapid lysis are not as obvious. In the absence of sodium in the medium, cells can not maintain the pH homeostasis, the intracellular pH will immediately be elevated to close to the medium pH, following passive transport of protons down a chemical gradient from inside to outside (Kitada et al., 1982; Krulwich et al., 1984). This could be one of the major reason causing rapid lysis, due to that alkaline pH can activate some alkaline proteases, while paralysing most cellular enzymes. It is also tempting to hypothesize that this rapid lysis may actually be driven by the deprivation of available inorganic carbon due to the lack of sodium-dependent transport. This could greatly increase the light-driven generation of active oxygen species through the Mehler reaction (Mehler, 1951), thus causing a lethal photooxidative effect in the light. This assumption is supported by our finding that no rapid lysis occurs when sodium deprived cultures of *Spirulina* were placed in the dark or treated by DCMU in the light (Schlesinger et al., 1996). In contrast, when a *Spirulina* culture was deprived of sodium at pH 8.5, no lysis was observed, nor did any changes in pigment composition occur during the first 2 hours after the onset of the deprivation.

Ammonia utilization. In the alkaliphilic cyanobacterium *S. platensis*, an active mechanism to translocate ammonium is probably not needed. Our data suggest that net uptake of ammonia to support optimal growth could be explained by a pH driven diffusion process (Boussiba et al., 1989; Boussiba and Gibson, 1991). It is perhaps significant in terms of cellular efficiency for ammonia utilization that cyanobacteria are generally very tolerant to high pH, and many can thrive in environments where the pH is 11 or above (Boussiba, 1989; Boussiba and Gibson, 1991).

Effects of uncouplers. The deleterious effects of uncouplers (carbonyl cyanide m-chlorophenyl hydrazone [CCCP], and nigericin) known to dissipate the pH gradient across biological membranes (Gaensslen and McCarty, 1971) were examined in the presence and absence of sodium on oxygen evolution *by S. platensis* at pH 10. Similar to in the case of ammonia, lower concentrations of these compound were needed to abolish oxygen evolution in the absence of sodium than in its presence (50 and 500 µM, respectively). The phenomena observed in *S. platensis* at pH 10 in the absence of sodium are partially due to the inability of cells to maintain an appropriate internal pH, probably as a result of inactivation of the sodium/proton antiporter by lack of sodium.

4. Summary

Cyanobacteria are extraordinarily interesting basic and applied research subjects, showing promising adaptability to extreme environments (Boussiba, 1991, 1997; Kerby and Rowell, 1992; Utkilen and Gjolme, 1992; Boussiba and Wu, 1999; Belkin and Boussiba, 1991; Schlesinger et al., 1996). Cyanobacteria are generally alkaliphilic (some of them being capable of growing at external pH up to 11.5), rather than acidiphilic (most cyanobacteria growing at pH above 5). Knowledge of the evolved strategies by alkaliphilic cyanobacteria to cope with harsh alkaline conditions will lead us to a better understanding of those microorganims found in extreme environments.

Considerable progress in the understanding of alkaliphilic cyanobacteria has been made:

a). The cytoplasma membrane of a few strains of alkaliphilic cyanobacteria has been purified and studied (Norling et al., 1997). A plasma membrane P-type ATPase (Xu et al., 1994; Gabbay-Azaria et al., 1994) has been detected in *Spirulina* with a pH optimum near 8.5, in accordance to the internal pH measured in *Spirulina* growing at pH 10 (Belkin and Boussiba, 1991).

b). Involvement of a Na^+/H^+ antiporter in the adaptation to high pH has been suggested. Sodium is essential for the growth and survival of alkaliphilic cyanobacteria in a pH-dependent manner, the higher the external pH, the higher the minimal Na^+ concentration to support growth (Miller et al., 1984; Belkin and Boussiba, 1991; Buck and Smith, 1995; Schlesinger et al., 1996).

c). The presence of an ATP driven sodium pump was proposed (Brown et al., 1990; Skulachev, 1994; Schlesinger et al., 1996). So far, little information is available concerning the activity of this kind of protein in cyanobacteria. Our finding that Na^+ can protect against the deleterious effect of uncouplers suggests that at pH 10, this system may operate to cope with the extrusion of sodium (Schlesinger et al., 1996).

d). The existence of Na^+/solute symporters which cotransport Na^+ and solutes is well documented (Kaplan et al., 1994; Ritchie, 1992; Lara et al., 1993; Fernández Valiente and Avendaño, 1993). This facilitates the completion of the Na^+/H^+ cycle for grown in alkaline medium.

e). Cytoplasmic or thylakoid pH of alkaliphilic cyanobacteria grown at alkali medium was accurately measured by various methods (Falkner et al., 1976; Peschek et al., 1985; Ritchie, 1991, 1992; Belkin and Boussiba, 1991; Dwivedi et al., 1994; Buck and Smith, 1995).

Further basic and applied research will be necessary to broaden the spectrum of biotechnological applications. It is not yet clear what affects the inability of obligatory alkaliphilic cyanobacteria to grow at neutral pH. Studies should aim to provide a clear indication as to which proteins are involved in mechanisms that allow *Spirulina* and other alkaliphilic cyanobacteria to cope with highly alkaline pH and of how sodium is involved in these mechanisms. Knowledge of cyanobacterial physiology in general and the mechanism involved in adaptation to high alkaline environment specifically may thus be meaningfully enriched. This information is also useful for commercial mass cultivation of *Spirulina*, which is receiving increasingly attention as a commercial source of single-cell protein (Cifferi, 1983), a rather encouraging sign. Alkaliphilic cyanobacteria may enable in the future the transformation of highly alkaline desert and wastelands into productive regions.

5. References

Batterton, J.C. and van Baalen, C. (1971) Arch. Microbiol. 76, 151-165.

Becker, D. and Brand, J. (1985) Plant Physiol. 79, 552-558.

Becker, D.W. and Brand, J.J. (1982) Biochem. Biophys. Res. Commun. 109, 1134-1139.

Belkin, S., Mehlhorn, R. and Packer, L. (1987) Plant Physiol. 84, 25-30.

Belkin, S. and Boussiba, S. (1991) Plant Cell Physiol. 32, 953-958.

Blumwald, E., Wolosin, J.M. and Packer, L. (1984) Biochem. Biophys. Res. Commun. 122, 452-459.

Booth, I. R. (1985) Microbiol. Rev. 49, 359-378.

222

Boussiba, S. (1989) Plant Cell Physiol. 32, 303-314.

Boussiba, S. (1991) Plant Soil 137, 177-180.

Boussiba, S. (1997) In: A.K. Rai (ed.), Cyanobacterial nitrogen metabolism and environmental biotechnology. Ellis Harwood Publishers, New Delhi, pp. 36-72.

Boussiba, S. and Gibson, J. (1991) FEMS Microbiol. Rev. 88, 1-14.

Boussiba, S., Resch, C.M. and Gibson, J. (1984) Arch. Microbiol. 138, 287-293.

Boussiba, S. and Wu, X. (1999) In: S. Kannaiyan (ed.), Biofertilizer technology. Wiley & Sons, N.Y.

Brock, T.D. (1978) Thermophilic microorganisms and life at high temperatures. Springer, New York.

Brown, I.I., Fadeyev, S.I., Gerasimenko, L.M., Kirik, I.I., Pushenko, M.Y. and Severina, L.I. (1990) Arch. Microbiol. 153, 409-411.

Brown, L. (1959) The mystery of the flamingoes. Hamlyn, London.

Buck, D. and Smith, G. (1995) FEMS Microbiol. Lett. 128, 315-320.

Buikema, W. and Haselkorn, R. (1993) Annu. Rev. Plant Physiol. 44, 33-52.

Ciferri, O. (1983) Microbiol. Rev. 47, 551-578.

Cole, G.A. (1968) in G.W. Brown (ed.), Desert biology. Academic Press, New York, Vol. 1, pp. 423-486.

Dwivedi, A., Srinivas, U., Singh, H.N. and Kumar, H.D. (1994) J. Gen. Appl. Microbiol. 40, 261-263.

Espie, G., Miller, A. and Canvin, D. (1988) Plant Physiol. 88, 757-763.

Espie, G.S., Miller, A.G. and Canvin, D.T. (1991) Can. J. Bot. 69, 936-944.

Espie, S.G. and Kandasamy, R.A. (1994) Plant Physiol. 104, 1419-1428.

Eugster, H.P. (1970) Mineralogical Society of America, Special Publication 3, 215-235.

Eugster, H.P. and Hardie, L.A. (1978) In: A. Lerman (ed.), Lakes: chemistry, geology and physics. Springer, New York, pp. 237-293.

Falkner, G., Horner, F., Werdan, K. and Heldt, H.W. (1976) Plant Physiol. 58, 717-718.

Fernández Valiente, E. and Avendaño, M.C. (1993) Plant Cell Physiol. 34, 201-207.

Florenzano, G., Sili, C., Pelosi, E. and Vincenzini, M. (1985) Arch .Microbiol. 140, 301-307.

Fogg, G.E. (1956) Bacteriol. Rev. 20, 148-165.

Fogg, G.E., Stewart, W.D.P., Fay, P. and Walsby, A.E. (1973) The blue-green Algae. Academic Press, London.

Gabbay-Azaria, R., Pick, U., Ben-Hayyim, G. and Tel-Or, E. (1994) Physiol. Plant. 90, 692-698.

Gabbay-Azaria, R., Schonfeld, M., Tel-Or, S., Messinger, R. and Tel-Or, E. (1992) Arch. Microbiol. 157, 183-190.

Gaensslen, R.E. and McCarty, R.E. (1971) Arch. Biochem. Biophys. 147, 55-65.

Garcia, M.L., Guffanti, A.A. and Krulwich, T.A. (1983) J. Bacteriol. 156, 1151-1157.

Gerasimenko, L.M., Dubinin, A.V. and Zavarzin, G.A. (1996) Microbiology 65, 736-740.

Grant, W.D. and Tindall, B.J. (1986) In: R.A. Herbert and G.A. Codd (eds.), Microbes in extreme environments. Academic Press, Orlando, pp. 25-54.

Grant, W.D., Mwatha, W.E. and Jones, B.E. (1990) FEMS Microbiol. Rev. 75, 255-270.

Guffanti, A.A., Susman, P., Blanco, R. and Krulwich, T.A. (1978) J. Biol. Chem. 253, 708-715.

Hardie, L.A. and Eugster, H.P. (1970) Mineralogical Society of America, Special Publication 3, 273-290.

Harold, F.M., Pavlasova, F. and Baarda, J.R. (1970) Biochim. Biophys. Acta 196, 235-244.

Hecky, R.E. and Kilham, P. (1973) Limnol. Oceanogr. 18, 53-71.

Horikoshi, K. (1986) In: R.A. Herbert and G.A. Codd (eds.), Microbes in extreme environments. Academic Press, Orlando, pp. 297-315.

Horikoshi, K. and Akiba, T. (1982) Alkalophilic microorganisms. Springer, Berlin.

Hsung, J.C. and Haug, H. (1975) Biochim. Biophys. Acta 389, 477-482.

Imhoff, J.F., Hashwa, F. and Trüper, H.G. (1978) Archiv für Hydrobiologie 84, 381-388.

Imhoff, J.F., Sahl, H.G., Soliman, G.S.H. and Trüper, H.G. (1979) Geomicrobiol. J. 1, 219-234.

Ivey, D.M., Guffanti, A.A., Bossewitch, J.S., Padan, E. and Krulwich, T.A. (1991) J. Biol. Chem. 266, 23483-23489.

Ivey, D.M., Guffanti, A.A., Shen, Z., Kudyan, N. and Krulwich, T.A. (1992) J. Bacteriol. 174, 4878-4884.

Jannasch, H.W. (1957) Archiv für Hydrobiologie 53, 425-433.

Kallas, T. and Castenholz, R.W. (1982a) J. Bacteriol. 149, 229-236.

Kallas, T. and Castenholz, R. W. (1982b) J. Bacteriol. 149, 237-246.

Kallas, T. and Dahlquist, F.W. (1981) Biochemistry 20, 5900-5907.

Kaplan, A., Scherer, S. and Lerner, M. (1989) Plant Physiol. 89, 1220-1225.

Kaplan, A., Schwarz, R., Lieman-Hurwitz, J., Ronen-Tarazi, M. and Reinhold, L. (1994) In: D.A. Bryant (ed.), The molecular biology of cyanobacteria. Kluwer Academic Publishers, Dordrecht, pp. 469-485.

Kaplan, A., Volokita, M., Zenvirth, D. and Reinhold, L. (1984) FEBS Lett. 176, 166-168.

Karazanov, V.V. and Ivanovsky, R.N. (1980) Biochim. Biophys. Acta 598, 91-99.

Kazmierczak, J., Coleman, M.L., Gruszczynski, M. and Kempe, S. (1996) Acta Palaeontologica Polonica 41, 319-338.

Kerby, N.W. and Rowell, P. (1992) In: N.H. Mann and N.G. Carr (eds.), Photosynthetic prokaryotes. Biotechnology Handbooks, vol 6. Plenum Press, New York, pp. 233-265.

Kitada, M. and Horikoshi, K. (1977) J. Bacteriol. 131, 784-788.

Kitada, M. and Horikoshi, K. (1992) J. Bacteriol. 174, 5936-5940.

Kitada, M., Guffanti A.A. and Krulwich, T.A. (1982) J. Bacteriol. 152, 1096-1104.

Koyama, M., Kiyomiya, A. and Nosoh, Y. (1976) FEBS Lett. 72, 77-78.

Krulwich, T.A. (1986) J. Membr. Biol. 89, 113-125.

Krulwich, T.A. (1995) Mol. Microbiol. 15, 403-410.

Krulwich, T.A. and Guffanti, A.A. (1983) Adv. Microb. Physiol. 24, 173-213.

Krulwich, T.A. and Guffanti, A.A. (1989a) Ann. Rev. Microbiol. 43, 435-463.

Krulwich, T.A. and Guffanti, A.A. (1989b) J. Bioenerg. Biomembr. 21, 663-677.

Krulwich, T.A. and Guffanti, A.A. (1992) J. Bioenerg. Biomembr. 24, 587-599.

Krulwich, T.A., Cheng, J. and Guffanti, A.A. (1994) J. Exp. Biol. 196, 457-470.

Krulwich, T.A., Federbush, J.G. and Guffanti, A.A. (1984) J. Biol. Chem. 260, 4055-4058.

Krulwich, T.A., Ito, M., Gilmour, R., Sturr, M.G., Guffanti, A.A. and Hicks, D.B. (1996) Biochim. Biophys. Acta 1275, 21-26.

Krulwich, T.A., Mandel, K.G., Borstein, R.F. and Guffanti, A.A. (1979) Biochem. Biophys. Res. Commun. 91, 58-62.

Kuroda, T., Shimamoto, T., Inaba, K., Kayahara, T., Tsuda, M. and Tsuchiya, T. (1994) J. Biochem. 115, 1162-1165.

Lara, C., Rodrigues, R. and Guerrero, M.G. (1993) J. Phycol. 29, 389-395.

Maeso, E.S., Pinas, F.F., Gonzalez, M.G. and Valiente, E.F. (1987) Plant Physiol. 85, 585-587.

McLachlan, J. and Gorham, P.R. (1962) Can. J. Microbiol. 8, 1-11.

McLaggan, D., Selwyn, M.J. and Dawson, A.P. (1984) FEBS Lett. 165, 254-258.

Mehler, A.H. (1951) Arch. Biochem. Biophys. 33, 65-77.

Melack, J.M. (1978) Arch. Hydrobiol. 84, 430-453.

Melack, J.M. and Kilham, P. (1974) Limnol. Oceanogr. 19, 743-755.

Miller, A.G., Espie, G.S. and Canvin, D.T. (1988) Plant Physiol. 86, 677-683.

Miller, A.G., Turpin, D.H. and Canvin, D.T. (1984) J. Bacteriol. 159, 100-106.

Mizushima, S., Machida, Y. and Kitahara, K. (1964) J. Bacteriol. 86, 1295-1300.

Molitor, V. and Peschek, G.A. (1986) FEBS Lett. 195, 145-150.

Molitor, V., Erber, W.W.A. and Peschek, G.A. (1986) FEBS Lett. 204, 251-256.

Nicholls, P., Obinger, C., Niederhauser, H. and Peschek, G.A. (1992) Biochim. Biophys. Acta 1098, 184-190.

Nitschmann, W.H. and Packer, L. (1992) Arch. Biochim. Biophys. 294, 347-352.

Nitschmann, W.H. and Peschek, G.A. (1982) FEBS Lett. 139, 77-80.

Nitschmann, W.H., Schmetterer, G., Muchl, R. and Peschek, G.A. (1982) Biochim. Biophys. Acta 682, 293-296.

Norling, B., Zarka, A. and Boussiba, S. (1997) Physiol. Plant. 99, 495-504.

Ogawa, S., Shulman, R. G., Glynn, P., Yamane, T. and Navon, G. (1978) Biochim. Biophys. Acta 502, 45-50.

Paschinger, H. (1977) Arch. Microbiol. 113, 285-291.

Padan, E., Zilberstein, D. and Schuldiner, S. (1981) Biochim. Biophys. Acta 650, 151-166.

Padan, E. and Schuldiner, S. (1994) Biochim. Biophys. Acta 1185, 129-151.

Padan, E. and Vitterbo, A. (1988) Meth. Enzymol. 167, 561-572.

Painter, T. (1993) Carbohydr. Polymers 20, 77-86.

Peschek, G.A., Crenzy, T., Schmetterer, G. and Nitschmann, W.H. (1985) Plant Physiol. 79, 278-282.

Peschek, G.A., Nitschmann, W.H. and Crenzy, T. (1988) Meth. Enzymol. 167, 361-380.

Ritchie, R.J. (1991) J. Plant Physiol. 137, 409-418.

Ritchie, R.J. (1992) J. Plant Physiol. 139, 320-330.

Rottenburg, H., Grunwald, T. and Avron, M. (1972) Eur. J. Biochem. 25, 54-63.

Scherer, S., Stürzl, E. and Böger, P. (1984) J. Bacteriol. 158, 609-614.

Schlesinger, P., Belkin, S. and Boussiba, S. (1996) J. Phycol. 32, 608-613.

Sharp, R.J. and Munster, M.J. (1986) In: R.A. Herbert and G.A. Codd (eds.), Microbes in extreme environments. Academic Press, Orlando, pp. 297-315.

Shimamoto, T., Inaba, K., Thelen, P., Ishikawa, T., Goldberg, E.B., Tsuda, M. and Tsuchiya, T. (1994) J. Biochem. 116, 285-290.

Skulachev, V.P. (1994) J. Bioenerg. Biomem. 26, 589-598.

Stanier, R.Y. and Cohen-Bazire, G. (1977) Ann. Rev. Microbiol. 31, 225-274.

Sugiyama, S., Matsukura, H. and Imae, Y. (1985) FEBS Lett. 182, 265-268.

Talling, J.F., Wood, R.B., Prosser, M.V. and Baxter, R.M. (1973) Freshwater Biol. 3, 53-76.

Tandeau de Marsac, N. (1991) In: L. Bogorad and I.K. Vasil (eds.), Cell culture and somatic cell genetics of plants, Vol 7b. Academic Press, New York, pp. 417-446.

Tandeau de Marsac, N. and Houmard, J. (1993) FEMS Microbiol. Rev. 104, 119-189.

Thomas, J. and Apte, K. (1984) J. Biosci. 6, 771-794.

Thomas, J.A., Cole, R.E. and Langworthy, T.A. (1976) Fed. Proc. Fed. Amer. Soc. Exper. Biol. 35, 1455.

Tindall, B.J. (1980) Ph.D Thesis, Leicester University.

Utkilen, H. and Gjolme, N. (1992) Appl. Environ. Microbiol. 58, 1321-1325.

Vonshak, A. (1997) *Spirulina platensis* (Arthrospira): physiology, cell-biology and biotechnology. Taylor & Francis, London.

Waser, M., Hess-Bienz, D., Davies, K. and Solioz, M. (1992) J. Biol. Chem. 267, 5396-5400.

Willey, J.M., Waterbury, J.B. and Greenberg, E.P. (1987) J. Bacteriol. 169, 3429-3434.

Wolk, P.C. (1973) Bacteriol. Rev. 37, 32-101.

Xu, C.H., Nejidat, A., Belkin, S. and Boussiba, S. (1994) Plant Cell Physiol. 35, 737-741.

Zhao, J. and Brand, J.J. (1988) Arch. Biochem. Biophys. 264, 657-664.

Zhao, J. and Brand, J.J. (1989) Plant Physiol. 91, 91-100.

VI

HALOPHILES

Biodata of **Aharon Oren** contributor of *"Life at High Salt Concentration: Possibilities and Limitatios"* and the co-author (with J. Seckbach) of *"A Vista into the Diverse Microbial Worlds: An Introduction to Microbes at the Edge of Life."*

Dr. **Aharon Oren** is a Professor of Microbial and Molecular Ecology at the Institute of Life Sciences, the Hebrew University of Jerusalem. He earned his Ph.D. at the Hebrew University in 1978 and spent a post-doctoral period with R.S. Wolfe and C.R. Woese at the University of Illinois at Urbana-Champaign. His research interests are in ecology, physiology and biochemistry of halophilic microorganisms and the microbiology of hypersaline environments. Recently he edded a book on "Microbiology and Biogeochemistry of Hypersaline Environments", published in 1999 by CRC Press, Boca Raton.

E-Mail: orena@shum.cc.huji.ac.il

J. Seckbach (ed.), Journey to Diverse Microbial Worlds, 227-238.

LIFE AT HIGH SALT CONCENTRATIONS: POSSIBILITIES AND LIMITATIONS

A. OREN
Division of Microbial and Molecular Ecology, Institute of Life Sciences, and the Moshe Shilo Minerva Center for Marine Biogeochemistry, The Hebrew University of Jerusalem, Jerusalem 91904, Israel

1. Metabolic diversity in hypersaline environments – field observations and pure culture studies

Microbial life exists at salt concentrations up to NaCl saturation (approximately 5.2 M) The bright red color of saltern crystallizer ponds and many hypersaline lakes world-wide shows that such environments may support development of dense microbial communities. However, the number of different microorganisms able to grow at these extremely high salinities is small.

With respect to adaptation to salt, microorganisms can be classified in groups, from non-halophiles through marine organisms and moderate halophiles (generally defined as growing best in the range of 0.5-2.5 M NaCl) to extreme halophiles such as *Halobacterium* and other members of the *Halobacteriaceae*. Halophilic behavior is found all over the phylogenetic tree, both in the archaeal, the bacterial and the eukaryal domain. The main primary producer in hypersaline aquatic environments all over the world is the eukaryal green alga *Dunaliella*. Within the archaeal domain halophiles are not only found within the family *Halobacteriaceae* – which contains the halophilic microorganisms par excellence, but also in the methanogenic family *Methanosarcinaceae*. Within the bacterial domain moderate halophiles are widespread. They can be found in the Proteobacteria (notably in the γ-subdivision), the Gram-positive bacteria, the cyanobacterial branch, the Spirochetes, and in the *Flavobacterium* branch.

Much more interesting than a survey of phylogenetic diversity within the halophile world is the examination of functional metabolic diversity in microorganisms able to live in hypersaline environments. It appears that many metabolic types commonly found in freshwater and marine environments are missing in environments with salt concentrations exceeding 100-150 g/l. Figure 1 presents a summary of the salt concentration limits of some important types of dissimilatory metabolism encountered in low-salt environments that are essential for the proper functioning of freshwater and marine ecosystems. While photosynthesis (both oxygenic with water as electron donor and anoxygenic on reduced sulfur species), aerobic respiration and anaerobic respiration with nitrate as electron acceptor can proceed up to the highest salt concentrations,

230

fermentation appears to have its upper salinity boundary at about 250 g/l salt. Methanogenesis using methanol or methylated amines as energy sources can function up to quite high salt concentrations (260 g/l being the maximum reported), but methanogenic bacteria have never yet been grown on hydrogen + CO_2 above 120 g/l. The upper salt limit for methanogenesis from acetate is even lower. Similarly, dissimilatory sulfate reducing bacteria that perform complete oxidation of acetate seem to be missing above 130 g/l salt. Another function that appears to be lacking at high salt concentrations is autotrophic nitrification.

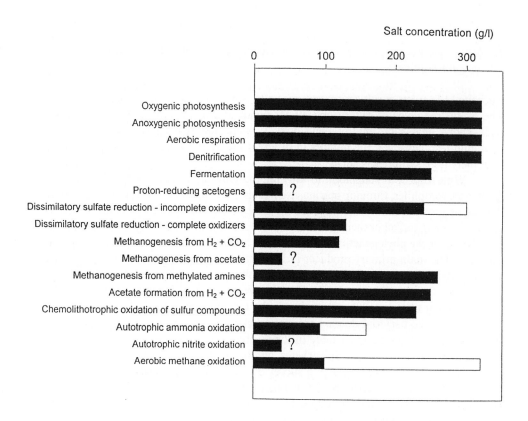

Figure 1. Approximate upper salt concentration limits for the occurrence of selected microbial processes. Values presented are based in part on laboratory studies of pure cultures (black bars) and on activity measurements of natural communities in hypersaline environments (white bars). Data were derived in part from Brandt and Ingvorsen (1997), Ollivier et al. (1998), Oremland and King (1989), Oren (1988, 1999), Rubentschik (1929), Sokolov and Trotsenko (1995), Zhilina and Zavarzin (1990), and many other sources. Reproduced from Oren (1999), with permission.

This chapter will investigate the possibilities and limitations of life in hypersaline environments, in an attempt to understand the reasons behind the reduced metabolic diversity encountered at high salt concentrations.

2. The hypothesis: energetic constraints determine what types of metabolism can function at elevated salt concentrations

Adaptation to life at high NaCl concentrations is energetically costly. All halophilic and halotolerant microorganisms have to maintain their cytoplasm at least isoosmotic with the external medium (Brown 1990). The intracellular milieu always differs greatly from that of the medium surrounding the cell. Some microorganisms synthesize organic osmotic solutes to provide the necessary osmotic balance; others use inorganic ions for the purpose. In the latter case the intracellular ionic composition differs from that of the outer medium, with K^+ rather than Na^+ being the dominant intracellular cation. In all cases the cell has to invest large amounts of energy to keep the intracellular Na^+ concentration low and to accumulate K^+ or organic osmotic solutes. The higher the salinity of the medium, the more energy is needed for haloadaptation.

Calculations of the energetic cost of adaptation to life at high salt shows that cells often have to spend a very large percentage of the energy gained in their dissimilatory metabolism to activate the ion pumps and in many cases to synthesize organic solutes. This poses severe problems for those groups of microorganisms that make a living by using dissimilatory processes that yield little energy only.

As documented below, an understanding of the energetic cost of adaptation to high salt, using the different strategies that Nature has devised ("high salt-in" or "high organic solute, low-salt in"), provides a plausible explanation for the apparent lack of many metabolic types of microorganisms in hypersaline environments. It is therefore proposed that energetic constraints determine the limits of life at high salt.

The ideas described here in brief have been documented in full in a recent review (Oren 1999). Readers are referred to that paper for a more detailed account.

3. The energetic cost of life at high salt concentrations – two strategies

3.1. THE "HIGH SALT-IN" OPTION

Two groups of halophilic prokaryotes use inorganic salts, mainly KCl, to provide osmotic balance with the high NaCl concentrations in their external environment: the archaeal order of the Halobacteriales, family *Halobacteriaceae*, and the order Haloanaerobiales (low G+C branch of the Gram-positive Bacteria) (families *Haloanaerobiaceae* and *Halobacteroidaceae*). K^+ and Cl^- concentrations of up to 4-5 M were found inside the cells (Kushner 1985; Oren 1986, 1992; Oren et al. 1997; Pérez-Fillol and Rodriguez-Valera 1986; Rengpipat et al. 1988). The presence of high concentrations of KCl requires adaptation of the whole intracellular enzymatic machinery. An in-depth survey of the special properties of the salt-adapted enzymes is beyond the framework of this chapter; details can be found in review papers by Lanyi (1974) and Dennis and Shimmin (1997).

The huge K^+ gradient over the cytoplasmic membrane (often up to three orders of magnitude) can only be generated and maintained at the expense of energy. The mechanism used to establish the Na^+ and K^+ gradients have been investigated in-depth in the Archaeon *Halobacterium salinarum*. As in most bioenergetic processes involving membranes, the primary energy source that serves as the driving force for the formation of ion gradients is the proton electrochemical gradient across the membrane. This gradient can be formed as a result of respiratory electron transport, by the hydrolysis of ATP, and/or by action of the light-driven proton pump bacteriorhodopsin. The H^+ gradient is then used to drive the outward transport of Na^+ ions, mediated by electrogenic Na^+/H^+ antiporters (Lanyi and McDonald 1976). A stoichiometry of 1 $Na^+/2$ H^+ was established for the main Na^+/H^+ antiporter of *H. salinarum* (Lanyi and Silverman 1979). K^+ probably enters the cell passively via a uniport system in response to the membrane potential (Lanyi and Hilliker 1976; Wagner et al. 1978). In addition, an active K^+ transport system that requires ATP for activation was identified in *Haloferax volcanii* (Meury and Kohiyama 1989). Chloride is accumulated via symport with Na^+ (Duschl and Wagner 1986) and/or using the light-driven chloride pump halorhodopsin (Schobert and Lanyi 1982).

Estimates of the energy requirement of the "high-salt-in" strategy showed that 1 ATP equivalents should suffice for the accumulation of 1.5-2 molecules of KCl [for details on the calculation see Oren (1999)]. Discussions on the energy burden of halophiles should also take into account the cost of maintenance of the ion gradients, once established. Passive permeability of the membranes to Na^+ is probably low (van de Vosseberg et al. 1995), and therefore it may be assumed that the amount of energy needed to meet the specific demands of members of the Halobacteriales and the Haloanaerobiales to life at high salt concentrations is moderate in comparison to the "high organic solute, low-salt in" option, discussed below.

3.2. THE "HIGH ORGANIC SOLUTE, LOW-SALT IN" OPTION

Microorganisms that use organic solutes to provide osmotic balance do not require specialized proteins in their cytoplasm. The "compatible" solutes do not harm protein structure and enzymatic activities of conventional enzymes. However, the energetic price the cells have to pay in this mode of osmotic adaptation is huge, as the biosynthesis of organic osmotic solutes is energetically much more expensive than building up high intracellular KCl concentrations.

A great variety of solutes have been shown to occur in different microorganisms. These include glycerol (found only in Eukarya, such as the halophilic alga *Dunaliella* and certain fungi) (Ben-Amotz and Avron 1981; Brown 1990), glycine betaine (in halophilic cyanobacteria, in anoxygenic photosynthetic Bacteria, and also in halophilic methanogenic Archaea), ectoine (1,4,5,6-tetrahydro-2-methyl-4-pyrimidine carboxylic acid) in many heterotrophic and phototrophic Proteobacteria, glucosylglycerol in cyanobacteria with a moderate salt tolerance, and the disaccharides sucrose and trehalose, found e.g. in many cyanobacteria (Mackay et al. 1984). The above list is far from exhaustive, for more complete lists see e.g. the reviews by Galinski (1993, 1995).

The amount of ATP molecules needed for the biosynthesis of the above-mentioned compounds varies with the size and complexity of the solutes. Simple calculations based on the known biosynthetic pathways show that for an autotroph the production of one molecule of solute costs between 30 ATP equivalents for glycerol to 109 ATP equivalents for the disaccharides sucrose and trehalose. For a heterotrophic microorganism that has to produce the solutes from glucose between 23 and 79 potential ATP molecules are "wasted" that could otherwise have been used for energy generation and growth. Glycerol, being the cheapest compatible solute to produce from a bioenergetic point of view, is not widely used, and its distribution is limited to the Eukarya. Most biological membranes are highly permeable to glycerol. Use of glycerol as compatible solute therefore requires special adaptations of the membrane structure, which are still incompletely understood, but are only found in a limited number of Eukarya (see Brown et al. 1982; Degani and Avron 1982; Gimmler and Hartung 1988). The cheapest organic osmotic solutes synthesized by prokaryotes are ectoine and glycine betaine, both at a cost of 54 and 40 ATP equivalents for autotrophic and heterotrophic microorganisms, respectively (see Oren 1999). Calculations show that the biosynthesis of osmotic solutes can increase the amount of energy required per unit of structural cell biomass manyfold (Oren 1999).

Cells that produce organic osmotic solutes not only have to spend much energy in producing them, but they also have to cope with loss of these solutes through a not completely impermeable membrane. Small amounts of glycerol and glucosylglycerol were shown to be lost from *Dunaliella* and halophilic cyanobacteria to the medium (Hagemann et al. 1997; Wegmann et al. 1980). Most halophiles therefore possess transport proteins in the membrane that enable the uptake of at least part of the lost solute from the medium. Glucosylglycerol producing cyanobacteria can accumulate glucosylglycerol from the medium (Hagemann et al. 1997). Many Bacteria and also methanogenic Archaea can take up glycine betaine (Moore et al. 1987; Proctor et al. 1997; Ventosa et al. 1998). Though the uptake of solutes from the medium against a concentration gradient also costs energy, the ability to accumulate solutes from outside means a significant energy saving as compared to biosynthesis. This also explains why many halophilic and halotolerant Bacteria, most of which are unable to produce glycine betaine, accumulate the compound from the medium when available, rather than spending large amounts of energy for the biosynthesis of e.g. ectoine (Imhoff and Rodriguez-Valera 1984; Ventosa et al. 1998).

The presence of high concentrations of organic solutes does not relieve the need for high activities of Na^+/H^+ antiporters to prevent the Na^+ ions leaking in, either due to passive permeability of the membrane or as a result of cotransport with e.g. amino acids, from damaging enzymatic activity. This again adds to the amount of energy required for osmotic adaptation in microorganisms that use organic solutes to provide osmotic balance while maintaining a low-salt cytoplasm.

4. Do energetic constraints explain the absence of certain metabolic functions at high salt?

We can now examine the above-presented idea that certain metabolic types of bacteria are missing at high salt concentrations due to energetic constraints. In order to be able to live at high salt concentrations a microorganism needs large amounts of energy to be able to cope with the high osmotic pressure outside. The "high-salt-in" strategy it preferable for strongly energy-limited organisms, as it is energetically much less costly than the production of organic osmotic solutes, as documented above. When insufficient energy is generated to supply the demands of cell growth, maintenance and osmotic adaptation, such a microorganism has no possibility to thrive in hypersaline environments. As the following examples show, there appears to be an excellent correlation between the amount of energy generated in the course of the different dissimilatory processes, the strategy ("high-salt-in" or "high organic solute, low-salt in") used to provide osmotic balance, and the salinity range at which the process occurs (see Fig. 1 for the upper salinity limits at which the different processes have been found, and Table 1 for thermodynamic data on some of the reactions discussed):

- Aerobic chemoorganotrophic bacteria have no problem coping with high salt, as large amounts of energy are available from respiration during complete oxidation to CO_2 (e.g. Glucose + 6 O_2 → 6 CO_2 + 6 H_2O; $\Delta G^{o'}$ = -2872 kJ = -239.3 kJ per 2 electrons). Thus, both aerobic halophilic Archaea of the family *Halobacteriaceae* that use the "high-salt-in" option and Bacteria such as members of the *Halomonadaceae* that synthesize ectoine and other organic solutes are found up to the highest salt concentrations.

- Dissimilatory nitrate reduction also can occur up to NaCl saturation, as the amount of energy obtained is not much lower than when molecular oxygen serves as electron acceptor in respiration.

- Phototrophic microorganisms are generally not energy-limited, and thus both oxygenic (*Dunaliella*, cyanobacteria) and anoxygenic types (*Halorhodospira*, *Thiohalocapsa*, *Halochromatium*), all producers of organic osmotic solutes, can be found up to very high salt concentrations.

- Fermentation of sugars, amino acids, etc. to products such as acetate, ethanol, hydrogen and CO_2 yields small amounts of energy only (typically around 3 ATP per molecule of glucose fermented). The finding that fermentative Bacteria can still be found up to 250 g/l salt can be explained by the fact that the members of the *Haloanaerobiales*, in contrast to all other halophilic representatives of the bacterial domain, use the "high-salt-in" option, which is energetically much more favorable than the production of organic solutes.

Table 1. Standard free energy change of different reactions discussed in the text.

Methanogenic reactions:

$H_2 + CO_2$:
    $4 H_2 + H^+ + HCO_3^- \rightarrow CH_4 + 3 H_2O$ -135.9 kJ
                            per mol of substrate –34.0 kJ

Acetate:
    $Acetate^- + H_2O \rightarrow CH_4 + HCO_3^-$ -31.1 kJ
                            per mol of substrate -31.1 kJ

Methanol:
    $4 Methanol \rightarrow 3 CH_4 + HCO_3^- + H_2O + H^+$ -314.7 kJ
                            per mol of substrate -78.7 kJ

Monomethylamine:
    $4 Methylamine + 3 H^+ + 3 H_2O \rightarrow 3 CH_4 + HCO_3^- + 4 NH_4^+$ -368.3 kJ
                            per mol of substrate -92.1 kJ

Dimethylamine:
    $2 Dimethylamine + H^+ + 3 H_2O \rightarrow 3 CH_4 + HCO_3^- + 2 NH_4^+$ -286.5 kJ
                            per mol of substrate -143.3 kJ

Trimethylamine:
    $4 Trimethylamine + 9 H_2O + H^+ \rightarrow 9 CH_4 + 3 HCO_3^- + 4 NH_4^+$ -764.5 kJ
                            per mol of substrate -191.1 kJ

Homoacetogenic bacteria:

$H_2 + CO_2$:
    $4 H_2 + H^+ + 2 HCO_3^- \rightarrow Acetate^- + 4 H_2O$ -104.6 kJ
                            per mol of substrate –26.1 kJ

Sulfate reducing bacteria:

$H_2 + SO_4^{2-}$:
    $4 H_2 + SO_4^{2-} + H^+ \rightarrow HS^- + 4 H_2O$ -152.3 kJ

Lactate + SO_4^{2-}:
    $2 Lactate^- + SO_4^{2-} \rightarrow 2 Acetate^- + HCO_3^- + HS^- + H^+$ -160.1 kJ

Acetate + SO_4^{2-}:
    $Acetate^- + SO_4^{2-} \rightarrow 2 HCO_3^- + HS^-$ -47.7 kJ

Aerobic chemoautotrophs:

$NH_4^+ + 1.5 O_2 \rightarrow NO_2^- + 2 H^+ + H_2O$ -274.6 kJ
                          per 2 electrons: -91.5 kJ

$NO_2^- + 0.5 O_2 \rightarrow NO_3^-$ -74.1 kJ
                          per 2 electrons: -74.1 kJ

$H_2S + 2 O_2 \rightarrow SO_4^{2-} + 2 H^+$ -796.3 kJ
                          per 2 electrons: 199.1 kJ

Aerobic methanotrophs:

$CH_4 + 2 O_2 \rightarrow CO_2 + 2 H_2O$ -818.0 kJ
                          per 2 electrons: -204.5 kJ

- The halophilic methanogenic bacteria produce osmotic solutes such as glycine betaine, β-glutamine, Nε-acetyl-β-lysine and others (Lai et al. 1991). In this respect they differ from the aerobic halophilic Archaea of the family *Halobacteriaceae*, which accumulate KCl. In view of the high energy cost of solute synthesis it is understandable that methanogenesis from $H_2 + CO_2$ or from acetate cannot occur at high salt concentrations, as the amount of energy generated during methanogenesis is very limited (see Table 1). However, methane is generated in hypersaline environments from substrates such as trimethylamine and other methylated amines (Oremland and King 1989). This is in agreement with the higher energy yield of the methanogenic reactions involved (Table 1, for a more in-depth discussion see Oren 1999).

- If methanogenesis on $H_2 + CO_2$ at high salt concentrations is not feasible for thermodynamic reasons, it should be expected that the formation of acetate from the same substrates by homoacetogenic bacteria would be still less favorable as the reaction yields even less energy (-26.1 vs. −34.0 kJ/mol H_2). However, it must be remembered that the halophilic homoacetogens known (*Acetohalobium arabaticum, Natroniella acetigena*) belong to the bacterial order Haloanaerobiales (Rainey et al. 1995; Zavarzin 1994; Zhilina and Zavarzin 1987, 1990; Zhilina et al. 1996). As discussed above, the Haloanaerobiales use the energetically cheap "high-salt-in" strategy of osmotic adaptation, and this probably allows homoacetogens to grow up to 240 g/l salt.

- The dissimilatory sulfate reducers accumulate organic osmotic solutes (trehalose, glycine betaine) (Welsh et al. 1996). In view of this energetically expensive mode of life it is understandable that dissimilatory sulfate reduction with complete oxidation of acetate does not appear to occur above 120 g/l salt (Brandt and Ingvorsen 1997): acetate oxidation with reduction of sulfate yields little energy only (Table 1). Much more energy can be generated by dissimilatory sulfate reduction with hydrogen or lactate as electron donors, and these processes were indeed shown to occur up to much higher salt concentrations (Caumette et al. 1990; Krekeler et al. 1997; Ollivier et al. 1994).

- Autotrophic nitrification, both the oxidation of ammonia to nitrite and nitrite oxidation to nitrate, yield little energy only, as the substrates are relatively oxidized. Moreover, a large fraction of the little energy obtained has to be used to drive uphill electron transfer for the production of NADPH, the electron donor needed for autotrophic CO_2 reduction. Being members of the Proteobacteria, nitrifying bacteria are expected to use the "high organic solute, low-salt in" mode of osmoregulation. Therefore it is understandable that nitrification was never shown to occur above 150 g/l salt (Koops et al. 1990; Post and Stube 1988; Rubentschik 1929).

- Oxidation of reduced sulfur compounds such as sulfide and elemental sulfur yields much more energy than oxidation of ammonia or nitrate. Therefore the existence of halophilic sulfur-oxidizing autotrophs is feasible (Wood and Kelly 1991).

- Aerobic oxidation of methane yields large amounts of energy (Table 1). Therefore the apparent lack of methane oxidizing activity in most hypersaline environments is puzzling (Giani et al. 1984; Slobodkin and Zavarzin 1992). However, recent studies of hypersaline environments in Ukraina have demonstrated the existence of highly halotolerant methanotrophic communities, able to function at salt concentrations exceeding 300 g/l (Khmelenina et al. 1996, 1997; Sokolov and Trotsenko 1995). Therefore the apparent lack of aerobic methane oxidizing activity in many hypersaline environments much have other reasons, not related with energetic constraints.

5. Conclusions

Field observations and pure culture studies show a decreasing diversity in microbial dissimilatory processes as the salinity increases up to saturation of NaCl. Energetic considerations related to the cost associated with adaptation to high salt provide a plausible explanation why certain metabolic types cease to occur even at relatively low salt concentrations, while others are found up to the highest salinities.

One might ask why Nature has not used the energetically more favorable "high-salt-in" strategy more extensively, beyond the two groups known thus far to accumulate high KCl concentrations: the Halobacteriales (living by aerobic respiration, in some cases with the option of denitrification) and the Haloanaerobiales (anaerobic fermenters and homoacetogens). It is possible that the far-reaching adaptation of the intracellular enzymatic machinery to the presence of high salt (Lanyi 1974; Dennis and Shimmin 1997) is not easy to achieve, and has thus far evolved only in the two specialized groups mentioned above.

A continued search for halophilic microorganisms performing the "missing" types of dissimilatory metabolism is to be recommended, as it may well yield novel types of Bacteria and Archaea with unusual modes of haloadaptation, thus increasing both the phylogenetic and the metabolic diversity known from hypersaline environments.

6. References

Ben-Amotz, A. and Avron, M. (1981) Trends Biochem. Sci. 6, 297-299.
Brandt, K.K. and Ingvorsen, K. (1997) Syst. Appl. Microbiol. 20, 366-373.
Brown, A.D. (1990) Microbial water stress physiology. Principles and perspectives. John Wiley & Sons, Chichester, UK.
Brown, F.F., Sussman, I., Avron, M. and Degani, H. (1982) Biochim. Biophys. Acta 690, 165-173.
Caumette, P., Cohen, Y., and Matheron, R. (1990) Syst. Appl. Microbiol. 14, 33-38.
Conrad, R., Frenzel, P. and Cohen, Y. (1995) FEMS Microbiol. Ecol. 16, 297-305.
Degani, H. and Avron, M. (1982) Biochim. Biophys. Acta 690, 174-177.
Dennis, P.P. and Shimmin, L.C. (1997) Microbiol. Mol. Biol. Rev. 61, 90-104.
Duschl, A. and Wagner, G. (1986) J. Bacteriol. 168, 548-552.
Galinski, E.A. (1993) Experientia 49, 487-496.
Galinski, E.A. (1995) Adv. Microb. Physiol. 37, 273-328.
Giani, D., Giani, L., Cohen, Y. and Krumbein, W.E. (1984) FEMS Microbiol. Lett. 25, 219-224.
Gimmler, H. and Hartung, W. (1988) J. Plant Physiol. 133, 165-172.

238

Hagemann, M., Richter, S. and Mikkat, S. (1997) J. Bacteriol. 179, 714-720.

Imhoff, J.F. and Rodriguez-Valera, F. (1984) J. Bacteriol. 160, 478-479.

Khmelenina, V.N., Starostina, N.G., Tsvetkova, M.G., Sokolov, A.P, Suzina, N.E. and Trotsenko Y.A. (1996) Mikrobiologiya 65, 696-703 (in Russian).

Khmelenina, V.N., Kalyuzhneya, M.G., Starostina, N.G., Suzina, N.E. and Trotsenko, Y.A. (1997) Curr. Microbiol. 35, 257-261.

Koops, H.-P., Böttcher, B., Möller, U., Pommerening-Röser, A. and Stehr, G. (1990) Arch. Microbiol. 154, 244-248.

Krekeler, D., Sigalevich, P., Teske, A., Cypionka, H. and Cohen, Y. (1997) Arch. Microbiol. 167, 369-375.

Kushner, D.J. (1985) In: C.R. Woese and R.S. Wolfe (eds.) The bacteria. A treatise on structure and function. Vol. VIII. Archaebacteria. Academic Press, Orlando, FL, pp. 171-214.

Lai, M.-C., Sowers K.R., Robertson, D.E., Roberts, M.F. and Gunsalus, R.P. (1991) J. Bacteriol. 173, 5352-5358.

Lanyi, J.K. (1974) Bacteriol. Rev. 38, 272-290.

Lanyi, J.K. and Hilliker, K. (1976) Biochim. Biophys. Acta 448, 181-184.

Lanyi, J.K., and R.E. McDonald. (1976) Biochemistry 15, 4608-4614.

Lanyi, J.K. and Silverman, M.P. (1979) J. Biol. Chem. 254, 4750-4755.

Mackay, M.A., Norton, R.S. and Borowitzka, L.J. (1984) J. Gen. Microbiol. 130, 2177-2191.

Meury, J. and Kohiyama, M. (1989) Arch. Microbiol. 151, 530-536.

Moore, D.J., Reed, R.H. and Stewart, W.D.P. (1987) Arch. Microbiol. 147, 399-405.

Ollivier, B., Hatchikian, C.E., Prensier, G., Guezennec, J. and Garcia, J.-L. (1991) Int. J. Syst. Bacteriol. 41, 74-81.

Ollivier, B., Caumette, P., Garcia, J.-L. and Mah, R.A. (1994) Microbiol. Rev. 58, 27-38.

Ollivier, B., Fardeau, M.-L, Cayol, J.-L., Magot, M., Patel, B.K.C., Prensier, G. and Garcia, J.-L. (1998) Int. J. Syst. Bacteriol. 48, 821-828.

Oremland, R.S. and King, G.M. (1989) In: Y. Cohen and E. Rosenberg (eds.) Microbial mats. Physiological ecology of benthic microbial communities. American Society for Microbiology, Washington, D.C. pp. 180-190.

Oren, A. (1986) Can. J. Microbiol. 32, 4-9.

Oren, A. (1988) Antonie van Leeuwenhoek 54, 267-277.

Oren, A. (1992) In: A. Balows, H.G. Trüper, M. Dworkin, W. Harder and K.-H. Schleifer (eds.) The prokaryotes. A handbook on the biology of bacteria: ecophysiology, isolation, identification, applications. 2nd ed. Springer-Verlag, New York, N.Y. pp. 1893-1900.

Oren, A. (1999) Microbiol. Mol. Biol. Rev. 63, 334-348.

Oren, A., Heldal, M. and Norland, S. (1997) Can. J. Microbiol. 43, 588-592.

Pérez-Fillol, M. and Rodriguez-Valera, F. (1986) Microbiología 2, 73-80.

Post, F.J. and Stube, J.C. (1988) Hydrobiologia 158, 89-100.

Proctor, L.M., Lai, R. and Gunsalus, R.P. (1997) Appl. Environ. Microbiol. 63, 2252-2257.

Rainey, F.A., Zhilina, T.N., Boulygina, E.S., Stackebrandt, E., Tourova, T.P. and Zavarzin, G.A. (1995) Anaerobe 1, 185-199.

Rengpipat, S., Lowe, S.E. and Zeikus, J.G. (1988) J. Bacteriol. 170, 3065-3071.

Rubentschik, L. (1929) Zentralbl. Bakteriol. II Abt. 77, 1-18.

Schobert, B. and Lanyi, J.K. (1982) J. Biol. Chem. 257, 10306-10313.

Slobodkin, A.I. and Zavarzin, G.A. (1992) Mikrobiologiya 61, 294-298 (in Russian).

Sokolov, A.P. and Trotsenko, Y.A. (1995) FEMS Microbiol. Ecol. 18, 299-304.

van de Vosseberg, J.L.C.M., Ubbink-Kok, T., Elferink, M.H.L., Driessen, A.J.M. and Konings, W.N. (1995) Mol. Microbiol. 18, 925-932.

Ventosa, A., Nieto, J.J. and Oren, A. (1998) Microbiol. Mol. Biol. Rev. 62, 504-544.

Wagner, G., Hartmann, R. and Oesterhelt, D. (1978) Eur. J. Biochem. 89, 169-179.

Wegmann, K., Ben-Amotz, A. and Avron, M. (1980) Plant Physiol. 66, 1196-1197.

Welsh, D.T., Lindsay Y.E., Caumette, P., Herbert, R.A. and Hannan, J. (1996) FEMS Microbiol. Lett. 140, 203-207.

Wood, A.P. and Kelly, D.P. (1991) Arch. Microbiol. 156, 277-280.

Zavarzin, G.A., Zhilina, T.N. and Pusheva, M.A. (1994) In: H.L. Drake (ed.) Acetogenesis. Chapman & Hall, N.Y. pp. 432-444.

Zhilina, T.N. and Zavarzin, G.A. (1987) Dokl. Akad. Nauk. SSSR 293, 464-468 (in Russian).

Zhilina, T.N. and Zavarzin, G.A. (1990) FEMS Microbiol. Rev. 87, 315-322.

Zhilina, T.N., Zavarzin, G.A., Detkova E.N. and Rainey, F.A. (1996) Curr. Microbiol. 32, 320-326.

Biodata of the authors of the chapter "*Newly Discovered Halophilic Fungi in the Dead Sea (Israel).*"

Dr. **Asya Buchalo** is a chief scientist at the Kholodny Inst. of Botany of the Nat. Acad. Sci. Kiev, Ukraine. She obtained her Ph.D. (1986) from the Kholodny Institute and received the Sci. Academy of Ukraine and the Ukraine State Sci. and Tech. awards.
E-mail: abuch@botan.kiev.ua

Dr. **Solomon P. Wasser**: is Head of International Center for Cryptogamic Plants and Fungi Institute of Evolution at the University of Haifa (Israel). He also serves as the Chair of the Department of Cryptogamic Plants, at Kholodny Institute of Botany, National Academy of Sciences, Kiev, and Ukraine. Dr. Wasser obtained his Ph.D. (1973) and D.Sc. (1982) from Kholodny Institute of Botany, Kiev, Ukraine. There he was elected a member of the Ukraine National Academy of Sciences (1988) and Professor of Botany and Mycology (1991). Among his publications are over 350 scientific articles (as author and co-author), including 32 books. He is the chief editor of 3 International Journals (*Algologia* [Ukraine], *Intl. J. Medicinal Mushrooms* [USA] and *Intl. J. Algae* [USA]) and has a dozen patents.
E-mail: spwasser@research.haifa.ac.il

Dr. **Eviatar Nevo** is currently a Professor and Chair of Biology in the University of Haifa (Israel) and chair of the Evolutionary Biology. He earned his Ph.D. from the Hebrew University in Jerusalem in 1964. His publication list has over 700 scientific articles and he is an editor, co-editor and author of several books dealing with various fields of evolutionary Biology. Dr. Nevo is the recipient of several awards and fellowships obtained from various sources and serves in the Editorial boards of a few Journals.
E-mail: nevo@research.haifa.ac.il

J. Seckbach (ed.), Journey to Diverse Microbial Worlds, 239-252.
© 2000 *Kluwer Academic Publishers. Printed in the Netherlands.*

NEWLY DISCOVERED HALOPHILIC FUNGI IN THE DEAD SEA (ISRAEL)

ASYA S. BUCHALO [1], EVIATAR NEVO [2], SOLOMON P. WASSER [1,2], AND PAUL A. VOLZ [3]

[1]*Kholodny Institute of Botany, National Academy of Sciences of Ukraine, Tereshchenkivska str. 2, 252601 Kyiv-MSP-1, Ukraine*
[2]*Institute of Evolution, University of Haifa, Mt Carmel, 31905 Haifa, Israel*
[3]*East Michigan Mycology Associates, 1805 Jackson Ave., Ann Arbor MI 48103-4039, U.S.A.*

1. Introduction

Organisms living in extreme environments are called extermophiles because they live under extreme conditions at the edge of life (Madigan and Marrs, 1997). These organisms (thermophiles, alkalophiles, acidophiles, halophiles) do not merely tolerate their extreme living conditions, which are detrimental to most organisms. Remarkably, they do best in these extreme habitats, and in many cases require one or more extremes for reproduction. In other words, they evolved unique adaptive evolutionary strategies to these extreme living conditions, thereby becoming narrow extreme specialists.

The number of catalogued fungi species is presently about 72,000 (Hawksworth et al. 1995; Hawksworth 1997), which is possibly only a small fraction of the existing fungal species. In particular, little is known about the fungi living in extreme environments, especially in extremely hypersaline habitats (Blomberg and Adler 1993; Adler 1996). In recent decades, researchers have paid attention to the underestimated but potentially important role of fungi in the degradation of organic material in marine and hypersaline ecosystems. The level of degradative processes caused by fungi in the sea is not understood. Jones (1988) showed that in marine ecosystems, fungi are much more active biodegraders than was previously thought. To date about 800 species of obligate marine fungi have been described, including species of Basidiomycotina, Ascomycotina, lichen-forming fungi, Deuteromycotina and yeasts (Kohlmeyer and Kohlmeyer 1979; Hawksworth et al. 1995).

Reports of fungi from extremely hypersaline environments are very few. However, lack of representation in the literature may reflect not the inability of fungi to colonize these extreme environments but rather the little effort that has been spent to identify them (Javor 1989). The only report known to us on the isolation of a halophilic filamentous fungus from a hypersaline lake is the description of a *Cladosporium* sp. (Hyphomycetes, anamorphic Mycosphaerellaceae) on a submerged piece of pine wood in the Great Salt Lake, Utah, USA (290 - 360 g l^{-1} salinity) (Cronin and Post 1977). In the same lake, nonfilamentous fungi from the genus *Thraustochytrium* were also found (Amon 1978; Brown 1990). In addition, a number of salt-tolerant fungi have been isolated from such sources as salted fish, seawater, and desert soils (Andrews and Pitt 1987; Blomberg and Adler 1993; Adler 1996). Two species that are commonly encountered on salted dried fish, *Polypaecilum pisce* and *Basipetospora halophila*, grow vigorously in saline media and can be cultured in salt-saturated conditions (Wheeler, Hocking and Pitt, 1988). Yeast species capable of growth in near saturated brines are found in several genera such as *Hansenula*, *Pichia*, *Zygosaccharomyces* and *Debaryomyces* (Onishi, 1963; Tokuoka, 1993).

Many fungi grow in the presence of very low water activities: *Aspergillus* sp. can grow down to a water activity (a_w) of 0.70 and *Xeromyces bisporus* L. R. Fraser may grow at an a_w as low as 0.61 (Kushner 1978; Brown 1990). The physiology of adaptation of salt-tolerant yeasts and other fungi to concentrated environments has been recently reviewed (Blomberg & Adler, 1993; Clipson & Jennings, 1993; Tokuoka, 1993). For a discussion of the molecular biology of osmostress responses of the yeast *Saccharomyces cerevisiae*, the reader is referred to the review by Mager and Varela (1993).

The Dead Sea, located in the Syrian-African rift valley, on the border between Israel and Jordan, is one of the most saline lakes on earth (salinity about 340 g l^{-1}). The lake differs from other hypersaline lakes in the unique ionic composition of its waters, with concentrations of divalent cations (Mg - 40.7 g l^{-1}; Ca - 17 g l^{-1}) exceeding those of monovalent cations (Na - 39.2 g l^{-1}; K - 7 g l^{-1}). The major anions are Cl (212 g l^{-1}) and Br (5 g l^{-1}). The water activity in undiluted Dead Sea water was reported to be about 0.669 in 1979 (Krumgalz and Millero, 1982), and today is probably even lower.

The Dead Sea is a natural laboratory for evolution toward adaptation to extremely high salt concentrations. Since the discovery of life in the Dead Sea by Wilkansky (Benjamin Elazari-Volcani) in 1936 (Wilkansky 1936), the lake is known to be inhabited by several types of microorganisms. These include archaeal and eubacterial prokaryotes, the unicellular green alga (*Dunaliella parva* Lerche), and possibly even protozoa (Elazari-Volcani 1940; Volcani 1944). Quantitative studies performed since 1980 have shown the lake to be a dynamic ecosystem. There are periods of dense blooms of algae (*D. parva*) and red halophilic Archaea, triggered by dilution of the upper water layers by fresh water floods during rainy winters. Blooms alternate with long periods characterized by an almost total absence of life forms (Oren 1988, 1992, 1993, 1999). The paucity in biodiversity of the Dead Sea biota is probably determined by the high concentration of magnesium and calcium and the unique composition of the Dead Sea waters

Before we initiated our studies, no fungi have been recorded in the Dead Sea, with the possible exception of an osmophilic yeast, reported by Kritzman (1973), able to grow in a medium containing 15% glucose and 12% salt. Further details are lacking, and no cultures have been preserved. We have recently described the first filamentous fungi from the Dead Sea (Buchalo et al., 1998a,b, 1999). Here we present data on the taxonomic position of 26 species of filamentous fungi isolated by our group from the Dead Sea. Growth characteristics of these fungi were determined at different temperatures and salinities in order to investigate the adaptation of these fungi to high salt concentrations and to assess their possible role in this extremely hypersaline and chemically unique habitat, at the edge of life.

2. Materials and Methods

Surface water samples were obtained from different sites of the Israeli Dead Sea shore in 1995-1998. For the isolation of fungi, portions of 2 ml of Dead Sea water were poured into Petri dishes and mixed by rotation with molten agar media. The following media were used: malt extract agar, malt extract agar with 20-50% (by volume) of Dead Sea water, and Czapek agar. Tetracycline and streptomycin (100 µg/ml each) were added to all media to inhibit bacterial growth. After solidification of the agar, plates were incubated both at 26 – 28°C and 37°C for 30 days. The fungal colonies obtained were transferred to tubes with agar media for identification and storage. All isolates were grown on glucose-peptone-yeast extract (GPY) agar (Molitoris and Schaumann 1986; Rohrmann et al. 1992), artificial seawater medium (RILA products, Teaneck, NJ, USA), salinities between 0 and 175‰ or on media prepared in different dilutions of Dead Sea water. The cultures were incubated at different temperatures (15, 22, 27, 35 and 37°C).

Fungi were also recovered from the deeper waters of the Dead Sea. Samples were collected on October 8, 1996 from the center of the lake, about 8 km north-east of Ein Gedi, by means of Go-Flo sampling bottles, or by pumping through a hose. Samples were transferred to sterile bottles. Triplicate portions of 2 ml of water from different depths were added to 20-ml glass scintillation vials containing 0.2 g malt extract and 1 g sucrose in 8 ml distilled water, sterilized by autoclaving, to give a final Dead Sea water concentration of 20% (by volume). The bottles were incubated at 25°C, and the occurrence of growth was followed for 30 days.

A number of fungal cultures were isolated from pieces of immersed wooden constructions. Pieces of wood (1-2 cm in diameter) on which fungal hyphae were observed under a binocular microscope were placed in Petri dishes on filter paper wetted with sterile Dead Sea water, and were incubated at 28°C for 30-40 days. The filter paper was regularly wetted with sterile distilled water. Mycelia developing were transferred to tubes with nutrient media for identification and storage.

In studies on the growth of Dead Sea fungi at different salinities and temperatures, colony diameter on Petri dishes (diameter 9 cm) with GPY and malt extract (ME) media was determined twice weekly for 7 weeks for *Ulocladium chlamydosporum* (DS-09), *Penicillium westlingii* (DS-13a; DS-13b), and *Gymnascella marismortui* (DS-02, DS-06, DS-07). Deionized water (I), authentic Dead Sea water (D), commercial Dead Sea Salt (S), commercial artificial sea water (RILA=R) and artificial seawater (A) were used with salinities between 0 ‰ (I) and 175 ‰ (half concentration of Dead Sea water) at temperatures of 22, 27 and 35°C.

3. Recent Discovery of Fungal Species in the Dead Sea (Israel)

Twenty six species representing 13 different genera of Zygomycotina, Ascomycotina and Mitosporic fungi (Hawksworth et al., 1995) were isolated from the Dead Sea. Only one species of Zygomycotina (*Absidia glauca*) was registered. Mitosporic fungi were represented by four species belonging to three genera (*Acremonium* Link, *Stachybotrys* Corda and *Ulocladium* Preuss). Species of Ascomycotina belonging to the genera *Aspergillus* Link (anamorphic Trichocomaceae) and *Chaetomium* Kuntze were the most numerous, and the genera *Penicillium* Link (anamorphic Trichocomaceae), *Cladosporium* Link and *Eurotium* Link were represented by 2 species, respectively. The other five genera of Ascomycotina found in the Dead Sea were represented by one species each. A new species of the genus *Gymnascella* Peck (Onygenales) has been found: *G. marismortui* Buchalo et al. (Buchalo et al., 1988a, 1999).

The list of fungi isolated from the Dead Sea

Zygomycotina

1. *Absidia glauca* Hagem, 1908
Description: Ellis and Hesseltine, 1965
Distribution: Europe (Norway, Denmark, the British Isles, Germany, Poland, Austria, Ukraine, Hungary, Italy); Asia (Israel, Turkey, India, Japan); Africa (Egypt); N. America (Canada, USA, Mexico); Australia, New Zealand.
These species were also found in soil around the Dead Sea (Steiman et al., 1995).

Ascomycotina

2. *Aspergillus caespitosus* Raper et Thom, 1994
Description: Raper and Fennell, 1965
Distribution: Europe (Russia); Asia (Russia, Israel); N. America (USA).

3. *A. carneus* (Tiegh.) Blochwitz, 1993
   = *Sterigmatocystis carnea* V. Tiegh, 1877
Descriptions: Raper and Fennell, 1965; Subramanian, 1971
Distribution: Europe (Poland, Ukraine, Italy); Asia (India, Hong Kong, Uzbekistan, Kuwait, Israel); Africa (S.A. Republic, Somalia, Egypt); N. America (the southern States of the USA, Canada).

4. *A. fumigatus* Fres., 1863
Description: Raper and Fennell, 1965; CM I Descriptions, N92, 1966; Subramanian, 1971
Distribution: a thermotolerant fungus with worldwide distribution.
A. fumigatus was registered earlier in soil around the Dead Sea. It was isolated only using a hyperosmotic medium (Guiraud et al., 1995).

5. *A. niger* Van Tiegh., 1867
Descriptions: Raper and Fennell, 1965; CM I Descriptions, N94, 1966;
Subramanian, 1971
Distribution: is documented from across the planet. It was also isolated from soil
around the Dead Sea (Steiman et al., 1995)

6. *A. phoenicis* (Cda) Thom, 1926
Descriptions: Raper and Fennell, 1965
Distribution: is documented from across the planet.

7. *A. terreus* Thom, 1918
Descriptions: Raper and Fennell, 1965; CM I Descriptions, N95, 1966;
Subramanian, 1971
Distribution: Europe (Spain, France, Italy, Greece, Czech and Slovak Republics,
Ukraine, the British Isles); Asia (China, Malaysia, India, Japan, Pakistan, Bangladesh,
Syria, Kuwait, Israel, Turkey); Africa (Libya, Egypt, Chad, Somalia, Zaire, S.A.
Republic, equatorial West Africa); N. America (the southern states of the USA); S.
America (Peru, Uruguay, Brazil, Argentina, Honduras); Australia.
A. terreus was isolated from soil around the Dead Sea (Steiman et al., 1995).

8. *A. ustus* (Bain.) Thom et Church
    = *Sterigmatocystis usta* Bain.
Descriptions: Raper and Fenell, 1965; Subramanian, 1971
Distribution: Europe (The British Isles, Austria, Czech and Slovak Republics, Russia,
Poland, Italy, Spain, France); Asia (Turkey, Pakistan, India, Nepal, China, Japan,
Israel, Syria, Kuwait); Africa (Egypt, Libya, Chad, Somalia, Zaire, South Africa,
Namibia); N. America (the USA); Central America (Jamaica, Hawaii, the Bahamas),
S. America (Peru, Argentina, Chile, Brazil).
A. ustus was isolated from the Dead Sea area (Steiman et al., 1995, 1997).

9. *Chaetomium aureum* Chivers, 1912
Descriptions: Skolko and Groves, 1953; Ames, 1963.
Distribution: Europe (Poland, Ukraine); N. America (Canada, USA, Mexico).

10. *Ch. flavigenum* van Warmelo, 1966
Description: Mycologia, 1966, 58, N6: 846-854.
Distribution: Africa (South Africa), Asia (Israel).

11. *Ch. funiculosum* Cooke, 1873
    = *Ch. setosum* Ellis et Everh., 1897
Descriptions: Udagawa, 1960; Ames, 1963; Seth, 1970.
Distribution: Europe (Germany, the British Isles); Asia (India, Nepal, Pakistan, Israel,
Japan); Africa (Central Africa); N.America (Canada, USA); Central America; S.
America (Brazil); Australia, New Zealand.

12. *Ch. nigricolor* Ames, 1950
Description: Ames, 1963
Distribution: Asia (India, Israel).

13. *Cladosporium cladosporioides* (Fres.) de Vries
  = *Penicillium cladosporioides* Fres.
  = *Hormodendrium cladosporioides* (Fres.) Sacc.
  = *Cladosporium hypophyllum* Fuckel
Descriptions: Elles, 1971, Subramanian, 1971
Distribution: worldwide distributed.
C. cladosporioides was isolated from the Dead Sea area (Guiraud et al., 1995; Steiman et al., 1997).

14. *C. macrocarpum* Preuss, 1948
Descriptions: de Vries, 1952; Subramanian, 1971; Ellis, 1971
Distribution: Europe (Germany, the British Isles, Czech and Slovak Republics, France, Cyprus); Asia (Turkey, Iraq, India, Nepal); Africa (Egypt, Libya, Kenya, S.A. Republic); N. America (USA); Australia.

15. *Emericella nidulans* (Eidam) Vuill., 1927
  = *Sterigmatocystis nidulans* Eidam, 1883
Anamorph: *Aspergillus nidulans* (Eidam) Winter, 1884
Descriptions: Raper and Fennell, 1965; CM I Descriptions, N93, 1966;
Subramanian, 1971
Distribution: Europe (Germany, the British Isles, France, Italy, Poland, Yugoslavia, Ukraine, Russia); Asia (Israel, Kuwait, Syria, Turkey, Pakistan, Bangladesh, Nepal, China, Japan); Africa (Egypt, Tunisia, Somalia, S.A. Republic); N.America (USA, Canada); S. America (Argentina, Peru).
E. nidulans was isolated earlier from soil around the Dead Sea (Steiman et al., 1995).
E. nidulans cultures were obtained only using a hyperosmotic medium.

16. *Eurotium amstelodami* Mangin
Anamorph: *Aspergillus amstelodami* (Mangin) Thom et Church
Descriptions: Raper and Fenell, 1965; Subramanian, 1971
Distribution: Europe (the British Isles, Italy, Russia); Asia (Turkey, Kuwait, Syria, India, Pakistan, Japan, Israel), Africa (South Africa); N. America (the USA); S. America (Brazil, Peru, Argentina); Australia.
E. amstelodami was isolated from the Dead Sea area (Steiman et al., 1995; 1997).

17. *Eurotium herbariorum* (Wiggers) Link et Gray
  = *E. repens* DB;
  = *E. rubrum* Koenig, Spieckermann et Bremer;
  = *Mucor herbariorum* Wiggers et Merat;
  = *Aspergillus sejunctus* Bain. et Sart;

= *A. umbrosus* Bain. et Sart.;

= *A. mangini* Thom et Raper

Anamorph: *Aspergillus glaucus* Link et Gray

Descriptions: Raper and Fenell, 1965; Subramanian, 1971

Distribution: Europe (the British Isles, Belgium, Germany, Russia, Italy); Asia (Israel, Turkey, Kuwait, Bangladesh, Pakistan, Japan, India); Africa (South Africa, equatorial West Africa); N. America (USA); S. America (Brazil, Argentina); Australia.

E. herbariorum was isolated from the Dead Sea area (Steiman et al., 1995, 1997).

18. *Gymnascella marismortui* Buchalo et al.

G. marismortui presently considered to be endemic to Israel, was described in 1998 from the Dead Sea (Buchalo et al., 1998a).

G. marismortui differs from all known species of the genus *Gymnascella*, including *G. punctata* (Dutta et Ghosh) Currah, by the shape of its ascospores, which have a very broad rim (Currah 1985). The shape of ascospores is an important taxonomic characteristic in the genus *Gymnascella*. *G. marismortui* also differs from *G punctata* by the presence of arthroconidia and chlamydospores, the presence of nodulose hyphae, the color and much smaller size of the ascomata (*ca.* 250 μm in *G. punctata* compared with 40-70 μm in *G. marismortui)*, and obligate halophily. (Buchalo et al., 1998a).

19. *Paecilomyces farinosus* (Holm ex Gray) A.H.Brown et G.Sm., 1957

= *Corynoides farinosa* Holm ex Gray, 1821

= *Isaria farinosa* (Holm ex Gray) Fr., 1832

= *Penicillium alboaurantium* G. Smith, 1957

Descriptions: Skou, 1967; Samson, 1974.

Distribution: Europe (Czech and Slovak Republics); Asia (Israel, Nepal, Russia, Japan); Africa (S.A. Republic); N. America (USA); S. America (Brazil).

P. farinosus was isolated from soil around the Dead Sea (Guiraud et al., 1995).

20. *Penicillium variabile* Sopp, 1912

Description: Raper and Thom, 1949

Distribution: Europe (Ireland, Belgium, France, Italy, Ukraine, Russia); Asia (Israel, Syria, Kazakhstan, India, Nepal, Pakistan, Russia); Africa (Egypt, Somalia, Central Africa, Namibia); N. America (USA, Canada); S. America (Brazil, Peru); Australia.

P. variabile was isolated from soil around the Dead Sea (Steiman et al., 1995).

21. *Penicillium westlingii* Zalessky, 1927

= *P. paxilli* Bain

Description: Samson and Pitt, 1990

Distribution: Europe (Austria, Poland); Asia (Israel).

22. *Thielavia terricola* (Gilman et Abbott) C. W. Emmons, 1930

= *Coniothyrium terricola* Gilman et Abbott, 1927;

= *Anixiopsis japonica* Saito; Minoura, 1948
Descriptions: Booth, 1961; Udagawa, 1963; von Arx, 1975.
Distribution: Europe (Italy, the British Isles, the Netherlands); Asia (Israel, Kuwait, Pakistan, Nepal, India, Japan, Turkey); Africa (Sudan, Sierra Leone, Nigeria, Central Africa, S.A. Republic); N. America (USA); Central America.

Mitosporic fungi

23. *Acremonium* sp.

24. *A. persicinum* (Nicot) W. Gams, 1971
  = *Cephalosporium asperum* E. Marchal, 1885;
  = *Paecilomyces persicinum* Nicot, 1958;
  = *Cephalosporium purpurascens* Sukap et Thirum, 1966
Description: W. Gams, 1971
Distribution: Europe (the British Isles, Germany, Belgium, Ukraine); Asia (Armenia, India, Thailand, New Guinea); Africa (Nigeria); N. America (Canada, USA).
A. persicinum was isolated from soil around the Dead Sea (Steiman et al., 1997).

25. *Stachybotrys chartarum* (Ehrenb. ex Link) Hughes, 1958
  = *Stilbospora chartarum* Ehrenb., 1818;
  = *Oidium chartarum* Ehrenb. ex Link, 1824;
  = *Stachybotrys atra* Corda, 1837
Descriptions: Ellis, 1971; Subramanian, 1971; Matsushima, 1975.
Distribution: almost all Europe and N. America; Asia (Israel, Kuwait, Iraq, Pakistan, Uzbekistan, Tajikistan, Japan, India, Nepal, Russia); Africa (Egypt, Libya, Zaire, central Africa, S.A. Republic); Central America; S. America (Brazil, Peru).
S. chartarum was isolated from soil around the Dead Sea (Guiraud et al., 1995).

26. *Ulocladium chlamydosporum* Mouchacca, 1971
Description: Revue de Mycologie, 1971, 36: 114.
Distribution: Africa (Egypt); Asia (Israel).
Ul. chlamydosporum was isolated from soil around the Dead Sea (Guiraud et al., 1995).

4. Growth at Different Salinities and Temperatures
In order to check the adaptation of the fungi isolated from the hypersaline Dead Sea and their role in this habitat, we investigated the influence of temperature and salinity of the growth of *Gymnascella marismortui* and the Dead Sea isolates of *Penicillium westlingii* and *Ulocladium chlamydosporum*. At all temperatures tested, *U. chlamydosporum* showed less growth with increasing salinity, the *G. marismortui* strains showed better growth with increasing salinity, whereas the *P. westlingii* strains grew equally well at all salinities (Figure 1).

The temperature optimum for growth on different media and salinities in most cases was 22°C for *U. chlamydosporum* and 27°C for *G. marismortui* (partially 35°C at 175‰ salinity). Growth of *P. westlingii* was generally good and only little influenced by temperature.

Whereas *U. chlamydosporum* seems less well adapted to temperatures and salinities of the Dead Sea, *G. marismortui* appears much better adapted and may represent authentic obligatory-halophile members of this habitat. The cosmopolitan *P. westlingii* strains show little growth differences and they may inhabit the Dead Sea, at least in a dormant state.

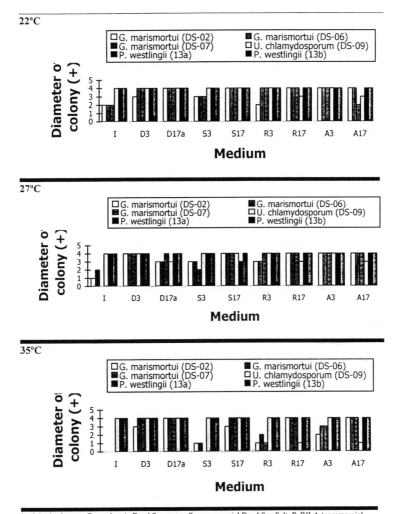

I - deionized water, D - authentic Dead Sea water, S - commercial Dead Sea Salt, R-RILA (commercial artificial sea water). A - artificial sea water (A). Salinities D3, R3, A3 - 35 ‰; D17, S17, R17, A17 - 175‰;. Growth (diameter of colony in cm): 0 - no growth, 1 - up to 0.5 cm, 2 - up to 2.0 cm, 3 - up to 4.5 cm. 4 - > 4.5 cm.

Figure 1. The influence of different media, temperatures and salinities on growth of fungi isolated from the Dead Sea (from Buchalo et al., 1999).

5. Discussion and Conclusion

Most fungal species isolated from the Dead Sea represent common soil fungi of panworld distribution (Mouchacca 1971; Ellis 1976; Pitt 1979; Ramirez 1982; Currah 1985; Gams 1993). Studies on the mycobiota of saline soils and saline estuarine sediments around the Dead Sea (Domsch et al. 1993; Steiman et al. 1995; Guiraud et al. 1995) showed that many species of filamentous fungi belonging to different taxa inhabit these soils. The authors reported that most of the fungi isolated were halotolerant.

Some of these fungi could have entered the Dead Sea with river water (Jordan river and all tributaries draining to the Dead Sea) or from the air, and may have adapted to the hypersaline environment. The occurrence of viable spores, the ability of isolated strains to grow and sporulate on nutrient media with high concentrations of Dead Sea, water, and the finding of growing hyphae on immersed wood of filamentous fungi in the Dead Sea, all suggest that fungi may be able to live in the Dead Sea, at least as dormant spores, and even possibly as vegetative mycelium. They may occur on substrates such as plant residues and wood, at least during the rainy period or in diluted areas near fresh water springs.

The foregoing findings suggest that filamentous fungi may play a hitherto overlooked role in the food web of the Dead Sea. The relatively low pH of the Dead Sea water (around 6) makes it a potentially suitable habitat for fungal life, but the salinity of the water is inhibiting growth. Noteworthy, the same is true for other forms of life inhabiting the lake, such as the green alga *Dunaliella parva* (Oren and Shilo 1985), and even the most magnesium-tolerant species among the halophilic Archaea (Mullakhanbhai and Larsen 1975; Oren, 1999). Only a significant dilution of the upper water layers of the Dead Sea enables the biota to overcome inhibition by the exceedingly high salt concentrations (and specifically the high concentrations of the divalent cations magnesium and calcium). To what extent the blooms of algae and Archaea that occurred in the lake in 1980 and in 1992 (Oren 1988, 1993) were accompanied by an increase in fungal biomass and activity is unknown. Final confirmation of vegetative growth of fungal species in the Dead Sea awaits future experiments *in situ* and the laboratory, using optimized liquid media with undiluted Dead Sea water. Agar media proved unsuitable for this purpose, because of crystal formation when the Dead Sea water concentration exceeded 50%. The finding that some of our fungal isolates grew in the presence of 50% Dead Sea water, as do the other microorganisms isolated from the lake, demonstrates that they may be adapted to life in the extremely stressful hypersaline Dead Sea water.

Since all species investigated could grow and produce polymer- and dye-degrading enzymes also at higher salinities and temperatures (Molitoris et al., 1998; Rawal et al., 1998; Buchalo et al., 1999) they probably represent inhabitants of the hypersaline Dead Sea, at least in a dormant state. However, *Gymnascella marismortui* may represent an authentic member of this habitat. In a recent review on fungi and salt stress Adler (1996) concluded that accumulation of polyols, mainly glycerol, is a major osmoregulatory strategy of saline fungi coupled with an effective exclusion of Na^+. The recent unravelling of the osmosensing signal transduction of *Saccharomyces cerevisiae* was also reviewed by Adler (1996), and that of Archea in the Dead Sea by Oren (1999). How relevant are these strategies to the Dead Sea Fungi?

The future challenge in studying fungal life in the hypersaline Dead Sea is to highlight the adaptive strategies and the genetic and physiological bases enabling fungi to live under these hostile and stressful conditions. Likewise, the entire life cycles of these fungi and their relation to brackish and freshwater types as well as soil species need to be elucidated. The question should be asked whether these fungal species display adaptive convergent molecular evolution to the Archaea and algae in the Dead Sea, or rather display an extreme version of salt tolerance strategies characteristic to saline fungi (Adler, 1996)? What are their physiological mechanisms to cope with high salinity? What are their ecological genomic reorganizations? Are their gene structure, regulation and control of transcription, translation and protein function unique? Are their genetic diversity levels, mutation, recombination (sexual) and DNA repair mechanisms adapted to the extreme hypersaline conditions? How useful could the Dead Sea fungi be to biotechnology and industry? The finding of so many hitherto undetected filamentous fungi in the Dead Sea opens new research opportunities, and challenges to understanding the genetic basis of adaptation and speciation of fungi to extreme hypersaline stress at the edge of life.

6. Acknowledgments

The authors thank Prof. H. P. Molitoris (Germany), Prof. A. Oren (Israel), Dr. I. Ellanskaja (Kyiv, Ukraine), Prof. D. L. Hawksworth and Dr. P. F. Cannon (UK) for invaluable comments, and Skipper M. Gonen and his crew for enabling sampling of Dead Sea water profiles. Financial support was obtained from the Israel Discount Bank Chair of Evolutionary Biology and the Ancell-Teicher Research Foundation for Genetics and Molecular Evolution (to E. N.), and USAID Grant No. TA-MOU-96-CA16-014 (to S. P. W.).

7. References

Adler, L. (1996). In:. Frankland, J. C., Magan, N. and. Gadd, G. M. (Eds). Fungi and Environmental Change, Cambridge: Cambridge University Press, pp. 217-234.

Amon, J. P. (1978). Mycologia 70: 1299-1301.

Andrews, S. and Pitt, J. L. (1987). Journal of General Microbiology 133: 233-238.

Blomberg, A. and Adler, L. (1993). In: Jennings, D. H. (Ed.). Stress Tolerance of Fungi. New York: Marcel Dekker, Inc., pp. 209-232.

Brown, A. D. (1990). Microbial Water Stress Physiology. Principles and Perspectives. Chichester: Wiley.

Buchalo, A. S.,. Nevo, E., Wasser, S. P., Oren, A. and Molitoris, H. P.(1998a). Proceedings of Royal Society London B 265: 1461-1465.

Buchalo, A. S.,. Nevo, E., Wasser, S. P., Oren, A. Molitoris, H. P. and Volz, P.R. (1998b), (Abstracts), VI International Mycological Congress, Jerusalem, Israel, 23-28 August, p. 180.

Buchalo, A. S.,. Wasser, S.P., Molitoris, H.P., Volz, P.A., Kurchenko, I., Lauwer, I. and. Rawal, B. (1999). In: Wasser, S. P. (Ed.) Kluwer Academic Publishers, Dortrecht, pp. 283-300.

Christensen, W. B. (1946). Journal of Bacteriology 52, 461-466.

Clipson, N. J. W. and Jennings, D. H. (1993). Canadian Journal of Botany 70: 2097-2105.

Cronin, A. E. and Post, F. J. (1977). Mycologia 69: 846-847.

Currah, R. S. (1985). Mycotaxon 24: 1-216.

Domsch, K. H., Gams, W. and Anderson, T. H. (1993). Compendium of soil fungi. Eching: IHW-Verlag.

Elazari-Volcani, B. (1940). Studies on the microflora of the Dead Sea. Ph.D. thesis, The Hebrew University of Jerusalem, Jerusalem (In Hebrew).

Ellis, M. B. (1976). More Dematiaceous Hyphomycetes. Kew: Commonwealth Mycological Institute.

Gams, W. (1993). Supplement and Corrigendum to the Compendium of Soil Fungi. Eching: IHW-Verlag.

Guiraud, P., Steiman, R., Seigle-Murandi, F. and Sage, L. (1995). Systematic and Applied Microbiology 18: 318-322.

Hawksworth, D.L. (1997). Mycologist 11: 18-22.

Hawksworth, D. L., Kirk, P. M., Sutton, B. C. and. Pegler, D. N. (1995). 8th edition. Wallingford, UK: CAB International, pp. 1-616.

Javor, B. (1989).. Berlin: Springer-Verlag.

Jones, E. B. G. (1988). The Mycologist 2: 150-157.

Kohlmeyer, J. and Kohlmeyer, E. (1979). Marine Mycology. The Higher Fungi. New York: Academic Press.

Kritzman, G. (1973). Observations on the microorganisms in the Dead Sea, M. Sc. thesis. The Hebrew University of Jerusalem (In Hebrew).

Krumgalz, B. S. and Millero, F. J. (1982). Marine Chemistry 11: 209-222.

Kushner, D. J. (1978). In:. Kushner, D. J. (Ed.). Microbial Life in Extreme Environments. New York: Academic Press, pp. 171-215.

Madigan, M. T. and Marrs, B.L. (1997). Extremophiles. Scientific American, pp. 66-71.

Mager, W. H. and Varela, J. C. S. (1993). Molecular Microbiology 10: 253-258.

Molitoris, H. P. and Schaumann, K. (1986). In:. Moss, S. T. (Ed.) The Biology of Marine Fungi. Cambridge: Cambridge University Press, pp. 35-47.

Molitoris, H. P., Lauer, I., Buchalo, A. S., Oren, A., Nevo, E. and Wasser, S. P. (1998). (Abstracts), VI International Mycological Congress, Jerusalem, Israel, August, 23-28, p. 169.

Mouchacca, D. J. (1971). Ulocladium Preuss. Review in Mycology 36: 114-122.

Mullakhanbhai, M. F. and H. Larsen, H. (1975). Archives of Microbiology 104: 207-214.

Onishi, H. (1963). Advances in Food Research 12: 53-90.

Oren, A. (1988). In: Marshall, K. C. (Ed.) Advances in Microbial Ecology. New York: Plenum Publishing Company, pp. 193-229.

Oren, A. (1992). In: Vreeland, R. H. and. Hochstein, L. I. (Eds.) The Biology of Halophilic Bacteria. Boca Raton: CRC Press, pp. 25-53.

Oren, A. (1993). Experientia 49: 518-522.

Oren, A. (1999). In: Wasser, S.P. (Ed.) Evolutionary Theory and Processes: Modern Perspectives. Kluwer Academic Publishers, Dordrecht, pp. 345-363.

Oren, A. and Shilo, M. (1985). Microbiology and Ecology 31: 229-237.

Pitt, J. I. (1979). The Genus Penicillium and its Telemorphic States Eupenicillium and Talaromyces. London: Academy Press.

Ramirez, C. (1982). Manual and Atlas of the Penicillia. Amsterdam: Elsevier Biomedical.

Rawal, B. S., Kurchenko, I., Buchalo, A. S., Wasser, S.P., Oren, A., Nevo, E. and Molitoris, H.P. (1998). (Abstracts), VI International Mycological Congress, Jerusalem, Israel, August, 23-28, p. 106.

Rohrmann, S.0., Lorenz, R. and .Molitoris, H. P. (1992). Canadian Journal of Botany 70: 2106-2110.

Steiman, R., Guiraud, P., Sage, L., Seigle-Murandi, F. and Lafond, J.-L. (1995). Systematic and Applied Microbiology 18: 310-317.

Steiman, R., Guiraud, P., Sage, L. and Seigle-Murandi, F. (1997). Antonie van Leeuwenhoek 72: 261-270.

Tokuoka, K. (1993). Journal of Applied Bacteriology 74: 101-110.

Volcani, B. (1944). The microorganisms of the Dead Sea. In: Papers Collected to Commemorate the 70th Anniversary of Dr. Chaim Weizmann, Daniel Sieff Research Institute, Rehovoth, pp. 71-85.

Wheeler, K., Hocking, A. and Pitt, J. I. (1988). Journal of General Microbiology 134: 2255-2261.

Wilkansky, B. (1936). Nature 138, 467.

VII

SYMBIOSES

Biodata of **Burkhard Büdel** contributor of *"Symbiosis (Living One inside the other)"*

Dr. Burkhard Büdel is a Professor in the Department of Biology at the University of Kaiserslautern, Germany. He obtained his Ph.D. from the University of Marburg in 1986. Subsequently he joined the University of Würzburg (Otto-Ludwig Lange's group) for 3.5 years. Then he spent a couple of years at the University of Darmstadt, the Senckenberg Research Institute at Frankfurt and at the University of Rostock. Professor Büdel's main interests are in biodiversity of microorganisms (cyanobacteria, algae and lichen), including systematic, taxonomy, phylogeny, ecology and ecophysiology. Among his research excursions abroad are places like South Africa, Australia, New Zealand, USA, Mexico, Costa Rica, Panama, Venezuela and French Guyana.

E-mail: **buedel@rhrk.uni-kl.de**

J. Seckbach (ed.), Journey to Diverse Microbial Worlds, 255-266.

SYMBIOSES (LIVING ONE INSIDE THE OTHER)

BURKHARD BÜDEL

General Botany
University of Kaiserslautern, Biology Department
P.O.Box 3049, D-67653 Kaiserslautern, Germany

1. Introduction

The nature of symbioses has been a matter of academic discussion for more than hundred years (e.g. DeBary 1887). Especially the lichen symbioses, varying from parasitism to mutualism, were treated several times in this context (e.g. Henssen and Jahns 1974; Nash 1996). In this chapter, I will apply the wide concept of DeBary (1887; Table 1).

Tab. 1: Types of symbioses according to DeBary (1887), + and - indicate the use for the organism (from Kappen et al. 1998).

		Organism 1		
		+	0	-
	+	mutualism	commensalism	parasitism
Organism 2	0	commensalism	neutralism	amensalism
	-	parasitism	amensalism	antagonism

Recently, symbiosis has been defined as a source of new metabolic activities for at least one of the partners (Douglas 1994). Symbiosis allows certain groups of fungi or animals like for example the Cnidaria to make indirectly use of atmospheric CO_2, or was an overwhelming successful step in evolution towards the development of plants via endocytobiosis. The resulting consortium is able to use new substrates or to live in extreme environments, and thus has an important influence on the dynamics and functioning of ecosystems (Kappen et al. 1998).

2. Cyanobacteria, Algae, and Fungi

Sitte and Eschbach (1992) defined cytosymbiosis as an intimate and long-lasting association of cells belonging to different taxa. Algae and cyanobacteria are very successful in the establishment of cytosymbiotic systems, as it is demonstrated in the excellent compilation edited by Reisser (1992) and the review of Friedl and Maier (1998).

Our knowledge of endocytobiotic systems, a special case of cytosymbiosis - one organism living inside the cell of another - is increasing rapidly. But, interestingly

258

enough, we actually do not know very much on how this process is initiated (Mollen-hauer and Mollenhauer 1991). The ecological situation must make the two partners meet and also allow survival of the incorporated part. In addition, endocytobiotic events have caused dramatic changes in taxonomy and phylogeny (Mollenhauer 1994). In the following two examples shall be presented representatively, where the process of the establishment was studied in more detail.

2.1. ENDOCYTOBIOSIS OF *GEOSIPHON PYRIFORME*

The coenocytic soil fungus *Geosiphon pyriforme* is to date the only example of a fungus living in endocytobiotic association with a cyanobacterium, i.e. *Nostoc punctiforme* (Kluge et al. 1997). The consortium as a whole lives C- and N-autotrophically (Kluge et al. 1991; Kluge et al. 1992). However, the *Geosiphon-Nostoc* cytosymbiosis can be experimentally separated and unified and was therefore regarded as rather „primitive" (Kluge et al. 1997). The taxonomic position of the fungus was proposed close to the zygomycete genus *Glomus* (Mollenhauer 1992). Striking similarities of the spore ultra-structure between various fungi of the Glomales and *Geosiphon* were found (Schüßler et al. 1994). Using SSU rDNA sequence fragments, the taxonomic position of *Geosiphon* could not be clarified finally, but the view was strongly supported, that *Geosiphon* rep-resents a very ancestral member of the Glomales and thus is closely related to AM-fungi (Gehrig et al. 1996).

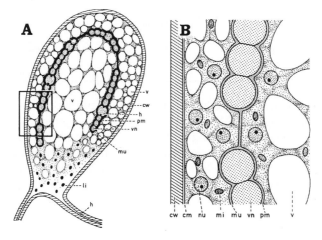

Figure 1: Inner structure of the *Geosiphon* bladder (from Mollenhauer and Kluge 1994, with permis-sion of the authors). A) total view. The fungal cytoplasm contains many vacuoles (v) in the upper part, whereas lipid droplets (li) and glycogen inclusions are located at the base of the bladder. A pericyano-bacterial membrane (pm) envelops the endosymbiont *Nostoc* (vn, h = heterocyst); B) detail of A; cw: cell wall of the bladder; cm: cell membrane (plasmalemma) of the fungus; mu: mucilage; nu: nucleus; mi: mitochondrion.

Inside the fungal coenocyte, the *Nostoc*-filaments are surrounded by a rather com-plex arrangement of vacuoles of different types (Fig. 1; Mollenhauer and Kluge 1994).

The symbiotic *Nostoc* is located in one tubulouse compartment, that forms a cup-shaped structure which is peripherally arranged (Fig.1). The *Nostoc*-surrounding membrane inside the bladder is of fungal nature (Schüßler et al. 1996).

The process of incorporation of the *Nostoc* filaments by *Geosiphon* is a phagocytosis, and it can only be performed if *Nostoc* primordia come into contact with growing hyphal tips of the fungus (Mollenhauer et al. 1996). The incorporation is summarized in Figs. 2 and 3. Only freshly formed *Nostoc* primordia can be incorporated, with the primary heterocytes of the primordium being excluded.

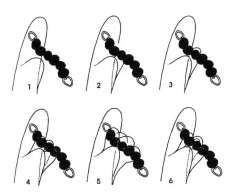

Figure 2: Schematic survey of the incorporation of a *Nostoc* filament (vegetative cells black, heterocytes with a double line) by protruding cytoplasm of *Geosiphon* (from Mollenhauer et al. 1996, with permission of the authors).

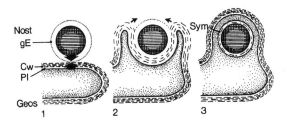

Figure 3: Three selected stages of the incorporation process as in Fig. 2 (from Mollenhauer at al. 1996, with permission of the authors). Longitudinal section of the feeding hypha (Geos); *Nostoc* (Nost) in cross section; Cw: fungal cell wall; gE: cyanobacterial gelatinous envelope; Pl fungal plasmalemma; Sym: pericyanobacterial membrane.

It was found that only the gelatinous envelope of the *Nostoc* primordium contains mannose, not that of its heterocytes (Schüßler et al. 1997). Specific partner recognition might therefore be based on the carbohydrate composition of the cyanobacterial envelope (Kluge et al. 1997).

260

2.2. DINOPHYCEAE

Dinoflagellates are described from many symbiotic associations and they may be either hosts or symbionts (Friedl and Maier 1998). The symbiosis can be fully synchronised with the intracellular symbiont depending on the host, or may be established new several times during the life span of the host. Even "stolen" plastids are known ("kleptoplastidy"), where the dinoflagellate takes the plastid by endocytotic uptake of another alga (e.g. Schnepf et al. 1989, Schnepf 1993, Lewitus et al. 1999).

Ultrastructural studies of the blue-green dinoflagellate *Gymnodinium aeruginosum* revealed, that it is a consortium of a dinophycean host and a vestigial, anucleate, cryptophycean endosymbiont (Schnepf et al. 1989). A single membrane (Fig. 4) surrounds the endosymbiont. Little is known about the endosymbiont aquisition of *Gymnodinium aeruginosum*, but it seems that a cell of *Chroomonas* is taken up by phagocytosis (Schnepf et al. 1989, Mollenhauer and Mollenhauer 1991).

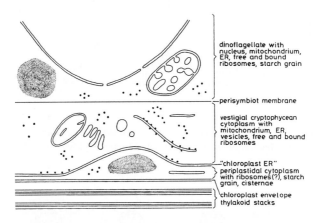

Figure 4: Membranes and compartments of host (*Gymnodinium aeruginosum*) and cryptophycean endocytosymbiont (from Schnepf et al. 1989, with permission of the authors).

2.3. LICHENS

4Lichens have been defined as the symbiotic phenotype of nutritionally specialized fungi that live as ecologically obligate biotrophs in symbiosis with algal and/or cyanobacterial photobionts (Honegger 1991, Palmqvist 1995). In a recent study it has been conclusively shown, that the lichen symbiosis occurred as early as in the Lower Devonian period (400 million years ago; Taylor et al. 1995). The lichen, named *Winfrenatia reticulata*, was composed of a fungus with aseptate hyphae, probably a member of the zygomycetes, and a unicellular cyanobacterium as the photobiont (=cyanobiont), showing typical stages of cyanobacterial cell division in thin sections (Taylor et al. 1997).

While morphology, taxonomy and phylogeny of the lichen symbiosis is quite well investigated, the initial process, triggering the dramatic morphogenetic change of the fungus, remains to be one the major biological enigmas. Since most lichen-forming fungi reproduce sexually, re-establishment of the symbiotic state is an essential step after each reproductive cycle. Unfortunately, very little is known about how this proceeds in nature (Honegger 1993). Most of the common green algal photobiont species are not abundant in the aposymbiotic stage (Tschermak-Woess 1988). Therefore, finding a compatible symbiont might be a major problem for all lichen-forming fungi reproducing mainly with aposymbiotic propagules. Dispersal by any kind of thallus fragmentation is an elegant way to overcome this problem. Concerning developmental biology of lichens, the interested reader is referred to the excellent review of Honegger (1993).

An accelerated evolution of mutualistic systems was recently demonstrated by Lutzoni and Pagel (1997), using nuclear ribosomal DNA of lichenized fungi and fungi associated with liverworts. Their results suggested an increased rate of evolution after the adoption of a mutualistic lifestyle.

2.3.1. *Mycobiont*

One out of five fungi is lichenized and we know from classical studies already that lichens are of polyphyletic origin (see also discussion by Jørgensen 1995). This view was recently confirmed and focused on, in an investigation of Gargas et al. (1995), where they found at least five independent origins, using the small subunit of rDNA for the reconstruction of the possible phylogeny.

Phylogenetic studies of higher ascomycetes (including the lichen forming ascomycetes) are enhanced by the introduction of molecular markers, with a main focus on 18S rDNA. Several groups, such as Plectomycetes and Pyrenomycetes, defined by their ascoma-type are supported, while others as the Discomycetes appear to be paraphyletic (Lumbsch in press). Re-evaluation of classical characters using molecular data led to the following results: 1) ascus types, often regarded as of major importance in ascomycete systematic, revealed for the prototunicate types to be of poor taxonomic value since prototunicate asci are polyphyletic; 2) the independence of Agyriales (include lichenized and non lichenized forms) supposed by morphological characters, is supported by sequence data, while the relationship to sister groups remains dubious (Lumbsch in press). Furthermore, recent molecular studies by Stenroos and DePriest (1998) on 27 representatives, indicated that the Lecanorales, representing the largest group of lichenized fungi, are polyphyletic in the broad circumscription and exclusion of some taxa, including the Pertusariales and Umbilicariacae, is suggested. The Pertusariales (Henssen 1976) and the Umbilicariaceae (Janex-Favre 1974) differ also in morphological features (ascus types and/or ascoma development) from the Lecanorales sensu strictu.

On the basis of the nuclear ribosomal repeat unit (ITS1, 5.8S, ITS2, and 25S), the lichen forming and non-lichen-forming ophalinoid mushrooms (basidiomycetes) of the genus *Omphalina* have been shown to be of monophyletic origin (Lutzoni 1997). This study also suggests, that there was a single transition to a mutualistic state during the evolution of *Omphalina*.

At present, molecularly inferred phylogenies stimulate evaluation of morphological character sets to a large and still increasing degree, and surely led and still lead to a renaissance of systematic biology in lichens.

2.3.2. *Photobiont*

Some forty genera of cyanobacteria and algae are known to occur as photobionts in lichens (Tschermak-Woess 1989), and overviews are given by several authors (Tschermak-Woess 1988, Büdel 1992, Gärtner 1992, Ettl and Gärtner 1995, Friedl and Büdel 1996). Three genera, *Trebouxia* (unicellular), *Trentepohlia* (filamentous), and *Nostoc* (filamentous) are the most frequent photobionts found in lichens. The geographically widespread, unicellular cyanobacterium *Chroococcidiopsis*, frequently occurring inside of rocks and/or underneath of quartz pebbles on soil, was recently found to be another frequent cyanobacterial photobiont in lichens living in the arid (micro- and macro-) environment (Büdel 1992, Friedl and Büdel 1996, Büdel et al. in press).

The phylogenetic relationship of symbiotic green algae, based on analyses of complete 18S rDNA resulted in the description of the class Trebouxiophyceae (Friedl 1995). This new circumscription includes almost all lichenized genera of autosporic, coccoid green algae, e.g. *Coccomyxa, Elliptochloris, Dictyochloropsis, Myrmecia, Stichococcus,* and *Trebouxia*, whereas the filamentous *Trentepohlia* is associated with the Ulvophyceae (Friedl 1997). Furthermore, the rDNA analyses indicate that the capacity to exist in lichen associations has multiple independent origins and that lichen algae are derived from non-symbiotic terrestrial green algae (Friedl 1997).

Diversity and specificity of the symbiotic cyanobacterial genus *Nostoc* from several lichens (*Nephroma arcticum, Peltigera aphthosa, P. membranacea* and *P. canina*) was investigated by Paulsrud and Lindblad (1998), using tRNA^{Leu} (UAA) introns. No evidence for sample heterogeneity was found. The symbionts seem to consist of only one cyanobacterium in each thallus. Even in lichens living close to each other, no variation was ever observed within a sample or within a single lichen thallus. This might be the result of high specificity, allowing only one or a few cyanobacterial strains to be accepted by one fungus (Paulsrud and Lindblad 1998).

Preliminary studies on the 16S rDNA of lichenized and free living species of the unicellular, baeocyte-producing genus *Chroococcidiopsis*, suggests multiple origin of the former group Pleurocapsales, where the genus had been placed in, and also close relationships between free living and lichenized members of the genus (Büdel and Friedl unpublished).

2.3.3. *Functional morphology*

Under physiological circumstances, gas filled internal spaces in the lichen medulla is never floated with water, even when the thallus is submerged for several hours. The technique of low temperature scanning electron microscopy (LTSEM) allowed new insights into the lichen thallus under almost natural hydration conditions (e.g. Honegger 1991, Honegger and Peter 1994, Scheidegger 1994). LTSEM observations suggest apoplastic water transport to the photobiont cells (Honegger 1991, Scheidegger 1994, Büdel 1999), resulting in an faster CO_2-diffusion within the lichen thallus (Fig. 5). CO_2-diffusion in water is 10^4 times slower than in air (Green at al. 1994). However, the thick fungal cell walls of the upper and/or lower cortices of many lichens are hydrophilic and can take up large amounts of water quite rapidly, thus decreasing CO_2-diffusion rates (Green at al. 1994).

Under low CO_2 availability, limited photosynthetic carbon gain might be compensated by the activity of a CO_2 concentrating mechanism (CCM), as it was demonstrated for free living green algae containing pyrenoids and many cyanobacteria, and also in lichens (Badger et al. 1993, Lange et al. 1994, Palmqvist 1995). Cyanobacteria are probably entirely dependent on their CCM, as their Rubisco has much lower affinity for CO_2 as compared to that of green algae (Palmqvist 1995). This again can be counterbalanced by a much more effective (up to 170-fold more in *Synechococcus leopoliensis*; Espie et al. 1991) CCM in cyanobacteria in comparison to green algae.

The importance of thallus morphology for photosynthetic carbon gain, related to thallus water content, was shown in several studies for cyanobacterial lichens of the genus *Peltula*, possessing the unicellular cyanobacterium *Chroococcidiopsis* as photobiont (Büdel 1999, Büdel et al. in press). This studies revealed the role of thallus morphologies, i.e. crustose-squamulose, peltate or semifruticose, resulting in quite different optima of thallus water contents for photosynthetic carbon gain, independently of the type of photobiont. As consequence, habitat occupation can be explained as being largely determined by the growth form of the *Peltula* species, adapted to the specific water regime of the habitat.

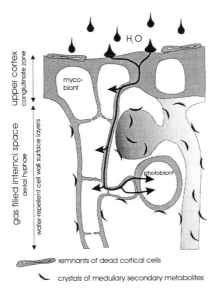

Figure 5: Apoplastic and symplastic water transport in lichen thalli (according to Honegger 1991, modified)

3. Plants and Cyanobacteria

3.1. HORNWORT AND LIVERWORT SYMBIOSIS

Some hornworts and liverworts are able to establish an extracellular symbioses in their gametophyte generation with the cyanobacterium *Nostoc*. Reconstitution experiments of the *Anthoceros-Nostoc* symbioses were successful not only with the original isolate of *Nostoc*, but also with several strains of *Nostoc*, originating from other symbioses, e.g. the angiosperm *Gunnera*, the cycads *Macrozamia* and *Cycas* and the lichen *Peltigera aphthosa* (Enderlin and Meeks 1983). All *Nostoc* strains shared the ability of dinitrogen-fixation and formation of hormogonia. In one *Nostoc* strain the heterocyte frequency reached a value of 44.5% in the symbiotic association compared with 8.9% in free living culture (Enderlin and Meeks 1983).

Genotypic (PCR amplification techniques) and phenotypic comparison of symbiotic cyanobacteria, coming from the hornwort *Phaeoceros*, with free living cyanobacteria, originating from several closely spaced locations, were investigated by West and Adams (1997) for their possible diversity. One of the symbiotic isolates turned out to be a species of *Calothrix*, a genus not known to form bryophyte symbioses so far. All other symbiotic isolates were *Nostoc* spp. All of the symbiotic and all but one of the free-living strains were able to reconstitute the symbioses with the hornwort *Phaeoceros* and the liverwort *Blasia*. From this studies it was implied by the authors, that many different strains can infect and establish a symbioses with a single hornwort or liverwort (West and Adams 1997).

3.2. AZOLLA-ANABAENA SYMBIOSIS

This is an extracellular symbioses between the dinitrogen-fixing cyanobacterium *Anabaena azollae* and the eukaryotic freshwater fern *Azolla*. The symbioses are maintained during sexual and asexual modes of reproduction. Up to now, it is not really clear whether or not there is uniformity or diversity in what we understand as *Anabaena azollae*. There is evidence for both, homogeneity but also for diversity of different strains of *A. azollae* (Schenk 1992).

3.3. CYCAD SYMBIOSIS

Cycads are an ancient group of plants that have a palm-like appearance, but belong to the gymnosperms. Besides the main root system, a coralloid root system located at the upper part of the hypocotyl, is also formed. Within the coralloid roots, nitrogen fixing cyanobacteria of the genera *Nostoc* or *Anabaena*, and probably two or three other genera live in an intercellular symbiosis (e.g. Lindblad 1987, Grilli-Caiola 1992). Restriction fragment length polymorphism of DNA from coralloid root cyanobacteria revealed that a mixture of *Nostoc*-strains can associate with a single cycad species, although a single *Nostoc*-strain might dominate in the coralloid root of one cycad species (Lindblad et al. 1989). Grobbelaar et al. (1987) determined cyanobacterial symbionts from 31 cycads from South Africa and identified the following species on a morphological basis: *Calothrix* sp., *Nostoc commune*, *N. punctiforme*, *N. ellipsosporum*, *N. paludosum*, *N. sphaericum* and *N muscorum*. However, since determinations were based on cultured material using keys based on morphological features of free living material with a known ecology, identifications are rather doubtful.

3.4. *GUNNERA* SYMBIOSIS

The angiosperm genus *Gunnera* is able to form an intracellular symbioses with the cyanobacterium *Nostoc*. The cyanobacterial specificity was tested recently by Johansson and Bergman (1994), using reconstitution experiments with several strains of heterocyte containing, filamentous cyanobacteria, originating from symbiotic systems and free-living ones, and *Gunnera manicata* as a test system. As already found for the bryophyte-cyanobacterium symbioses (se above), the infecting cyanobacteria were all symbiotic isolates, either from *Gunnera* itself, from the hornwort *Anthoceros*, the cycad *Macrozamia* or the lichen *Peltigera canina*. But no other genera than *Nostoc* were capable of infection.

4. Conclusions

From the above-mentioned results of recent investigations, it appears to be obvious that the diversity of symbiotic organisms at the side of the photosynthetically active inhabitant is probably much larger than previously thought. With the combination of modern molecular techniques and classical methods, new insights into the still fascinating biological phenomenon symbiosis can be gained.

5. References

Badger, M.R., Pfanz, H., Büdel, B., Heber, U. and Lange, O.L. (1993) Planta 191: 57-70.

Douglas, E.A. (1994) Symbiotic interactions, New York.

Büdel, B. (1992) in: W. Reisser (ed.) *Algae and symbioses: plants, animals, fungi, viruses, interactions explored*, Biopress Limited, Bristol, pp. 301-324.

Büdel, B. (1999) Bielefelder Ökologische Beiträge 14: 134-143.

Büdel, B., Becker, U., Sterflinger, K. and Follmann, G., in: S. Porembski and W. Barthlott (eds.), *The vegetation of inselbergs*, Ecological Studies, Springer-Verlag, Berlin, Heidelberg (in press).

Enderlin, C.S. and Meeks, J.C. (1983) Planta 158: 157-165.

Espie, G.S., Miller, A.G., Canvin, D.T. (1991) Plant Physiol. 97: 943-953.

Ettl, H. and Gärtner, G. (1995) Syllabus der Boden-, Luft- und Flechtenalgen, Gustav Fischer-Verlag, Stuttgart.

Friedl, T. (1995) J. Phycol. 31: 632-639.

Friedl, T. (1997) Pl. Syst. Evol. [Suppl.] 11: 87-101.

Friedl, T. and Büdel, B. (1996) in: T.H. III Nash (ed.) *Lichen biology*, Cambridge University Press, Cambridge, pp. 8-23.

Friedl, T. and Maier, U.G. (1998) Cytosymbiosis. Progress in Botany 59: 259-282.

Gargas, A., DePriest, P.T., Grube, M. and Tehlers, A. (1995) Science 268: 1492-1495.

Gärtner, G. (1992) in: W. Reisser (ed.) *Algae and symbioses: plants, animals, fungi, viruses, interactions explored*, Biopress Limited, Bristol, pp. 325-338.

Gehrig, H., Schüßler, A., and Kluge, M. (1996) Mol. Evol. 43: 71-81.

Green, T.G.A., Lange, O.L., and Cowan, I.R. (1994) Crypt. Bot. 4: 166-178.

Grilli Caiola, M. (1992) in: W. Reisser (ed.) *Algae and symbioses: plants, animals, fungi, viruses, interactions explored*, Biopress Limited, Bristol, pp. 231-254.

Grobbelaar, N., Scott, W.E., Hattingh, W. and Marshall, J. (1987) South African J. Bot. 53: 111-118.

Henssen, A. (1976) in: D.H. Brown, D.L. Hawksworth, and R.H. Bailey (eds.), *Lichenology: Progress and Problems*, Academic Press, London, pp. 107-138.

Henssen, A. and Jahns, H.-M. (1974) Lichenes, Georg Thieme Verlag, Stuttgart.

266

Honegger, R. (1991) Ann. Rev. Plant Physiol. Plant Mol. Biol. 42: 553-578.
Honegger, R. (1993) New Phytol. 125: 659-677.
Honegger, R. and Peter, M. (1994) Symbiosis 16: 167-186.
Janex-Favre, M.-C. (1974) Rev. Bryol. Lichenol. 40: 59-86.
Johansson, C. and Bergman, B. (1994) New Phytol. 126: 643-652.
Jørgensen, P.M. (1995) Int. Lichenol. Newslett. 28: 53-54.
Kappen, L., Sattelmacher, B., Dittert, K. und Buscot, F. (1998) in: *Handbuch der Umweltwissenschaften* 3. Erg. Lfg., pp. 3-28.
Kluge, M., Mollenhauer, D., and Mollenhauer, R. (1991) Planta 185: 311-377.
Kluge, M., Mollenhauer, D., Mollenhauer, R., and Kape, R. (1992) Bot. Acta 105: 343-344.
Kluge, M., Gehrig, H., Mollenhauer, D., Mollenhauer, R., Schnepf, E. and Schüßler, A. (1997) in: H.E.A. Schenk, R. Herrmann, K.W. Jeon, N.E. Müller, and W. Schwemmler, (eds.) *Eukaryotism and symbiosis*, Springer-Verlag Berlin, Heidelberg, pp. 469-476.
Lange, O.L., Büdel, B., Zellner, H., Zotz, G. and Meyer, A. (1994) Bot. Acta 107: 279-290.
Lewitus, A.J., Glasgow, H.B. and Burkholder, J.M. (1999) J. Phycol. 35: 303-312.
Lindblad, P. (1987) *Nostoc*-cycad symbiosis: with emphasis on the cyanobiont, Acta Univ. Ups., Comprehensive summaries of Uppsala dissertations from the Faculty of Science 70, Uppsala.
Lindblad, P., Haselkorn, R., Bergman, B., and Nierzwicki-Bauer, S.A. (1989) Arch. Microbiol. 152: 20-24.
Lumbsch, H.T. Naturwissenschaften (in press).
Lutzoni, F.M. (1997) Syst. Biol. 46(3): 373-406.
Lutzoni, F., Pagel, M. (1997) Proc. Natl. Acad. Sci. USA 94:11 422-11427.
Mollenhauer, D. (1992) in: W. Reisser (ed.) *Algae and symbioses: plants, animals, fungi, viruses, interactions explored*, Biopress Limited, Bristol, pp. 339-351.
Mollenhauer, D. (1994) Senckenberg-Buch 70: 339-364.
Mollenhauer, D. and Mollenhauer, R. (1991) Annali di Botanica 49: 149-167.
Mollenhauer, D. and Kluge, M. (1994) Endocytobiosis & Cell Res. 10: 29-34.
Mollenhauer, D., Mollenhauer, R. and Kluge, M. (1996) Protoplasma 193: 3-9.
Nash, T. (ed.; 1996) *Lichen Biology*. Cambridge University Press, Cambridge.
Palmqvist, K. (1995) Symbiosis 18: 95-109.
Paulsrud, P. and Lindblad, P. (1998) Appl. Environ. Microbiol. 64: 310-315.
Reisser, W. (1992) (ed.) Algae and symbioses: plants, animals, fungi, viruses interactions explored. Biopress, Bristol.
Scheidegger, C. (1994) Crypt. Bot. 4: 290-299.
Schenk, H.E.A. (1992) in: A. Balows, H.G. Trüper, M. Dworkin, W. Harder, and K.-H. Schleifer, *The Prokaryotes*, Springer-Verlag New York, Berlin, Heidelberg, pp. 3819-3854.
Schnepf, E. (1993) in: R.A. Lewin (ed.) *Origins of plastids*, Chapman and Hall, London, pp 53-76.
Schnepf, E., Winter, S. and Mollenhauer, D. (1989) Pl. Syst. Evol. 164: 75-91.
Schüßler, A., Mollenhauer, D., Schnepf, E. and Kluge, M. (1994) Bot. Acta 107: 36-45.
Schüßler, A., Bonfante, P., Schnepf, E., Molllenhauer, D. and Kluge, M. (1996) Protoplasma 190: 53-67.
Schüßler, A, Meyer, T., Gehrig, H., and Kluge, M. (1997) Eur. J. Phycol. 32: 233-239.
Sitte, P. and Eschbach, S. (1992) Prog. Bot. 53: 29-43.
Stenroos, S. and DePriest, P.T. (1998) Am J. Bot. 85: 1548-1559.
Taylor, T.N., Hass, H., Remy, W. and Kerp, H. (1995) Nature 378: 244.
Taylor, T.N., Hass, H. and Kerp, H. (1997) Am. J. Bot. 84: 992-1004.
Tschermak-Woess, E. (1988) in: M. Galun (ed.) *Handbook of lichenology*, CRC Press, Boca Raton, pp. 39-92.
West, N.J. and Adams, D.G. (1997 Appl. Environ. Microbiol. 63: 4479-4484.

VIII

BIOLUMINESCENCE

Biodata of **Margo G. Haygood** author of the chapter entitled "*Luminous Bacteria*" (with co-author A. Scott)

Dr. Margo Haygood is an associate Professor of Marine Biology in Scripps Institution of Oceanography (University of California, San Diego, La Jolla CA). After spending a couple of years at the University of Tokyo, she received her Ph.D. (1984) from the Scripps Institution of Oceanography where she has served as a faculty member since 1987.

Dr. Haygood's research interest is in: Environmental microbiology; bacterial evolution, marine biotechnology; role of microbial symbionts in the production of bioactive compounds; iron acquisition and regulation in marine bacteria; physiology, and evolution and development of light organ symbioses. She gained honors and awards from sources such the NSF Graduate Fellowship, Japanese Ministry of Education and the office of Naval Research.

E-Mail: <u>mhaygood@ucsd.edu</u>

J. Seckbach (ed.), Journey to Diverse Microbial Worlds, 269-285.
© 2000 *Kluwer Academic Publishers. Printed in the Netherlands.*

LUMINOUS BACTERIA

MARGO HAYGOOD and **SCOTT ALLEN**
Marine Biology Research Division
Scripps Institution of Oceanography
University of California, San Diego
La Jolla, CA 92093-0202. USA.

1. Introduction

Although their size and structural complexity are limited, microbes are the champions of the biosphere in metabolic and biochemical diversity. One of the most enchanting manifestations of the biochemical diversity of microbes is bacterial bioluminescence. In the past, bacterial bioluminescence was more familiar to the public than it is today. The invention of the refrigerator light may be responsible for this decline. Decaying marine fish occasionally develop a population of visible luminous bacteria, but unless the specimen is viewed in darkness, the light will be overlooked. The bioluminescence of fish and meat due to bacteria has undoubtedly been observed with interest, if not understood, throughout human history. Bacteriological study of luminous bacteria began in the mid-nineteenth century and has continued to the present (Harvey 1952).

Study of the luminous bacteria has made important contributions to other fields. The phenomenon of quorum sensing, first elucidated in the luminous bacteria, has emerged as a profoundly important paradigm for intercellular communication among bacteria (Fuqua et al. 1996). Luminous bacteria and their genes are widely applied for detecting toxic contaminants, for monitoring release of genetically engineered microorganisms, and as reporters for biological research (Simpson et al. 1998; Chatterjee and Meighen 1995).

2. Taxonomy

The luminous bacteria are united by a single biochemical property, not by a common evolutionary history. In general, bioluminescence systems are diverse and have multiple independent origins. In the bacteria however, the enzymes responsible for bioluminescence are related and clearly have a common evolutionary origin. In contrast, the bacteria themselves occur in three groups within the proteobacteria, suggesting that the luminescence system may have been laterally transferred. Table 1 shows the species of luminous bacteria that have been cultivated and described. Within each of the three

Table 1. Species of luminous bacteria

Species	Typical habitat	Symbiotic hosts	Temperature preference[1]	Reference
Vibrio				
cholerae[2]	Freshwater, estuarine	–	Warm	(Hada et al. 1985)
harveyi	Seawater, enteric	–	Warm	(Baumann and Baumann 1981)
splendidus	Seawater	–	(Eurythermal)	(Baumann and Baumann 1981)
orientalis	Seawater	–	Eurythermal	(Yang et al. 1983)
fischeri subgroup				(Ruimy et al. 1994)
fischeri	Seawater, symbiotic	Fish: Monocentridae; Squid: *Euprymna, Sepiola*	Temperate	(Baumann and Baumann 1981)
logei	Seawater, symbiotic	Squid: *Sepiola*	Cold	(Fidopiastis et al. 1998)
salmonicida[3]	Fish pathogen	–	Cold	(Fidopiastis et al. 1999)
Photobacterium				
leiognathi	Seawater, symbiotic	Fish: three families of Perciformes	Warm	(Baumann and Baumann 1981)
phosphoreum	Seawater, enteric, symbiotic	Fish: diverse, six families in four orders	Cold	(Baumann and Baumann 1981)
Shewanella				
hanedai	Seawater	–	Cold	(Jensen et al. 1980)
woodii	Seawater, squid ink	–	Cold	(Makemson et al. 1997)
Photorhabdus				
luminescens	Symbiotic	Nematodes: *Heterorhabditis*	Warm	(Szallas et al. 1997)

[1]Warm, grows at 35°C, but not at 4°C; temperate, grows at neither 35°C nor 4°C; cold, grows at 4°C, but not at 35°C; eurythermal, grows at both 4°C and 35°C; ( ), variable among strains. [2]A few strains are luminous. [3]Only luminous with addition of aldehyde or autoinducer

main groups, the marine vibrios (including both *Vibrio* and *Photobacterium* species), the *Shewanella* species, and the *Photorhabdus* species, the luminous species are more closely related to non-luminous relatives than to members of the other groups. There are many non-luminous species of the genus *Vibrio*, and several non-luminous *Photobacterium* (*angustum*, *damselae*, *histaminum*, *iliopiscarium* and *profundum*). Most *Shewanella* species are not luminous. The closest relatives of *Photorhabdus* species are non-luminous nematode symbionts in the genus *Xenorhabdus*. Non-luminous strains are not uncommon among the luminous species. *V. salmonicida* was only recently recognized as luminescent because light production requires the addition of exogenous substrates (Fidopiastis et al. 1999). It is not known whether this may be true of other *Vibrio* species.

The obligate light organ symbionts of flashlight fishes (family Anomalopidae) form a coherent group within the genus *Vibrio*. Likewise, the symbionts of two species of deep-sea anglerfishes (suborder Ceratioidei) have been shown to be most closely related to *Vibrio* species (Haygood and Distel 1993). Neither of these groups of symbionts have been cultivated or named.

3.   Habitats

Bioluminescence is predominantly a marine phenomenon. Much of the ocean is dark, or dim enough for bioluminescence to be perceived, and most of the organisms in midwater environments (200-1000 m) are bioluminescent (Young 1983). Luminous bacteria are also found mainly in marine habitats. Freshwater luminous isolates (a few strains of *Vibrio cholerae*) are rare (Hada et al. 1985). Only one genus of terrestrial luminous bacteria exists: *Photorhabdus*.

4.   Enumeration of luminous bacteria

Surveys of luminous bacteria in seawater have been done many times in numerous locations. Usually the method used is plating on agar medium and counting visibly luminous colonies in the dark. There are several drawbacks to this approach. First, Lee and Ruby showed conclusively that in *V. fischeri* in Kanehoe Bay, viable cells (cells that can colonize a juvenile squid light organ) exceeded colony-forming units by one to two orders of magnitude (Lee and Ruby 1995). This discrepancy between total and cultivable bacteria is a classic phenomenon in microbial ecology, and makes measurements based on colony formation unreliable. Furthermore, it is now clear that some strains of bonafide bioluminescent bacteria do not glow when grown under standard conditions. *V. fischeri* strains symbiotic with the squid *Euprymna scalopes* are luminescent in the light organ habitat, but not in laboratory culture (Boettcher and Ruby 1990). Likewise, *V. salmonicida* produces light only in the presence of exogenous substrates (Fidopiastis, et al. 1999). This means that some luminous cells that can form colonies may be counted as non-luminous bacteria. Strains that contain luminescence genes but do not produce visible light can be readily identified by gene probing or PCR amplification of luminescence genes. However, it is unclear whether a particular strain thus identified is amplification or probing of luminescence genes are beginning to be more frequently employed for enumerating luminous bacteria (Lee and Ruby 1992). These are more sensitive than rRNA surveys and avoid cultivation artifacts, but cannot be correlated with light production, and are subject to other artifacts such as PCR bias.

5. Marine Habitats

5.1. SEAWATER

Keeping in mind these major caveats, we can summarize what is known about the presence of luminous bacteria in seawater. Based on plating experiments, marine luminous bacteria are ubiquitous, but rarely abundant, in seawater, typically occurring at levels of 1-10 cfu/ml in most locations. Surveys based on direct counting of 16S rRNA genes in nucleic acids extracted from seawater samples without cultivation have radically changed our view of the bacterial populations dominant in seawater, but have been of little use in studying the luminous bacteria. Only *P. phosphoreum* has been detected in molecular surveys (Fuhrman et al. 1993). Based on plate count surveys, *P. phosphoreum* is not significantly more abundant than other species, but occurs in colder, deeper water where the total bacterial population is less, so it may rise above the detection limit in these samples. Of course, rRNA gene sequences cannot verify whether the cells detected are actually capable of luminescence. In any case, luminous bacteria are not ordinarily dominant members of seawater populations. One exception to this generalization is the milky seas phenomenon in which the surface of the ocean glows. This is believed to be due to development of a population of luminous bacteria following a plankton bloom (Lapota et al. 1988).

The distributions of luminous species in seawater can be predicted largely by their temperature optima for growth. The best example is Ruby's classic study of relative abundance of *V. fischeri* and *V. harveyi* in coastal San Diego waters (Ruby and Nealson 1978). In months when water temperatures are cooler, *V. fischeri* predominated, in warmer months, *V. harveyi* predominated, in agreement with their relative temperature optima for growth. Likewise, in open ocean environments, warm water species such as *V. harveyi* predominate in surface waters, while *P. phosphoreum*, a psychrophilic species, is found at greater depths (Ruby et al. 1980).

5.2. ENTERIC HABITATS

Like other marine vibrios, luminous bacteria in the genus *Vibrio* and *Photobacterium* are common members of the enteric flora of marine animals. They have been found in fishes and crustaceans, and may be present in other animals. *P. phosphoreum* can be the dominant cultivable bacterium in the gut tracts of some fishes, such as hatchetfishes, and it is a normal member of the gut flora of cod (*Gadus morua*) (Ruby and Morin 1979; Dalgaard et al. 1997). In marine fish packaged in a modified atmosphere, the growth of *P. phosphoreum* is favored by high CO_2 and the high levels of trimethlyamine-N-oxide present in fish tissue. Thus, since it is always present, *P. phosphoreum* is virtually the sole spoilage agent of marine fish wherever modified atmosphere packaging is used for fish processing (Dalgaard et al. 1997). The enteric habitat may be the most important for luminous marine vibrios; in seawater they may simply be dispersing from one enteric habitat to another. This is an area that deserves much greater study, because we cannot understand the ecology of luminous bacteria until we determine the populations in this habitat, their activities, and their impact on other environments, such as seawater.

5.3.   LIGHT ORGAN SYMBIOSES

One of the most fascinating qualities of the luminous bacteria is their propensity for forming symbiotic associations with higher organisms. In these cases an animal, such as a fish, develops a specialized organ for maintaining the bacteria and uses the light produced for a variety of purposes, such as predation, defense against predation, and intraspecies communication. Except in one group, the tunicates, the bacteria are extracellular, and excess bacteria are released into the surrounding environment, thus contributing to seawater populations.

There are two categories of light organ symbioses, facultative and obligate. These terms refer to the bacterial point of view; in all cases, the association is obligate for the animal host. Specifically, although artificially created aposymbiotic animals can survive indefinitely in captivity, in nature they have not been found, suggesting that they are at a disadvantage in the absence of symbionts. In the facultative symbioses, the symbiotic bacteria are members of species that also occur free-living. The bacteria can be isolated readily from the light organ and released symbionts can be isolated from the environment. In the obligate symbioses, the bacteria have not been successfully isolated and there is no evidence that they persist outside the host, or inhabit other ecological niches.

5.3.1.   *Types Of Light Organ Symbioses:*
Fishes.   Light organ symbioses are most common among fishes (Haygood 1993). Twenty families of fishes contain species that have symbiotic light organs. In addition, *Lumiconger arafura* (Congridae, Anguilliformes) has been reported to have a luminescent organ derived from the intestinal tract that might house bacterial symbionts (Castle and Paxton 1984). A report claiming the discovery of light organ symbionts in myctophid and stomiiform fishes was shown conclusively to be an artifact (Foran 1991; Haygood et al. 1994). *P. phosphoreum* is the most promiscuous symbiont, forming associations with fishes from six families in four orders. All are fishes that live in deep, cold water, where *P. phosphoreum* is the dominant species of luminous bacteria, and is a common enteric inhabitant. All of the light organs are connected to the gastrointestinal tract and probably evolved independently from enteric associations. *P. leiognathi* occurs in light organs of fishes in three families, all within the order Perciformes. All are warm water species, consistent with the temperature preferences of *P. leiognathi*, and all of the light organs are modifications of the gastrointestinal tract. *V. fischeri* is found in the light organs of only one family, the Monocentridae, but it also occurs as a symbiont of squid (see below) and is thus the only species of luminous bacteria to be found in symbiotic associations in different animal phyla. Deep-sea anglerfishes (suborder Ceratioidei) are a prominent group in the midwater and deep sea. All but two families in the suborder have luminescent lures containing luminous bacteria located on the illicium, a modified dorsal

spine. None of the symbionts have been cultivated and sequence data are available for only two species, *Melanocetus johnsoni* and *Cryptopsaras couei* (Haygood and Distel 1993). These two species have distinct, but related, symbionts affiliated with the genus *Vibrio*. In contrast to the typical case in which hosts control light emission by physical barriers such as shutters or chromatophores, the deep-sea anglerfishes *Haplophryne mollis* and *Linophyrne arborifera* appear to exert some form of biochemical control on light emission (Hansen and Herring 1977; Herring and Munk 1994). The anomalopids are a family of tropical reef fishes with suborbital light organs used for hunting prey and communication. The symbionts are obligate. They form a cluster within the genus *Vibrio*, with a distinct sequence type found in each genus of the family (Haygood and Distel 1993). This pattern is what would be expected in an obligate symbiont that diverged in concert with its host.

Cephalopods. Two families of squids, the Sepiolidae and the Loliginidae contain species with light organs inhabited by symbiotic luminous bacteria. All are believed to be facultative associations. In sepiolids, *V. fischeri* colonizes light organs of *Euprymna* species and *V. fischeri* and *V. logei* both occupy light organs of *Sepiola* species, sometimes as a mixed culture (Fidopiastis et al. 1998). Sepiolids, which recruit their symbionts from free-living populations at each generation, have *V. fischeri* symbionts that exhibit genetic differentiation correlated with relationships among hosts, suggesting parallel divergence (Nishiguchi et al. 1998). In addition, the relative efficiency of colonization of the symbiotic strains for different hosts suggests that phenotypic specialization according to host has occurred (Nishiguchi et al. 1998). The relationship between *V. fischeri* and *Euprymna scolopes* has been a fertile field for investigation of the events surrounding light organ colonization. The presence of *V. fischeri* induces development of the light organ (McFall-Ngai 1994). Colonization by *V. fischeri* requires motility, a mannose binding adhesin, iron acqusition capability, luminescence, autoinducer synthesis, amino acid synthesis and cyclic AMP utilization (Ruby 1996).

Tunicates. Pyrosomes are pelagic colonial tunicates that have long been admired for their spectacular luminescence. The observation of bacteria-like bodies in the light organs led to the suggestion that symbiotic bacteria might be responsible for the light (Mackie and Bone 1978). This was proven by assay of bacterial luciferase, but bacteria have not been successfully cultivated (Leisman 1980). The pyrosomes are a distinct contrast to all other luminescent symbioses. The bacteria are intracellular, rather than in a glandular organ. The pyrosomes flash in response to mechanical stimulation, and the light expands from the site of stimulation sweeping across the entire colony (Bowlby et al. 1990). In most other bacterial light organs, the light is emitted continuously by the bacteria, and is controlled physically by the host. Pyrosomes appear to be able to control light production, as opposed to emission of light from the animal. This obligate symbiosis represents the extreme in adaptation of symbionts to the host environment.

5.3.2. *growth rates*

In general, growth of bacteria in symbiotic light organs is slow relative to the maximum growth rates observed in rich culture media. In fishes, direct measurements have been made in monocentrids and anomalopids. This is feasible because the light organs open to the exterior and released bacteria can be counted. In these environments, the doubling time of the bacteria ranges from 8 to 24 hours (Haygood et al. 1984). In the squid, *Euprymna scalopes*, a distinctly different pattern prevails. Ninety percent of the symbionts are expelled each morning before the squid retires for the day into the sand. The remaining bacteria repopulate the organ by nightfall (Ruby and Lee 1998). These symbionts evidently have frequent opportunities to grow rapidly.

5.3.3. Evolution of Light Organ Symbioses

In fishes it seems likely that most of the symbioses arose when free-living marine vibrios developed a stable association with a host. This occurred many times independently. The obligate symbioses would have evolved from facultative ones. The obligate symbionts are very specialized for the symbiotic habitat and have lost the traits necessary to adapt to varying conditions, including those of laboratory cultivation. They represent the extreme of symbiotic adaptation. However, it is apparent that a continuum of adaptation exists. The symbiotic strains of *V. fischeri* associated with *Euprymna* sp. are genetically and phenotypically distinct from other strains of *V. fischeri*; clearly some degree of genetic isolation from strictly free-living strains has developed (Nishiguchi et al. 1998). More direct transmission from one generation of hosts to the next makes the symbiont less dependent on other habitats, and releases it from selection for free-living traits, as well as isolates it genetically from free-living strains. Strains of *P. phosphoreum* do not appear to be specialized according to host. The bacteria probably pass freely between seawater, enteric and light organ populations of several hosts in the same habitat and have little opportunity to diverge. *Euprymna* sp. rely upon free-living *V. fischeri* as a source of inoculum for the next generation, but they are able to dominate the *V. fischeri* population locally by releasing bacteria, resulting in genetic divergence with maintenance of free-living capability. The light organ symbioses of the fish families Monocentridae and Anomalopidae make a remarkable comparison. All members of both families have light organ symbionts, and thus probably arose from a light organ bearing ancestor in both cases. The families are closely related and are believed to have arisen at the same time, in the late Cretaceous. Both symbioses have had equal time to evolve, yet the monocentrids are facultative and the anomalopids are obligate. Perhaps seawater populations of the monocentrid strains of *V. fischeri* were reliable enough that no selective pressure existed to evolve another means of transmission. Maybe the anomalopids experienced unreliable availability of symbionts and experienced selective pressure for a more direct means of transmission rather than relying on free-living populations. This would relieve the selective pressure for the symbionts to maintain the metabolic adaptability required for free-living existence and allow them to become highly specialized for the symbiotic habitat. Certainly this comparison emphasizes that progression from facultative to obligate symbiosis is not inevitable, and that either type can be stable over long periods of time.

5.4. AQUACULTURE PATHOGENS

Like other marine vibrios, luminous bacteria can occur as pathogens. Luminous infections in crustaceans such as amphipods and shrimp are occasionally observed. *V. harveyi* is the luminous bacterium usually isolated. Recently, luminous bacteria, especially *V. harveyi,* have emerged as important pathogens in shrimp aquaculture (Lavilla-Pitogo and de la Pena 1998; Lavilla-Pitogo et al. 1998; Moriarty 1998). Near total collapse of the shrimp population in aquaculture ponds due to infection is not uncommon. The luminous vibrios are constantly present at low levels in the source water, and can reach high levels in the water and sediments of the ponds. Use of antibiotics has led to selection of antibiotic resistant strains. Some success in controlling these infections has been achieved by adding innocuous competing bacteria to ponds (Moriarty 1998). Intensive culture may select for increased pathogenicity, as in nosocomial strains of human pathogens that become established in hospitals. Crowded conditions select for increased virulence because the host need not survive long before the pathogen encounters another host.

5.5. FISH PATHOGENS

V. salmonicida, a pathogen that causes cold water vibriosis in salmon, trout and cod was recently found to be capable of bioluminescence (Fidopiastis et al. 1999). This had not been noticed in the past because the bacteria require exogenous aldehyde in order to produce detectable bioluminescence. Fidopiastis et al. suggested that the luciferase reaction may enhance virulence. This would occur because, in the absence of aldehyde, the luminescence reaction produces hydrogen peroxide from superoxide, both of which could have a toxic effect on host tissue. This effect has been demonstrated *in vivo* in *Escherichia coli* carrying plasmids bearing *V. harveyi* luciferase genes (Gonzalez-Flecha and Demple 1994).

5.6. SEDIMENT

Although luminous bacteria are not usually major members of sediment populations, in certain circumstances they can reach much higher levels in sediments than in overlying water. This is the case in habitats influenced by the *Euprymna scolopes/V. fischeri* symbiosis, where sediment counts are 70 fold higher than in the seawater (Lee and Ruby 1994). There are also substantial numbers of luminous bacteria in sediments of shrimp aquaculture ponds (Moriarty 1998).

6.   Terrestrial Habitats

The genus *Photorhabdus* contains the only terrestrial luminous bacteria. One species has been described, *P. luminescens,* but one or more others are believed to exist (Liu et al. 1997; Szallas et al. 1997). Like members of the genus *Xenorhabdus,* which are closely related to *Photorhabdus* but are not luminous, *Photorhabdus* strains are symbionts of nematodes that are in turn, pathogens of insects (Boemare et al. 1997). *P. luminescens* and other *Photorhabdus* species are specific symbionts of nematodes of the genus *Heterorhabditis. Xenorhabdus* species are symbionts of *Steinernema* nematodes. Although the nematode host is cosmopolitan, the bacteria are never isolated directly from soil, but are only found in the host. Historical accounts of glowing wounds are attributed to *Photorhabdus,* and clinical isolates are known. The bacteria are carried within the intestine of infective juveniles, a dispersal stage of the nematode. When the nematode penetrates the insect, the bacteria grow and provide food, a signal required to exit the dispersal stage, and other reproductive factors for the nematode. When the bacteria reach stationary phase in the insect, they begin to glow. The luminescence is hypothesized to aid in dispersal of the infective juveniles by attracting scavengers. If injected into an insect, the bacteria alone can kill the insect. Only *Photorhabdus* strains can support the growth of *Heterorhabditis* nematodes, or colonize the infective juvenile stage. *Photorhabdus* has an impressive arsenal of weapons with which to attack on two fronts. One is to overcome the defenses of the insect, the other is to inhibit competing saprophytic bacteria. Means to achieve the former include several insecticidal proteins and inhibition of the host immune system by a diffusible compound. For the latter, the bacteria produce four classes of antibiotics, as well as proteases and lipases. The bacteria also produce crystalline protein inclusion bodies high in methionine. Apparently these are not insect toxins, but may serve as a food for the nematode, since the nematode cannot develop in strains that have mutations in these proteins (Bintrim and Ensign 1998). The nematode/bacterial associations, as well as the insecticidal proteins derived from them, are targets for development for pest control.

7.   Biochemistry

The same biochemical reaction is responsible for luminescence in all luminous bacteria (Tu and Mager 1995; Wilson and Hastings 1998). Luciferase is the term used for light-emitting enzymes. Luciferases do not consititute an evolutionarily related enzyme family, rather they have multiple evolutionary origins. All luminescence reactions are oxidations requiring oxygen. Most luciferases act upon a luciferin, a substrate specific to the luminescence reaction that is sequestered by the organism until used for light production.

Bacterial luciferase catalyzes the simultaneous oxidation by oxygen of reduced flavin mononucleotide ($FMNH_2$) and a long-chain aldehyde (tetradecanal *in vivo*) with the emission of light at approximately 490 nm:

$$FMNH_2 + O_2 + RCHO \rightarrow FMN + RCOOH + H_2O + light$$

Thus bacterial bioluminescence is distinctive in several respects. It has two substrates, unlike a classical luciferin/luciferase reaction. Neither substrate is a unique, specialized structure as in other luciferins. Finally, rather than using a luciferin that is sequestered specifically for light production, $FNMH_2$ is an integral part of the redox economy of the cell. Consequently, when the enzyme system is induced, light is produced as a continuous glow rather than a flash.

Because this reaction is unique to bacteria, and distinct from all other bioluminescence systems, *in vitro* assay of bacterial luciferase is diagnostic for the presence of luminous bacteria. This assay has been used to confirm the bacterial origin of luminescence in symbioses in which bacteria cannot be cultivated, such as pyrosomes, anomalopid fishes, and deep-sea anglerfishes (Leisman 1980).

Bacterial luciferase is a heterodimer. The alpha subunit is about 40 KDa, and the beta subunit is about 35 KDa. The corresponding genes are *luxA* and *luxB*. The two genes are related, probably due to gene duplication. Bacterial luciferase has an extraordinarily high affinity for oxygen. The $FMNH_2$ is provided to the luciferase by flavin reductases that catalyze the reduction of FMN by NAD(P)H. The long-chain aldehyde is provided by recycling of the oxidized product by a fatty acid reductase complex. *Lux C, D* and *E* code for this enzyme complex.

Secondary emitters are common in bioluminescence. In bacteria, two secondary emitters have been well characterized: LumP and YFP. LumP occurs in *P. phosphoreum* and *P. leiognathi* and blue shifts the emission spectrum to 475 nm. LumP is a 21 KDa protein with a lumazine fluorophore. YFP (yellow fluorescent protein) occurs in one strain of *V. fischeri*. It shifts emission toward the red to 540 nm, and has FMN as its fluorophore. LumP and YFP are homologous proteins related to riboflavin synthase.

The core enzymes required for bioluminescence are luciferase and the fatty acid reductase complex (Meighen 1994). If the genes for these enzymes, *lux A, B, C, D,* and *E*, are cloned into another bacterium such as *Escherichia coli* and expressed, light production will ensue. $FMNH_2$ synthesis and recycling and aldehyde synthesis are provided by the host cell, so that *lux A, B, C, D* and *E* are sufficient to confer light production. In all luminous bacteria studied so far these core genes are homologous and are arranged in the same order: CDABE, forming a cassette conducive to lateral transfer of bioluminescence among bacteria. This suggests that a common evolutionary origin, followed by lateral transfer, accounts for the distribution of bioluminescence among bacteria today. Accessory genes for secondary emitters, riboflavin biosynthesis and regulation are often, but not always, linked to the core genes.

8.   Regulation

Bioluminescence in luminous bacteria is not constitutive; it is typically induced in late exponential phase. The increase in light can be 3-4 orders of magnitude, one of the most dramatic enzyme inductions known in bacteria. The fact that luminous bacteria produce light only when population densities are high enough to allow light to be perceived suggests that light itself has a selective advantage for the bacteria in certain circumstances. This is clearly the case for facultative light organ symbionts, but is less obvious for other luminous bacteria. Most luminous bacteria use an intercellular signaling system to sense population density and regulate luminescence. Although there are common themes in how this is accomplished, diversity is the rule for the detailed mechanisms.

Most luminous bacteria, with the conspicuous exception of *Photorhabdus*, secrete a small molecule, an acyl homoserine lactone, known as an autoinducer. When the concentration of this molecule rises due to population growth, a signaling cascade triggers light production. This phenomenon was originally described in the marine luminous bacteria as autoinduction. Recently we have learned that many Gram-negative bacteria use acyl homoserine lactone mediated population sensing for regulating cooperative activities (Hardman et al. 1998). The types of activites regulated are production of extracellular enzymes and antibiotics, conjugation, and virulence, all activites that are more successful when carried out by a group than by an isolated cell. In this context, the phenomenon has been labeled quorum sensing.

Quorum sensing systems of two bacteria, *V. fischeri* and *V. harveyi* are the best characterized (Bassler and Silverman 1995; Dunlap 1999; Freeman and Bassler 1999a; Freeman and Bassler 1999b; Surette et al. 1999). Neither of these species responds to the signals produced by the other; each has a distinct chemical channel for communication. Despite extensive study, they are not completely elucidated, but some useful comparisons can be made. The reader should bear in mind that these organisms may have additional quorum sensing systems that may or may not impact bioluminescence regulation. In addition, many other factors regulate bioluminescence and may interact with the systems described. Each species has two signaling pathways, one of which is similar between the two species. Both species use acyl homoserine lactone signals, and two-component regulators. Beyond these themes, most details are different. Table 2 lists genes involved in regulation of bioluminescence; Table 3 shows regulatory cascades.

V. fischeri has two acyl homoserine lactone autoinducers, VAI-1, β-ketocaproyl homoserine lactone, and VAI-2, octanoyl homoserine lactone. VAI-1 positively regulates luminescence and VAI-2 has a negative effect. *V. harveyi* has one acyl homoserine lactone autoinducer, HAI-1, β-hydroxybutyryl homoserine lactone, and another autoinducer of unknown structure, HAI-2. Both positively regulate luminescence.

In *V. fischeri luxR* is a required transcriptional activator of the *lux* operon containing the luciferase and fatty acid reductase genes. VAI-1 binds LuxR and facilitates DNA binding. A distinctive feature of *V. fischeri* not found in other luminous bacteria is that *luxI*, the gene responsible for VAI-1 synthesis, is located in the *lux* operon. This results in positive feedback; as increasing VAI-1 concentrations promote *lux* operon transcription, VAI-1 synthesis is concomitantly increased. Other luminous bacteria have induction responses similar to *V. fischeri*; there may be positive feedback systems involved that have yet to be discovered. VAI-2 is synthesized by the protein encoded by *ainS*, which is not located in the *lux* operon. VAI-2 negatively impacts LuxR activity either via a two component hybrid sensor kinase, AinR, or by direct competitive inhibition of VAI-1 binding to LuxR.

V. harveyi also has an obligatory transcriptional activator called LuxR, but it is not homologous to *V. fischeri* LuxR. In contrast to *V. fischeri*, in *V. harveyi* autoinduction signals impact transcription via LuxO, a transcriptional repressor of the two component response regulator family. LuxO repressor is active when phosphorylated and inactive when dephosphorylated. Signals from the two autoinducers are integrated by LuxU and relayed to LuxO. HAI-1 is sensed by LuxN, a two component hybrid sensor kinase somewhat similar to AinR. HAI-2 is sensed by LuxPQ, a two component hybrid sensor kinase. Thus changes in concentration of both HAI-1 and HAI-2 impact LuxO activity via a phosphorelay cascade integrated by LuxU. The two systems are functionally redundant; a mutation in either pathway does not eliminate population density-dependent luminescence induction. HAI-1 appears to be specific to *V. harveyi*, but HAI-2 is produced by other species of bacteria as well. Thus *V. harveyi* may be able to monitor both its own population and that of other bacteria that share its habitat.

Table 2. Genes involved in regulation of bioluminescence

Organism	Gene name	Homologous genes	Function	Class
V. fischeri	*luxI*		VAI-1 synthesis	luxI/LuxR
	luxR		VAI-1 binding Promoter binding Transcriptional activator	luxI/luxR
	ainS	*V. harveyi luxM*	VAI-2 synthesis	
	ainR	*V. harveyi luxN*	VAI-2 response?	Two component hybrid sensor kinase
V. harveyi	*luxLM*	*V. fischeri ainS*	HAI-1 synthesis	
	luxN	*V. fischeri ainR*	HAI-1 response	Two component hybrid sensor kinase
	luxS		HAI-2 synthesis	
	luxPQ		HAI-2 response	Two component hybrid sensor kinase
	luxU		Integrator of HAI-1 and HAI-2 signals, controlled by *luxU* phosphorylation	
	LuxO		Transcriptional repressor,	Two component response regulator
	LuxR	Not *V. fischeri luxR*	Required transcriptional activator	

Table 3. Regulatory cascades in *V. fischeri* and *V. harveyi*

Organism	Type	Autoinducer synthesis	Autoinducer (effect)		Response
V. fischeri	luxI/luxR	*luxI*	VAI-1 (+)	→	*luxR* → *lux* operon
	II	*ainS*	VAI-2 (−)	→ *ainR*	
V. harveyi	II	*luxLM*	HAI-1 (+)	→ *luxN*	*luxO* → *lux* operon
	III	*luxS*	HAI-2 (+)	→ *luxPQ* → *luxU*	*luxR*

Even though they do not appear in *V. harveyi*, homologs of *V. fischeri* luxI and luxR are found in many other Gram-negative bacteria that use acyl homoserine lactone quorum sensing. *V. harveyi* LuxS homologs are also found in other bacteria, suggesting that the HAI-2 class of autoinducers may be as widespread and important as the VAI-1 class. Just as *V. harveyi* may be able to monitor and influence its neighbors, higher organisms may intervene in bacterial quorum sensing to inhibit pathogens and fouling. Marine red algae produce furanones that specifically inhibit quorum sensing regulated activities in marine bacteria, including bioluminescence, swarming, motility and exoenzyme synthesis.

In addition to quorum sensing regulation, bioluminescence expression is affected by a multitude of other factors, including iron, oxygen, catabolite repression and others (Dunlap 1991). These factors could affect the known regulatory circuits or act independently.

9. Function and Evolution

Although it would seem obvious that the function of bacterial bioluminescence is light production, the question of whether light production was its original function, and how it evolved, is unresolved. Bioluminescence is an expensive function. In addition to the cost of enzyme synthesis, it consumes NAD(P)H and ATP in the luciferase reaction and aldehyde recycling reaction. The complex regulation dedicated to luminescence supports the notion that it has an important function. The prevalence of quorum sensing regulated bioluminescence suggests that production of perceptible light is currently important, but functions based on light perception could only have evolved fairly recently, after the evolution of vision in animals. In bioluminescent symbioses, the light itself is obviously critical, and symbionts such as *V. fischeri* have adapted regulation accordingly. However, if luminescence had no physiological function it seems likely that dark mutants would frequently take over light organ populations. Population density regulated light production might also aid in dispersal of enteric bacteria, which could achieve populations in fecal pellets sufficient to produce perceptible luminescence and aid re-entry into enteric habitats by attracting scavengers.

Many possible physiological functions of luminescence have been proposed. These are not necessarily mutually exclusive, and could be synergistic. These would allow evolution of luminescence prior to visual animals. If luciferase has a physiological function(s), perhaps light is not a required product. This suggests that non-luminous bacterial luciferases could exist. One of the most frequently suggested functions is scavenging of oxygen to reduce oxygen toxicity, based on the high affinity of luciferase for oxygen. A hint that this may have some validity has emerged from studies of the *V. fischeri/E. scalopes* symbiosis. The host makes a haloperoxidase that inhibits bacteria by producing superoxide (Small and McFall-Ngai 1999). Luminescence appears to be required for persistent colonization, possibly due to the protection conferred by oxygen scavenging by luciferase (Ruby 1996). In a similar vein, luminescence could aid in maintaining redox balance under iron-limited conditions, when electron transport mediated oxidation of NADH and $FMNH_2$ is inhibited. This could be advantageous to symbionts whose growth is controlled by host restriction of iron availability (Haygood and Nealson 1985).

A reasonable scenario for the evolution of bacterial luminescence is that the enzyme evolved to fill a physiological function. After the appearance of visual animals late in evolution, separately evolved quorum sensing systems were combined with bioluminescence genes in various combinations in different bacteria, as a selective advantage of light itself emerged. The core genes were laterally transferred at least once, to *Photorhabdus*

10. Conclusion

Luminous bacteria have long been viewed as a delightful oddity among bacteria. Research efforts of the past several decades have shown that the luminous bacteria can play a vital role in bacteriology, as excellent model systems for revealing mysteries that reach far beyond the luminous bacteria themselves. The mechanisms of bacterial intercellular communication, and the evolution and mechanisms underlying stable relationships between bacteria and symbiotic hosts .are areas in which these fascinating organisms are proving bountiful sources of knowledge.

9.   References

Bassler, B. L. and Silverman, M. R. (1995). in: J. A. Hoch and T. J. Silhavy (eds.) Two-Component Signal Transduction. Washington, D.C., ASM Press pp 431-445.

Baumann, P. and Baumann, L (1981) in: M. P. Starr, H. Stolp, H. G. Trüper, A. Balows and H. G. Schlegel.(eds) The Prokaryotes New York, Springer-Verlag pp 1302-1331.

Bintrim, S. B. and Ensign, J. C. (1998). J. of Bacteriol. 180: 1261-1269.

Boemare, N., Givaudan, A., Brehelin, M., Laumond, C. (1997). Symbiosis 22: 21-45.

Boettcher, K. J. and Ruby, E. G. (1990). J. Bacteriol. 172: 3701-3706.

Bowlby, M. R., Widder, E. A., Case, J.F. (1990). Biol. Bull. 179: 340-350.

Castle, P. H. J. and Paxton, J. R. (1984). Copeia 1984: 72-81.

Chatterjee, J. and Meighen, E. A. (1995). Photochem. and Photobiol. 62: 641-650.

Dalgaard, P., Manfio, G. P., Goodfellow, M.. (1997). Zentralblatt fuer Bakteriol. 285: 157-168.

Dunlap, P. V. (1991). Photochem. Photobiol. 54: 1157-1170.

Dunlap, P. V. (1999). J. Molec. Microbiol. Biotechnol. 1: 5-12.

Fidopiastis, P. M., Sorum, H., Ruby, E. G. (1999). Arch. of Microbiol. 171: 205-209.

Fidopiastis, P. M., Von Boletzky, S., Ruby, E. G. (1998). J. of Bacteriol. 180: 59-64.

Foran, D. (1991). J. Exp. Zool. 259: 1-8.

Freeman, J. A. and Bassler, B. L. (1999a). Mol. Microbiol. 31: 665-678.

Freeman, J. A. and Bassler, B. L. (1999b). J. Bacteriol. 181: 899-906.

Fuhrman, J. A., McCallum, K., Davis, A.A. (1993). Appl. Environ. Microbiol. 59: 1294-1302.

Fuqua, C., Winans, C., Greenberg, E.P. (1996). Ann. Rev. Microbiol. 50: 727-751.

Gonzalez-Flecha, B. and Demple, B. (1994). J. Bacteriol. 176: 2293-2299.

Hada, H. S., J. Stemmler, J., Grossbard, M.L., West, P.A., Potrikus, C.J., Hastings, J.W., Colwell, R.R.(1985). System. Appl. Microbiol. 6: 203-209.

Hansen, K. and Herring, P. J. (1977). J. Zool. Lond. 182: 103-124.

Hardman, A. M., Stewart, G. S. A. B., Williams, P. (1998). Antonie van Leeuwenhoek 74: 199-210.

Harvey, E. N. (1952). Bioluminescence. New York, Academic Press.

Haygood, M. G. (1993). Crit. Rev. Microbiol. 19: 191-216.

Haygood, M. G. and Distel, D. L. (1993). Nature 363: 154-156.

Haygood, M. G., Edwards, D. B., Mowlds, G., Rosenblatt, R. H. (1994). J. Exper. Zool. 270: 225-231.

Haygood, M. G. Nealson, and K. H. (1985). Symbiosis **1**: 39-51.

Haygood, M. G., Tebo, B.M., Nealson, K.H. (1984). Mar. Biol. **78**: 249-254.

Herring, P. J. and Munk, O. (1994). J. Mar. Biol. Assoc. UK **74**: 747-763.

Jensen, M. J., Tebo, B.M., Baumann, P., Mandel, M., Nealson, K.H. (1980). Curr. Microbiol. **3**: 311-315.

Lapota, D., Galt, C., Losee, J. R., Huddell, H. D., Orzech, J. K., Nealson, K. H.(1988). J. Exper. Mar. Biol. Ecol. **119**: 55-82.

Lavilla-Pitogo, C. R. and De la Pena, L. D. (1998). Fish Pathol. **33**: 405-411.

Lavilla-Pitogo, C. R., Leano, E. M., Paner, M. G. (1998). Aquaculture **164**: 337-349.

Lee, K. H. and Ruby, E. G. (1992). Appl. Environ. Microbiol. **58**: 942-947.

Lee, K. H. and Ruby, E. G. (1994). Appl. and Environ. Microbiol. **60**: 1565-1571.

Lee, K. H. and Ruby, E. G. (1995). Appl. and Environ. Microbiol. **61**: 278-283.

Leisman, G., Cohn, D.H. and Nealson, K.H. (1980). Science **208**: 1271-1273.

Liu, J., R. Berry, R., Poinar, G., Moldenke, A. (1997). Int. J. System. Bacteriol. **47**: 948-951.

Mackie, G. O. and. Bone, Q. (1978). Proc. R. Soc. (B) **202**: 483-495.

Makemson, J. C., Fulayfil, N. R., Landry, W., Van Ert, L. M., Wimpee, C. F., Widder, E. A., Case, J. F. (1997). Int. J. System. Bacteriol. **47**: 1034-1039.

McFall-Ngai, M. J. (1994). Amer. Zool. **34**: 554-561.

Meighen, E. A. (1994). Ann. Rev. Genetics **28**: 117-139.

Moriarty, D. J. W. (1998). Aquaculture **164**: 351-358.

Nishiguchi, M.K., Ruby, E.G. and McFall-Ngai, M.J. (1998) Appl. Environ. Microbiol. 64: 3209-3213.

Ruby, E. G. (1996). Ann. Rev. Microbiol. **50**: 591-624.

Ruby, E. G. and Lee, K. H. (1998). Appl. Environ. Microbiol. **64**: 805-812.

Ruby, E. G. and Morin, J. G. (1979). Appl. Environ. Microbiol. **38**: 406-411.

Ruby, E. G. and Nealson, K. H. (1978). Limnol. Oceanogr. **23**: 530-533.

Ruby, E.G., Greenberg, E.P. and Hastings, J.W. (1980) Appl. and Environ. Microbiol. **39**: 302-306.

Ruimy, R., V. Breittmayer, V., Elbaze, P., Lefay, B., Boussemart, O., Gauthier, M., Christen, R.(1994). Int. J. Sytem. Bacteriol. **44**: 416-426.

Simpson, M. L., Sayler, G. S., Applegate, B. M., Ripp, S., Nivens, D. E., Paulus, M. J., Jellison, G. E., Jr. (1998). Trends Biotech. **16**: 332-338.

Small, A. L. and McFall-Ngai, M. J. (1999). J. Cell. Biochem. **72**: 445-457.

Surette, M. G., Miller, M.B., Bassler, B.L.(1999). Proc. Natl. Acad. Sci. USA **96**: 1639-1644.

Szallas, E., Koch, C., Fodor, A., Burghardt, J., Buss, O., Szentirmai, A., Nealson, K. H., Stackebrandt, E. (1997). Int. J. of System. Bacteriol. **47**: 402-407.

Tu, S. C. and Mager, H. I. X. (1995). Photochem. and Photobiol. **62**: 615-624.

Wilson, T. and Hastings, J. W. (1998). Ann. Rev. Cell Dev. Biol. **14**: 197-230.

Yang, Y., Yeh, L.P., Cao, Y., Baumann, L., Baumann, P., Tang, S.T., Beaman, B. (1983). Curr. Microbiol. **8**: 95-100.

Young, R. E. (1983). Bull. Mar. Sci. **33**: 829-845.

IX

VERSATILE MICROBIAL LIFE ON THE EDGE

Biodata of **David L. Valentine** contributor of *"Diversity of Methanogens"* (with co-author David L. Boone).

David Valentine received his B.S. in chemistry and biochemistry (1995) and M.S. in chemistry (1996) from the University of California, San Diego. He also received an additional M.S. in Earth System Science from the University of California at Irvine, in 1997. He is currently a Ph.D. candidate in the Department of Earth System Science at the University of University of California, Irvine. His research interests include anaerobic microbial ecology, isotope biogeochemistry, and the geobiology of methane hydrates.

E-mail: **dvalenti@uci.edu**

Biodata of **David R. Boone** co-author of *"Diversity of Metanogens"* with author David Valentine.

Dr. David Boone is a Professor of Environmental Microbiology at Portland State University (Portland, Oregon). He received his Ph.D. in Microbiology and Cell Science from the University of Florida in 1977. Dr. Boone is a Fellow of the American Academy of Microbiology, a Trustee of Bergey's Manual Trust, and the chair of the ICSB Subcommittee for taxonomy of Methanogens. His research interests are physiology, ecology and phylogeny of methanogens and other organisms involved in anoxic degradation of organic matter.

E-mail: **booned@pdx.edu**

J. Seckbach (ed.), Journey to Diverse Microbial Worlds, 289-302.
© 2000 *Kluwer Academic Publishers. Printed in the Netherlands.*

DIVERSITY OF METHANOGENS

DAVID L. VALENTINE[1] and **DAVID R. BOONE**[2]
*[1]University of California, Department of Earth System Science
Irvine, CA 92697-3100, USA* and *[2]Portland State University
Department of Environmental Biology, P.O. Box 751
Portland, OR 97207-0751, USA*

1. Introduction

Methanogens are a phylogenetically distinct group of strictly anaerobic *Archaea*, characterized by the ability to produce methane as their major metabolic product. They are the only known organisms that produce a hydrocarbon as their catabolic end product. Methanogens thrive in habitats devoid of oxygen, where they play an important role in the degradation of organic matter. These organisms are further characterized by additional similarities: 1) their catabolic substrate range is limited to simple molecules, generally not exceeding two carbons in size, 2) their membranes contain novel lipids not found in most other microorganisms, 3) they lack peptidoglycan containing muramic acid, and 4) they all utilize a very similar catabolic mechanism.

Though all methanogens share several key characteristics, there is also exceptional diversity found within this group. Much of the diversity exists within the context of environmental conditions, including extremes of environmental conditions. Methanogens are capable of thriving at extremes of temperature, thriving in permanently cold Antarctic lakes and marine sediments, and in near-boiling hydrothermal vents. They thrive inside rocks, located miles below the surface of the planet, as well as in areas of extreme pH, salinity, and nutrient limitation (Table 1). Methanogens are diverse in other aspects as well. They exist in a range of shapes and sizes, occupying different environmental niches, and producing many unique compounds; they are also capable of performing many unusual functions.

Methanogens are important environmentally, economically, and scientifically. The atmospheric concentration of their catabolic product, methane, has been increasing rapidly over the last century due to human activities. Methane is an important greenhouse gas and also contributes to ozone depletion (Cicerone and Oremland, 1988). Around 75% of the methane released into the atmosphere is believed to have been produced by methanogens, leading to an interest in environmental aspects of methanogenesis. Methanogenesis also plays a key role in the global carbon cycle as the terminal step in the degradation of organic matter, as well as by supporting many aerobic methanotrophic communities. Economically, methane is a major component of natural gas and is an important source of heat and energy. In addition, an estimated 10% of the feedstock given to cattle is lost to methanogenesis within the rumen (Hungate, 1966). Scientifically, methanogens constitute a large portion of the *Archaea* (Woese, 1987), a domain of life which is considered to be distinct from Eukarya and true Bacteria. Methanogenesis is thought to be one of the most ancient forms of cellular metabolism, and may have been the original method of energy acquisition in the evolution of life.

2. Physiological Diversity

Methanogens occupy a suite of unusual environments on Earth, and are nearly ubiquitous in environments where oxygen is absent. Much of the diversity of methanogens is due to selective environmental pressures. Methanogens have adapted to almost every conceivable environment on Earth including environments with

Table 1. Habitats of methanogens

General Environment	Habitats Harboring Methanogens	Examples
Terrestrial	Swamps, Marshes, Bogs Tundra, Floodplains, Fens Soils, Trees	Everglades Alaskan North Slope Tree rot
Aquatic	Sediments (lake and ocean) Anoxic waters, Salt marsh, Ocean waters	Lake Constance Cariaco Trench Tropical North Pacific
Geological	Geothermal Hydrothermal Deep terrestrial subsurface	Yellowstone Deep sea vents Within basalt
Animals	Ruminal Termites, Human gut Symbionts	Cows, Sheep, etc. Termite hindgut Protozoa
Anthropogenic	Waste digestors Rice paddies Landfills	Sewage sludge Soil rhizosphere

extreme temperature, salinity, pH, and pressure. Methanogens can exist at all temperatures between freezing and boiling, at all pH's from acidic (below 5) to alkaline (above 10), and at all salinities from freshwater to saturated brine (Fig. 1). In addition to physical settings, methanogens have adapted to environments in an ecological sense. They are found to occupy many unusual niches, and their metabolism varies accordingly.

2.1. CATABOLISM

Catabolism is the portion of metabolism that generates energy for an organism. In methanogens, the catabolic pathway always involves the production of methane. The catabolic diversity of methanogens is best explained by examining their role in degradation of organic material. Organic material is generally degraded in a series of steps in which several types of organisms participate. Large organic molecules are first broken down by extracellular enzymes; the resulting compounds can be consumed by various fermenting organisms. Fermentation products like H_2 and organic acids are then oxidized to CO_2 and H_2O by respiring organisms. Respiring organisms rely on the presence of a terminal oxidant, and will generally consume the strongest available oxidant until fully depleted. Oxygen is utilized first because it yields the most energy, followed by NO_3^-, Fe^{3+}, Mn^{4+}, and then SO_4^{2-} (Cord-Ruwisch et al., 1988). Methanogenesis generally occurs only after SO_4^{2-} is fully depleted, as sulfate-reducing bacteria are able to outcompete methanogens for common substrates (Hoehler et al., 1998). Carbon dioxide acts as the terminal oxidant during methanogenesis. Unlike other oxidants, CO_2 is never depleted during the decomposition of organic matter because it is simultaneously produced by the same processes of degradation. The separation of oxidants is observed spatially in many sediments or temporally during the digestion of organic matter.

Though many methanogens perform similar functions in the overall decomposition of organic matter, the dynamics of the environment and the available metabolites may differ. Most methanogens have adapted by specializing in one form of metabolism (Table 2). The majority of methanogens are hydrogenotrophic and utilize a CO_2 reduction pathway with either H_2 or formate ($HCOO^-$) acting as the reductant. Some methanogens are aceticlastic and utilize a pathway in which they cleave the carbon-carbon bond of acetate to form CH_4 and CO_2 (also called acetate fermentation). In this case, organic matter is not oxidized fully to CO_2, but rather the methyl group of acetate is converted directly to methane. Other methanogens are methylotrophic and utilize methylated compounds like methanol, methylamines, and dimethylsulfide; many of these organisms oxidize some of their substrate to CO_2 in order to generate reducing equivalents for CH_4 production. A few CO_2 reducing methanogens are even capable of utilizing longer chain alcohols to generate reducing equivalents.

Methylated compounds are important methane precursors in saline environments and in SO_4^{2-} reducing environments. Otherwise, CO_2 reduction and acetate fermentation are the dominant methanogenic pathways. Both of these latter pathways are important in anoxic sediments, anaerobic digestors, marshes, and flooded soils, though one form of metabolism often dominates in any particular setting. Several environments tend to favor CO_2 reduction, especially geothermal and hydrothermal systems where H_2 is delivered from below. The guts of animals, especially the rumen,

Table 2. Catabolic diversity of methanogens

Primary Substrate	Catabolic Reaction/ Type of Methanogenesis	Representative Environments	Organisms	Comments
H_2 (Hydrogen)	$4H_2 + CO_2 \rightarrow CH_4 + 2H_2O$ CO_2 Reduction or Hydrogenotrophic	Most environments	Most methanogens	most common form of methanogenesis
$HCOO^-$ (formate)	$4HCOO^- + 4H^+ \rightarrow CH_4 + 3CO_2 + 2H_2O$ CO_2 Reduction	Rumen Waste digestor	Many methanogens	
CH_3CH_2OH (alcohols)	$2CH_3CH_2OH + CO_2 \rightarrow CH_4 + 2CH_3COOH$ CO_2 Reduction	?	*Methanomicrobiales*	environmental importance unknown
CO (carbon monoxide)	$4CO + 2H_2O \rightarrow CH_4 + 3CO_2$ CO Reduction	?	*Methanobacteriales* *Methanosarcinales*	environmental importance unknown
CH_3COO^- (acetate)	$CH_3COO^- + H^+ \rightarrow CH_4 + CO_2$ Aceticlastic or Acetate Fermentation	Marsh/ Waste digestor Freshwater	*Methanosarcinales*	environmentally important
$(CH_3)_3N$ (trimethyamine)	$4(CH_3)_3N + 6H_2O \rightarrow 9CH_4 + 3CO_2 + 4NH_3$ Methylotrophic or C_1	Marine sediments Saline waters	*Methanosarcinales*	non-competitive substrate
CH_3OH (methanol)	$4CH_3OH \rightarrow 3CH_4 + CO_2 + 2H_2O$ Methylotrophic or C_1	?	*Methanosarcinales*	environmental importance unknown
CH_3OH (methanol)	$CH_3OH + H_2 \rightarrow CH_4 + H_2O$ Methylotrophic or C_1	?	*Methanospheara* (genus)	environmental importance unknown
$(CH_3)_2S$ (dimethylsulfide)	$2(CH_3)_2S + 2H_2O \rightarrow 3CH_4 + CO_2 + 2H_2S$ Methylotrophic or C_1	Marine environments Oil fields	*Methanosarcinales*	

also favor CO_2 reduction over acetate fermentation. The fundamental reason for this dominance is that CO_2 reduction is more energetically favorable than acetate fermentation. The difference in energetics translates to a slower growth rate for aceticlastic methanogens, and thus they are unable to maintain active populations within the digestive tract. Most methanogens must compete with respiring organisms for acetate, formate, and H_2, which are often referred to as competitive substrates. Many methylotrophic methanogens consume non-competitive substrates such as trimethylamine. The ability to consume these substrates allows methylotrophic methanogens to thrive in environments where other terminal electron acceptors, like SO_4^{2-}, are still present. Such environments may include salt marshes as well as anoxic waters.

2.2. ANABOLISM

Anabolism is the portion of metabolism in which energy is utilized to synthesize complex molecules from simple ones. The anabolic diversity of methanogens is limited by the fact that most methanogens are autotrophic. Biosynthesis in most autotrophic methanogens occurs through a modified Ljungdahl-Wood pathway in which acetyl-CoA is formed by the action of acetyl-CoA synthase. Also called "carbon monoxide dehydrogenase," this enzyme is capable of methylating carbon dioxide with a methyl group originating within the catabolic CO_2 reduction pathway. Further biosynthesis occurs by use of an incomplete tricarboxylic acid cycle. Most methanogens utilize this cycle in the reductive direction with the exception of aceticlastic and methylotrophic methanogens which utilize the cycle in an oxidative direction (Sprott *et al.*, 1993). Further anabolic reactions form a diverse assortment of products including unusual lipids and cell envelope components.

Methanogens, like all *Archaea*, are characterized by membrane lipids which differ from those of both *Eukaryotes* and true bacteria (De Rosa *et al.*, 1986). Archaeobacterial lipids are based on ether linkages and are generally formed by the condensation of a glycerol or other polyols, with an isoprenoid alcohol (generally containing either 20, 25, or 40 carbons). Lipids found in true bacteria and in *Eukaryotes* are based on ester linkages, formed by the condensation between a fatty acid and an alcohol. Methanogens demonstrate a considerable diversity of ether-linked membrane lipids.

Methanogens synthesize a diverse assortment of cellular envelopes. The primary differences between envelopes are the various compounds used as fundamental building blocks. Cell envelopes found in methanogens can be categorized into three groups by their primary constituent: 1) pseudomurein, 2) protein or glycoprotein layers, and 3) heteropolysaccharides (Jones *et al.*, 1987). Pseudomurein envelopes are found in *Methanobacteriales*, and differ from bacterial murein in that L-talosaminuronic acid replaces muramic acid, glycosidic linkages differ, and amino acids differ in sequence and enantiomer. Protein containing envelopes are also common among methanogens, including *Methanococcales* and *Methanomicrobiales*, and can coexist with pseudomurein envelopes. Heteropolysaccharides are only found in a few methanogens, including *Methanosarcina*, where they form an unusually rigid envelope.

2.3. TEMPERATURE

As shown in Fig. 1, methanogens are able to grow within a wide range of temperatures. The current temperature extremes for the growth of pure cultures ranges from -2°C for *Methanogenium frigidum* (Franzmann *et al.*, 1997), to 110°C for *Methanopyrus kandleri* (Kurr *et al.*, 1991). Both of these extreme temperatures approach the limits within which any form of life is known to thrive. Many adaptations are necessary for growth in extreme conditions (van de Vossenberg *et al.*, 1998). Life in perennially cold oligotrophic environments may require an unusually slow growth rate, unsaturated membrane lipids, and a low enzymatic temperature optimum. In contrast, hyperthermophilic methanogens are characterized by a rapid growth rate, membrane-spanning lipids, enzymes with high temperature optima, and extra protection of their cellular envelopes. Due to the slow growth rate of psychrophilic and psychrotolerant methanogens, these species are underrepresented in both laboratory studies and culture collections.

2.4. SALINITY

Figure 1 shows that methanogens are capable of growing within a wide range of salinities. Most known methanogens prefer low salinities, though some sodium is required because of its importance in energy conservation. Many methanogens, particularly *Methanococcales*, require marine salinities for growth. A few methylotrophic species within the *Methanosarcinaceae* are extremely halophilic. The reason that all moderately and extremely halophilic methanogens are methylotrophic is, presumably, that several osmoprotectants utilized by other organisms in hypersaline environments are precursors of the methylated substrates utilized by methylotrophic methanogens.

2.5. pH

The pH range of cultured methanogens does not yet encompass the diversity known to exist in nature. As shown in Fig. 1, the pH range for growth of known methanogens ranges from pH = 4 to pH = 10. Many environments on Earth foster active methanogenic populations below this range (Williams and Crawford, 1984).

2.6. MORPHOLOGY

Many components of diversity are apparent as morphological differences between methanogens. Some key differences include shape, size, motility, and the presence of gas vesicles. In general, methanogens can take on many different shapes, and are limited in size from about 0.2 μm in diameter to filaments up to 100 μm long.

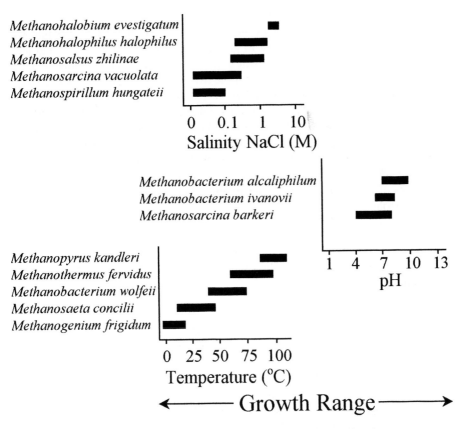

Figure 1. Temperature, salinity, and pH tolerance of selected methanogens

Aggregates of *Methanosarcina* can be up to 5 mm in diameter. Another morphological variation in methanogens is the presence or absence of flagella and gas vesicles. Many species of *Methanobacteriales* are rod-shaped while *Methanococcales* are roughly coccoid; other methanogens may appear as long, thin spirals, angular plate-shaped, or clumps of round cells. In some species, morphology even changes with the life cycle (Boone and Mah, 1987). Flagella are primarily found in the orders of *Methanococcales* and *Methanomicrobiales*, and are used for chemotaxis. Gas vesicles are found within some *Methanosarcinales*, and may function to provide flotation.

3. Taxonomy (phylogeny)

Until recently, bacterial classification relied almost exclusively on phenotypic characteristics; more recently the comparison of genetic sequences, especially the sequence of the major RNA molecule (or its gene) of the small subunit of the ribosome, has been used to unravel evolutionary relationships between organisms. This analysis has been extremely useful in guiding the taxonomy of bacteria, and has also illuminated the evolutionary placement of prokaryotes in the larger framework of all life on Earth. This analysis indicates that the two prokaryotic groups, *Bacteria* and *Archaea* are at least as distantly related to each other than either is to Eukarya. Analysis of evolutionary trees leads to the conclusion that life can be divided into three major branches, the *Archaea*, the *Bacteria*, and the *Eukarya* (Fig. 2), the latter group including the kingdoms of plants, animals, fungi, and protists. Thus, these three major branches must be assigned a new taxonomic level, called "domains," higher than the kingdom. Methanogens are one of the three major present-day representatives of *Archaea*, the other two groups comprising extreme halophiles and extreme thermophiles.

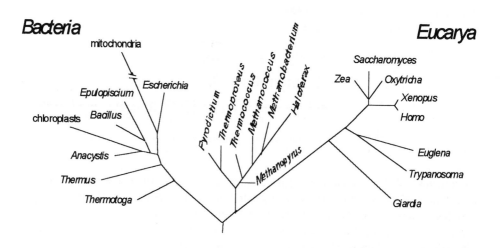

Figure 2. Universal phylogenetic tree based on 16S rRNA sequences

Methanogens are found in five orders in the kingdom *Euryarchaeota* (Fig. 3). There is not room in a review such as this for a detailed description of the taxonomy of methanogens, but such can be found in Bergey's Manual of Systematic Bacteriology, the first edition of which was published in four volumes (Holt *et al.*, 1994) and the second edition of which is in preparation for publication in 1999 or 2000.

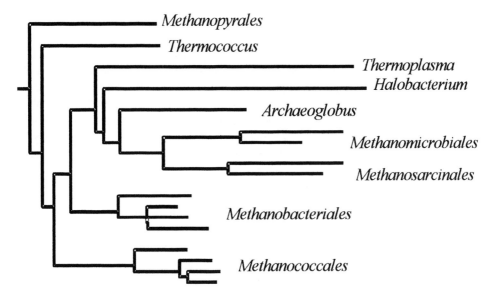

Figure 3. Phylogeny of methanogens

3.1. THE *METHANOBACTERIALES*

The order *Methanobacteriales* contains mainly rod-shaped cells. The ultrastructure of the cell wall viewed by electron microscopy is reminiscent of Gram-positive bacteria, and in fact these organisms appear Gram positive in Gram stains. However, as described above, the chemical structure of the cell wall differs from the murein of bacteria. All of the species in this order grow by reducing CO_2 with H_2, and many can also use formate to reduce CO_2. One genus, *Methanosphaera*, grows exclusively by reducing methanol with H_2. Members of *Methanobacteriales* are found in anaerobic digestors, sediments, anoxic soils, and in the rumen. One genus, *Methanobrevibacter*, is the most commonly encountered methanogen in the colon of mammals.

3.2. THE *METHANOCOCCALES*

The order *Methanococcales* includes mostly marine organisms. These organisms have cell walls composed of protein, so they are sensitive to detergents and usually osmotically sensitive. All grow by reduction of CO_2 to methane with H_2 as electron donor, and many can use formate as well. This group includes extreme thermophiles from hydrothermal marine vents (*Methanococcus jannaschii*) which grow at temperatures up to 94°C.

3.3. THE *METHANOMICROBIALES*

The *Methanomicrobiales* is perhaps the most diverse of the methanogenic orders. The habitat of the order is varied, but the source of many isolates is anaerobic digestors and anoxic sediments. The morphology is also highly diverse, including rods, cocci, irregular cocci, ring- or corpuscle-shaped organisms, plates, and spirals. Like the previously described orders, CO_2 reduction is the common catabolic activity, and all organisms use H_2 as reductant. Many can also use formate, and some can use alcohols (by oxidizing the alcohols to carboxylic acids or ketones).

3.4. THE *METHANOSARCINALES*

Whereas *Methanomicrobiales* is morphologically the most diverse order of methanogens, *Methanosarcinales* is physiologically the most diverse. This order comprises two families, *Methanosarcinaceae* and *Methanosaetaceae*. Members of this latter family are sheathed rods that grow strictly by cleaving acetate into CH_4 and CO_2. This order contains a single genus (*Methanosaeta*) with two species, the mesophilic *Methanosaeta concilii* and the thermophilic *Methanosaeta thermophila*.

All members of the family *Methanosarcinaceae* grow by dismutating methyl compounds such as methanol, methylamines, or methyl sulfides, with one methyl group oxidized to CO_2 for each three methyl groups reduced to methane. In addition, ammonia (from methylamines) or sulfide (from methyl sulfides) may be liberated. In addition to this methylotrophic growth, most organisms in the genus *Methanosarcina* can also grow by CO_2 reduction with H_2 or by acetate cleavage. Many members of this family are halophilic, growing at marine salinity (*Methanococcoides*, *Methanolobus*, and *Methanosalsum*), and others are even more halophilic, growing in saturated brines.

3.5. THE *METHANOPYRALES*

The fifth order of methanogens is the extremely thermophilic *Methanopyrales*. This order contains only a single, rod-shaped species, *Methanopyrus kandleri*. It grows optimally at 98°C by using H_2 to reduce CO_2 to CH_4.

4. Additional Aspects of Diversity

As with many other microbes, the diversity of methanogens is not completely known. Evidence from laboratory and field experiments implicates methanogens as participants in diverse phenomena including: 1) methane oxidation in anaerobic environments (Hinrichs *et al.*, 1999; Hoehler *et al.*, 1994), 2) reduction of partially oxidized sulfur compounds (Stetter and Gaag, 1983), 3) reduction of ferric iron (Vargas *et al.*, 1998), 4) formation of methylated mercury (Pak and Bartha, 1998), and 5) the oxidation of acetate. Evidence also indicates that methanogens can persist in a variety of seemingly hostile environments including aerated soils (Andersen *et al.*,

1998; Peters and Conrad, 1995), semi-oxic termite guts (Leadbetter and Breznak, 1996), ocean waters (Kiene, 1991), and the deep terrestrial subsurface (Stevens and McKinley, 1995). Continued studies will undoubtedly yield additional diverse behaviors.

It is generally believed that the vast majority of extant microorganisms have yet to be cultured. The increasing use of molecular biological tools in environmental settings is helping to constrain the problem of uncultured diversity. High-power molecular tools, including PCR-based phylogenetics, can determine the phylogeny of organisms in environmental samples, but indicate little about their physiology or importance. It is now possible to further determine the natural diversity of methanogens by coupling molecular-biological tools with biogeochemical analysis (including lipid biomarkers) and cultivation strategies. Such integrated approaches are likely to enhance our knowledge of methanogenic diversity.

The study of methanogenic diversity has been made possible, in large part, because of the existence of culture collections. Culture collections store most known species of methanogens and make them available for scientific purposes. Three culture collections harbor the majority of known methanogens: 1) the DSMZ (DSMZ = Deutsche Sammlung von Mikroorganismen und Zellkulturen GmbH - German Collection of Microorganisms and Cell Cultures; web address: http://www.dsmz.de/dsmzhome.htm) currently containing over 225 strains, 2) the OCM (OCM = Oregon Collection for Methanogens; web address: http://caddis.esr.pdx.edu/OCM/) currently containing over 175 strains, and 3) the ATCC (ATCC = American Type Culture Collection; web address: http://www.atcc.org/) currently containing over 40 strains. The existence and accessibility of these culture collections allows for both the maintenance and study of methanogenic diversity.

5. Conclusion

The methanogens are a group of strictly anaerobic *Archaea* that produce methane as their primary metabolic product. While metabolism is limited to methane production from small carbon-containing molecules, they demonstrate an amazing diversity. Methanogens are found to thrive in extreme environments, to produce many unusual compounds, and to participate in many novel activities. Methanogens are chemautotrophic by nature, able to live on the most fundamental compounds of life: H_2 and CO_2. Perhaps the most telling facet of methanogenic diversity is that the same form of metabolism that allows them to grow from volcanic gases also comprises the final step in the degradation of complex organic material.

6. Further Reading

The most comprehensive review of methanogenesis to date can be found in the book entitled "Methanogenesis: Ecology, Physiology, Biochemistry and Genetics" edited by J. G. Ferry (Ferry, 1993). Several additional reviews cover other aspects of

302

methanogenesis, including the biogeochemical cycle of methane (Cicerone and Oremland, 1988), energetics of syntrophic methanogenic degradation (Schink, 1997), and systematics of bacteria (Holt *et al.*, 1994).

7. Acknowledgements

We thank J. Y. King and J. A. Peterson for helpful comments on the manuscript. Preparation of this chapter was supported by the NSF Life in Extreme Environments special competition (to DLV through W.S. Reeburgh).

8. References

Andersen, B. L., Bidoglio, G., Leip, A., and Rembges, D. (1998) *Global Biogeochem Cycles* 12, 587-594.

Boone, D. R., and Mah, R. A. (1987) *Appl Environ Microbiol* 54, 1699-1700.

Cicerone, R. J., and Oremland, R. S. (1988) *Global Biogeochem Cycles* 2, 299-327.

Cord-Ruwisch, R., Seitz, H.-J., and Conrad, R. (1988) *Arch Microbiol* 149, 350-357.

De Rosa, M., Gambacorta, A., and Gliozzi, A. (1986) *Microbiol Rev* 50, 70-80.

Ferry, J. G. (1993) *Methanogenesis: ecology, physiology, biochemistry & genetics*. New York: Chapman & Hall.

Franzmann, P. D., Liu, Y., Balkwill, D. L., Aldrich, H. C., Conway de Macario, E., and Boone, D. R. (1997) *Int J Syst Bacteriol* 47, 1068-1072.

Hinrichs, K.-U., Hayes, J. M., Silva, S. P., Brewer, P. G., and DeLong, E. F. (1999) *Nature* 398, 802-805.

Hoehler, T. M., Alperin, M. J., Albert, D. B., and Martens, C. S. (1994) *Global Biogeochem Cycles* 8, 451-463.

Hoehler, T. M., Alperin, M. J., Albert, D. B., and Martens, C. S. (1998) *Geochim Cosmochim Acta* 62, 1745-1756.

Holt, J. G., Krieg, N. R., Sneath, P. H. A., Staley, J. T., and Williams, S. T. (1994) *Bergey's manual of determinative bacteriology.* . Baltimore: Williams Wilkins.

Hungate, R. E. (1966) *The rumen and its microbes*: Academic Press. . New York

Jones, W. J., Nagle, D. P., Jr., and Whitman, W. B. (1987) *Microbiol Rev* 51, 135-177.

Kiene, R. P. (1991) in: J. E. Rogers and W. B. Whitman (eds.) *Microbial Production and Consumption of Greenhouse Gases: Methane, Nitrogen Oxides, and Halomethanes,* , Washington, DC: Amer. Soc. Microbiol., pp. viii, 298

Kurr, M., Huber, R., König, H., Jannasch, H. W., Fricke, H., Trincone, A., Kristjansson, J. K., and Stetter, K. O. (1991) *Arch Microbiol.* 156, 239-247.

Leadbetter, J. R., and Breznak, J. A. (1996) *Appl Environ Microbiol* 62, 3620-3631.

Pak, K. R., and Bartha, R. (1998) *Appl Environ Microbiol* 64, 1987-1990.

Peters, V., and Conrad, R. (1995) *Appl Environ Microbiol* 61, 1673-1676.

Schink, B. (1997) *Microbiol Mol. Biol. Rev.* 61, 262-280.

Sprott, G. D., Ekiel, I., and Patel, G. B. (1993) *Appl Environ Microbiol* 59, 1092-1098.

Stetter, K. O., and Gaag, G. (1983) *Nature* 305, 309-311.

Stevens, T. O., and McKinley, J. P. (1995) *Science* 270, 450-454.

van de Vossenberg, J., Driessen, A. J. M., and Konings, W. N. (1998) *Extremophiles* 2, 163-170.

Vargas, M., Kashefi, K., BluntHarris, E. L., and Lovley, D. R. (1998) *Nature* 395, 65-67.

Williams, R. T., and Crawford, R. L. (1984) *Appl Environ Microbiol* 47, 1266-1271.

Woese, C. (1987) *Microbiol Rev* 51, 221-271.

Biodata of **Eliora Z. Ron** contributor of *"Microbial Life on Petroleum."*

Dr. Eliora Ron is a Professor for Microbiology at Tel-Aviv University and holds the Manja and Morris Chair for Biotechnology and Biophysics. Dr. Ron received her M.Sc. (1962) majoring in Microbiology, Genetics and Biochemistry, from the Hebrew University. She obtained her Ph.D. from Harvard University (1967). She served as the President of the Israeli Society of Microbiology (1995-1999). Dr. Ron's research interests include: 1. Use of microorganisms to treat environmental pollution-bioremediation of hydrocarbon and heavy metals pollution. 2. Microbial produced polymers with potential applications. 3. Molecular systems for stress response in microorganisms, and 4. Whole cells Biosensors. Dr. Ron has more than 100 publications, including chapters in books and review articles and is the inventor of several patents.

E-mail: eliora@post.tau.ac.il

J. Seckbach (ed.), Journey to Diverse Microbial Worlds, 303-315.

MICROBIAL LIFE ON PETROLEUM

Eliora Z. Ron
Department of Molecular Microbiology & Biotechnology
George S. Wise Faculty of Life Sciences
Tel Aviv University, Ramat Aviv, Israel 69978

1. Introduction

The subject of microbial life on petroleum has been studied by scientists with a broad spectrum of interests. The studies include basic research on the biochemistry of hydrocarbon oxidation, the genetic factors coding for the enzymes, the physiology of the oil degrading bacteria and the ecology of contaminated environments. In addition, oil degrading bacteria often produce a variety of surface active molecules that have numerous potential biotechnological and industrial uses. Another important aspect involves the potential of using oil degrading bacteria for combating oil pollution - bioremediation. Numerous recent reviews have been written on the subject (Korda et al., 1997; Leahy and Colwell, 1990; Morgan and Watkinson, 1989; Prince, 1993; Rosenberg, 1991; Rosenberg and Ron, 1996; Swannell et al., 1996).

Petroleum hydrocarbons produce energy when oxidized. Therefor, growth on petroleum is also dependent upon the presence of oxygenases and is often enhanced by emulsifiers that change the surface properties of the hydrocarbons and increase the bioavailability of the hydrophobic compounds, especially those with low water solubility. In addition, since hydrocarbons do not contain nitrogen, bacterial growth also depends on the availability of nitrogen. This chapter will discuss several aspects of microbial life on petroleum, the physiology and genetics of oil degrading bacteria, as well as interactions between various types of bacteria in the community.

2. Bacterial Growth on Petroleum

2.1. LIFE IN THE PRESENCE OF PETROLEUM

Petroleum is a mixture of a large variety of hydrocarbons, aliphatic and aromatic. These include inhibitory compounds as well as compounds that are highly carcinogenic. As a result, bacteria growing on petroleum have to be environmentally persistent and relatively resistant to the toxic effects of the hydrocarbons. Particularly toxic and the most carcinogenic compounds are the polycyclic aromatic hydrocarbons (PAHs), especially those having 4-5 rings (Cerniglia, 1992; Miller and Miller, 1981). The toxic effects of oil are largely due to dissolving the lipid portion of the cytoplasmic and other membranes. It has shown that hydrophobic organic solvents, such as toluene, are toxic for living organisms because they accumulate in and disrupt cell membranes. Recently the effect of these solvents on bacteria and potential resistance mechanisms have been elucidated. Several bacterial strains have been isolated that can adapt to the presence of organic solvents under conditions previously believed to be lethal. The mechanisms that contribute to the solvent tolerance of these strains involve alterations in the composition of the cytoplasmic and outer membrane, including accumulation of *trans* unsaturated fatty acids in the membrane instead of the *cis* isomers (Isken and de Bont, 1998; Weber et al., 1994).

2.2. UTILIZATION OF HYDROCARBONS AS SOURCES OF CARBON AND ENERGY

The various compounds present in petroleum vary in their biodegradability - the most rapidly biodegradable are the n-alkanes (C14-C18). Aromatic compounds, especially the PAHs, are more resistant to biodegradation, and among these only phenanthrene is degraded at a reasonable rate (Sepic et al., 1996).

Microorganisms - fungi and bacteria - capable of oxidizing hydrocarbons can use them as a source of energy and carbon. These microorganisms are ubiquitous in nature, and possess oxidative enzymes (oxygenases) capable of degrading the various types of hydrocarbons - short-chain, long-chain, and numerous aromatic compounds, including polycyclic aromatic hydrocarbons. They contain a wide variety of pathways for degrading and modifying aromatic compounds, including dioxygenase and monooxygenase ring attacks, and cleavage of catechols by both *ortho-* and *meta-*routes. The specificity of the process is determined by the genetic potential of the particular microorganism and is expressed in the hydrocarbon substrate specificity of the oxygenase and in the ability of the carbon source to induce the various enzyme activities necessary for its biodegradation. It should be noted that several enzymatic reactions for the degradation of hydrocarbon have been demonstrated only in bacteria, and were not shown in fungi. Anaerobic breakdown is also possible in certain Bacteria. For example, a reductive benzoate pathway is the central conduit for the anaerobic biodegradation of aromatic pollutants. The reduction of the benzene ring requires a large input of energy and this metabolic capability has, so far, been reported only in bacteria (*Rhodopseudomonas palustris*) (Egland et al., 1997). The reader is referred to several reviews for information on the enzymatic processes involved in the various oxidation reactions (Cerniglia, 1992; Ensley et al., 1982; 1990; Ensley et al., 1996, 1998; Gibson, 1967; Gibson et al., 1991; 1996; van Beilen et al., 1994; Zylstra and Gibson, 1991).

2.3. OIL DEGRADING BACTERIA

Bacteria growing on petroleum can adhere to the hydrophobic oil droplets, usually due to their hydrophobic surface or to the production of specific, fimbriae-like, appendages (Rosenberg et al., 1998; Rosenberg and Rosenberg, 1981). The bacteria growing on petroleum usually produce potent emulsifiers that help to disperse the oil and to detach the bacteria from it after the utilizable substrate has been depleted (Rosenberg and Ron, 1998).

The bacteria that have been repeatedly isolated as petroleum biodegraders include various strains of *Acinetobacter* (Rosenberg, 1991; Rosenberg and Rosenberg, 1981), several of which were shown to multiply in oil concentrations as high as 5% (Hanson et al., 1997). *Pseudomonas* strains are known to possess a variety of degradative capabilities, and it is therefore not surprising that many oil degrading strains were isolated, including the well studied *P. putida , P. oleovorans, P. fluorescens, P. aeruginosa* and *P. stutzeri* (Chakrabarty et al., 1973; Dagher et al., 1997; Frantz and Chakrabarty, 1986; Grimberg et al., 1996; Rheinwald et al., 1973; van Beilen, et al., 1994). Reports include a variety of *Sphingomonas* strains (Dagher, et al., 1997; Khan et al., 1996; Willumsen et al., 1998; Yrjala et al., 1997) and *Burkholderia* (Johnson and Olsen, 1997; Laurie and Lloyd-Jones, 1999; Shields et al., 1995). Recently strains of *Rhizobium* were implicated in recycling of aromatic compounds, suggesting the possibility of bioremediating oil near plant roots (Damaj and Ahmad, 1996). Among Gram positive bacteria most of the studies have been carried out on *Rhodococcus* (Asturias et al., 1995; Warhurst and Fewson, 1994; Whyte et al., 1998) and *Mycobacterium* (Churchill et al., 1999; Govindaswami et al., 1995; Wang et al., 1995).

The research on oil degrading bacteria has recently taken a completely new direction with the discovery of novel types of marine bacteria capable of utilizing aromatic and aliphatic hydrocarbons as a sole carbon source. These Gram-negative, aerobic, rod-shaped bacteria use a limited number of organic compounds, including aliphatic hydrocarbons, volatile fatty acids, and pyruvate and its methyl ether. They do not use amino acids, proteins, carbohydrates or polysaccharides. The 16S rRNA gene sequence analyses showed that these strains are all members of the γ-subclass of the *Proteobacteria*. However, they did not demonstrate a close phylogenetic relationship to any previously described species. Morphological, physiological and genotypic differences between the new strains and other bacteria justified the creation of novel genera and species. These include *Neptunomonas naphthovorans* (Hedlund et al., 1999) *Alcanivorax borkumensis* (Abraham et al., 1998; Ensley, et al., 1998; Yakimov et al., 1998), *Cycloclasticus* (Button et al., 1998; Dyksterhouse et al., 1995; Geiselbrecht et al., 1998, 1996; Wang et al., 1996), *Fundibacter* (Bruns and Berthe-Corti, 1999) *Marinobacter* (Gauthier et al., 1992; Huu et al., 1999), *Gelidibacter* and *Psychroserpens* (Bowman et al., 1997).

2.4. GENES AND PLASMIDS

The genes involved in hydrocarbon degradation are usually organized in inducible operons, often located on large, low-copy conjugative plasmids. However, there are demonstrations of constitutive operons and of operons located on the chromosome. The degradative plasmids and the gene clusters on the chromosomes undergo rearrangements and shuffling. The following paragraphs will briefly touch on some of the main findings concerning genes and plasmids involved in hydrocarabon degradation. It should be noted that the genes discussed here were described only from several bacterial groups that have been well studied. One study, of 43 bacterial strains from different sources (mostly *Pseudomonas* spp.) capable of degrading aromatic and PAHs indicated that only 14 strains showed homology to one of the previously described degradative genes (Foght and Westlake, 1988). Even less is known about genes and plasmids of hydrocarbon degrading Gram positive bacteria.

The initial reports of degradative genes date back to the beginning of the 1970s, with the discovery of the Oct and Tol plasmids in *Pseudomonas.* The catabolic plasmids can be divided into three general families, represented by the plasmids controlling the degradation of alkanes (such as the OCT plasmid), plasmids like the NAH plasmids for oxidation of naphthalene and salicylate, and plasmids such as the TOL plasmids for the degradation of toluene and xylene (Assinder and Williams, 1990; Chakrabarty et al., 1973; Franklin et al., 1981; Frantz and Chakrabarty, 1986; Furukawa et al., 1983). Bacteria can harbor one plasmid or multiple compatible degradative energy-generating plasmids (Chakrabarty, 1992; Whyte et al, 1997).

The genes coding for alkane oxidation in *Acinetobacter* are chromosomal (Singer and Finnerty, 1984) and are plasmid coded (Oct) in *Pseudomonas* (van Beilen et al., 1994). It is interesting to note that some of the genes that participate in alkane oxidation probably comprise indispensable constituents of the bacterial cell. For example, in *P. oleovorans*, one of the gene products that catalyses the initial oxidation of aliphatic substrates, the alkane hydroxylase, is an integral cytoplasmic membrane protein, and constitutes after induction 1.5-2% of the total cell protein (Nieboer et al., 1993).

The naphthalene catabolic genes are usually located on plasmids. These plasmids - called NAH plasmids - constitute a group of plasmids that vary in electrophoretic mobility and endonuclease restriction patterns, but are highly homologous as shown by Southern hybridization. It therefore appears that NAH is a family of highly homologous plasmids, that have accumulated differences probably due to deletions and insertions (Connors and Barnsley, 1982; Daly et al., 1997). In this group the best-studied plasmid is NAH7 of *P. putida* PpG7 (Assinder and Williams, 1990; Dunn and Gunsalus, 1973). It carries two operons, one of which enables the utilization of naphthalene and the other salicylate, that are turned on by the product of another NAH7 gene, *nahR* , in the presence of salicylate (Simon et al., 1993). The plasmid-encoded genes of the naphthalene degradation pathway of (NAH7) have recently been proven to be involved in degradation of PAHs other than naphthalene (Sanseverino et al., 1993). *P. fluorescens* strain 5RL carries a plasmid with regions highly homologous to upper and lower NAH7 catabolic genes and has been shown capable of mineralization of phenanthrene and anthracene, as well as naphthalene.

The family of TOL plasmids (such as pWW0 of *P. putida*) specifies enzymes for the oxidative catabolism of toluene and xylenes (Franklin, et al., 1981; Frantz and Chakrabarty, 1986) and their synthesis is regulated by the product of *xylR* (Harayama et al., 1986). The first enzyme in this pathway, xylene monooxygenase, is composed of two subunits. It is interesting to note that the sequence of one of the subunits - XylM - has a 25% identity with alkane hydroxylase, which catalyzes a similar reaction (the omega-hydroxylation of fatty acids and the terminal hydroxylation of alkanes) (Suzuki et al., 1991).

The TOL plasmids, like the NAH plasmids, also constitute a family of related plasmids, with differences that can be attributed to DNA rearrangements, such as duplication of catabolic genes that occurs in many TOL plasmids (Osborne et al., 1988). Another type of gene rearrangement results in localization of the TOL genes on the chromosome. In *P. putida* MW1000 (Sinclair et al., 1986) and in *P. putida* TMB (Favaro et al., 1996) the chromosomal genes are almost identical to those found in the TOL plasmid pWW0, suggesting that the chromosomal location of these TOL genes is probably a result of transposition (Favaro et al., 1996; Tsuda et al., 1989). This suggestion is compatible with the finding that many of the catabolic genes involved in hydrocarbon degradation have been shown to be associated with insertion sequences. The presence of these insertion sequences results in a rapid dissemination of the degradative gene clusters among the bacterial populations, greatly expands the substrate range of the microorganisms in the environment, and aids the evolution of new and novel degradative pathways (Tan, 1999).

Comparison of different plasmids with overlapping activities - such as the lower pathways encoded by the TOL and NAH plasmids - indicates a considerable similarity of gene organization and DNA sequence, and suggest a common origin (Williams and Sayers, 1994). The genes coding for degradation of aromatic hydrocarbons in Gram positive bacteria, such as *Rhodococcus* and *Mycobacterium* show homology to similar genes of *P. putida,* suggesting they may have originated in Gram negative microorganisms, probably *Pseudomonas*, and later transferred to the Gram positive bacterium (Asturias, et al. 1995; Churchill, et al. 1999).

2.5. MICROBIAL COMMUNITIES

Recently, studies have begun to focus on the ecology of hydrocarbon degradation by microbial populations in the natural environment. The studies involve isolation and enumeration of bacteria, as well as culture-independent molecular approaches. These include hybridization with probes of hydrocarbon degradation genes and direct amplification of the small-subunit rRNA genes by PCR with universally conserved or specific primers. These molecular studies make it possible to determine the biodiversity and distribution of microorganisms and genes in sites contaminated with petroleum and can be used to monitor changes in microbial populations and microbial biodiversity following exposure to petroleum and during the degradation period. The results lead the way to understanding the relationships between the various bacteria and the effect of environmental factors on these relationships and on the community as a whole. This approach is highly relevant for the understanding of bacterial ecology in contaminated sites as well as for planning bioremediation schemes.

The results of these experiments clearly show that in chronically contaminated sites the half lives for mineralization of chemicals such as naphthalene is significantly shorter than in sites that had no previous exposure. Microbiological analysis of sediments indicated that hydrocarbon-utilizing microbial populations also varied among ecosystems, and were 5 to 12 times greater in sediment after chronic petrogenic chemical exposure than in sediment from an uncontaminated ecosystem. These results provide useful estimates for the rates of hydrocarbon mineralization in different natural ecosystems and on the degradative pathway for microbial metabolism in various environments (Heitkamp et al., 1987; Herman et al., 1994). The types of relationships between the various microorganisms are illustrated in the following example: PAHs with low water solubility tend to adsorb on soil particles, making them even less bioavailable to degrading microorganisms. It has been shown that in communities containing the white rot fungi, for instance, the conversion of PAHs is faster, probably because the fungus converts the PAHs to more water soluble and bioavailable products, that are then readily mineralized by the indigenous bacteria (Meulenberg et al., 1997). Another example of a successful consortium is the enhanced biodegradation of aromatic pollutants in cocultures of anaerobic and aerobic bacteria. Electrophilic aromatic pollutants with multiple chloro, nitro and azo groups have proven to be persistent to biodegradation by aerobic bacteria. These compounds are readily reduced by anaerobic consortia to lower chlorinated aromatics or aromatic amines that are then subject to aerobic bacterial attack. The aerobic bacteria also help in reducing the oxygen, providing the anaerobic microniches required for the initial reductive steps (Field et al., 1995).

3. Bioremediation

Bioremediation involves the acceleration of natural biodegradative processes in contaminated environments by improving the availability of materials (e.g., nutrients, or oxygen) or microorganisms. Thus, bioremediation usually consists of application of nitrogenous and phosphorous fertilizers, supplying air and often adding bacteria and possible emulsifiers. The addition of bacteria and emulsifiers is advantageous when bacterial growth is slow - low temperatures, high concentrations of pollutants - or when the pollutants consist of compounds that are difficult to degrade - such as PAHs. The added microorganisms are usually natural or engineered bacteria that have a high capability of degrading the pollutants present in the site. However, the reintroduction of indigenous microorganisms isolated from the contaminated site after culturing is generally more effective (Korda et al., 1997; Prince, 1993; Rosenberg and Ron, 1996; Swannell et al., 1996).

3.1. RATE LIMITING FACTORS

3.1.1. Nitrogen
The major limitation in bioremediation of hydrocarbons and halogenated compounds in open systems (large bodies of soil and water) is the availability of nitrogen, which is not present in sufficient amounts in these pollutants. In theory, approximately 150 mg of nitrogen are consumed in the conversion of 1 g of hydrocarbon to cell material. The nitrogen requirement for maximum growth of hydrocarbon oxidizers can generally be satisfied by salts such as ammonium sulfate, ammonium nitrate and ammonium chloride. All of these compounds have a high water solubility, which reduces their effectiveness in open systems because of rapid dilution. In addition, excess of soluble nitrogen is by itself a severe environmental problem (Rosenberg et al., 1997, 1998; Rosenberg and Ron, 1996).

The biologically-correct solution to the problem is to use bacteria that are capable of degrading organic pollutants while using atmospheric nitrogen - nitrogen fixers. However, there is an internal incompatibility of the conditions required for the two processes. While nitrogen fixation is essentially anaerobic, hydrocarbon utilization requires oxygen. A possible solution would involve bacteria able to carry out the two processes in "compartments" of some kind, such as cyanobacteria in which nitrogen fixation is limited to the heterocysts (Kuritz and Wolk, 1995).

One solution that appears to be effective is the use of fertilizers that have an affinity for hydrocarbons and adhere to the oil. The most extensively studied example is the oleophilic fertilizer Inipol EAP 22, an oil-in-water microemulsion containing a N:P ratio of

7.3:2.8 (Button et al., 1991; Glaser, 1991; Lindstrom et al., 1991). Inipol EAP 22 is composed of oleic acid, lauryl phosphate, 2-butoxyl-1-ethanol, urea and water. The fertilizer was used in large quantities to treat the oil-contaminated shoreline following the 1989 oil spill in Prince William Sound, Alaska. There are at least three problems with Inipol EAP 22. First, it contains large amounts of oleic acid, which serves as an alternative carbon source, thereby further increasing the C:N ratio in the environment. Second, it contains an emulsifier which could be harmful to the environment. Third, as soon as the fertilizer comes into contact with water, the emulsion breaks, releasing the urea to the water phase where it is not available for the microorganisms. A newly developed controlled release (not slow release) nitrogen fertilizer, that depends on microbial-catalyzed hydrolysis of an oleophilic polymer, has been used successfully in oil spills on beaches (Rosenberg et al., 1997; Rosenberg and Ron, 1996). The effectiveness of this fertilizer is mostly due to the fact that it remains adsorbed to the oil in aqueous environments and is, therefore, available to the oil-degrading bacteria that adhere to the oil droplets.

3.1.2. Air
Oxygen becomes limiting especially in depth, and its limitation results in a considerable slowing down of the bioremediation process. Supply of oxygen is achieved by processes like bioventing. Recent studies suggest the use of peroxides (H_2O_2) as a supplemental oxygen source (Fiorenza and Ward, 1997).

3.1.3. Bioavailability and the use of surfactants
The low water solubility of many hydrocarbons, especially the polycyclic aromatic hydrocarbons (PAHs) is believed to limit their availability to microorganisms, which is a potential problem for bioremediation of contaminated sites. It has been assumed that surfactants would enhance the bioavailability of hydrophobic compounds. Several surfactants have been studies and both negative and positive effects of the surfactants on biodegradation were observed. For example, the addition of the surfactant Tergitol NP-10 increased the dissolution rate of solid-phase phenanthrene and resulted in an overall increase in the growth of a strain of *P. stutzeri* (Grimberg et al., 1996). A similar effect was obtained by the addition of Tween 80 to two *Sphingomonas* strains: the rate of fluoranthene mineralization was almost doubled. In contrast, the same surfactant inhibited the rate of fluoranthene mineralization by two strains of *Mycobacterium* (Willumsen, et al., 1998) and no stimulation was observed in other studies using several surfactants (Bruheim et al., 1997; Bruheim and Eimhjellen, 1998).

Another approach is to use biosurfactants that are produced by many bacteria. These are more effective, selective, environmentally friendly, and more stable than many synthetic surfactants. The biosurfactants include low molecular weight compounds, such as glycolipids in which carbohydrates are attached to a long-chain aliphatic acid and high molecular weight compounds, as reviewed by Desai and Banat (1997) and by Rosenberg and Ron (1997, 1998). The high molecular weight surfactants are less effective in reducing the interfacial tension, but coat the oil droplets and prevent their coalescence. A high molecular weight bioemulsifier - Alasan - that consists of a protein-bound heteropolysaccharide (Navon-Venezia et al., 1995) was recently shown to significantly increase the rate of biodegradation of several PAHs (Rosenberg et al.,

1998). Clearly, optimization of this process would involve selecting the best microorganisms and most suitable surfactants for PAH bioremediation.

3.2. MONITORING THE EFFICACY OF BIOREMEDIATION

The efficacy of bioremediation is determined chemically, by measuring the change in total petroleum hydrocarbons, usually complemented by chromatographic results (gas chromatography, or gas chromatography-mass spectrometry). Recently, there are attempts to introduce biosensors, especially microbial whole cell biosensors, to monitor the rate of elimination of the pollutants. These biosensors consist of promoters from genes that are induced by the presence of the pollutant of interest. These promoters are fused to reporter genes whose expression can be easily monitored. An example of such biosensors is the strain of *Escherichia coli* K-12 that carries the promoter of the *xylS* gene - induced by aromatic hydrocarbons - from *P. putida* that has been fused to a promoterless luciferase operon (from *Vibrio harveyi* or firefly). The expression of luciferase is measured by light emission. It was induced in the presence of aromatic compounds. The lower detection limit for *m*-xylene was 5 mM (Kobatake et al., 1995). Similar biosensors were constructed that can monitor the level of additional environmental pollutants, such as toluene or octane (Applegate et al., 1998; Sticher et al., 1997). Recently new whole cell biosensors have been constructed in which the monitoring of gene expression is electrochemical. These promoter-based, whole cell biosensors are suitable for on line and *in situ* monitoring of pollutants (Biran et al., 1999).

One of the advantages of biosensors is that they discriminate between the total hydrocarbons and the bioavailable hydrocarbons, the latter being at least as important as the value for total hydrocarbons. Biosensors constitute a cost-effective, convenient solution to monitoring the progress of bioremediation treatments. It is expected that the use of biosensors will become widespread, especially if further developments improve the capabilities for in situ and on line monitoring.

3.3. FUTURE PROSPECTS

Bioremediation is a promising technology for treatment of petroleum pollution. Most of the components of crude oils are biodegradable, and are totally mineralized. Optimization of the bioremediation processes is still required, especially in the case of hydrocarbons with low availability (such as PAHs), in cold areas, or in sites when oxygen is limiting. Yet, because bioremediation is ecologically friendly and cost effective it will probably become the technology of choice to treat the majority of pollutants.

4. Conclusions

Life on petroleum involves microbial communities that are resistant to the toxicity of the hydrocarbons and have a broad spectrum of degradative abilities. These microorganisms produce a broad spectrum of enzymes and surfactants with potential applications in several industries - home care, cleaning, cosmetics, food and pharmaceutical industries. The study of bacteria living in environments contaminated with petroleum is interesting with respect to the genetics, physiology and biochemistry of the bacteria, as well as the structure and function of the communities that they form. These studies will constitute the basis for using bacteria and bacterial consortia for bioremediation of pollution and creating pollutant-free environments.

5. Acknowledgements

This work was supported by a grant of the Ministry of Science, Israel, and by the Manja and Morris Leigh Chair for Biophysics and Biotechnology.

6. References

Abraham, W. R., Meyer, H., and Yakimov, M. (1998) Biochim. Biophys. Acta. **1393**:57-62.

Applegate, B. M., Kehrmeyer, S. R., and Sayler, G. S. (1998) Appl. Environ. Microbiol. **64**:2730-2735.

Assinder, S. J., and Williams, P. A. (1990) Adv. Microb. Physiol. **31**:1-69.

Asturias, J. A., Diaz, E., and Timmis, K. N. (1995) Gene. **156**:11-18.

Biran, I., Klementy, E., Hengge-Aronis, R., Ron, E. Z., and Rishpon, J. (1999) Microbiol. (in press).

Bowman, J. P., McCammon, S. A., Brown, J. L., Nichols, P. D., and McMeekin, T. A. (1997) Int. J. Syst. Bacteriol. **47**:670-677.

Bruheim, P., Bredholt, H., and Eimhjellen, K. (1997) Can. J. Microbiol. **43**:17-22.

Bruheim, P., and Eimhjellen, K. (1998) Can J Microbiol. **44**:195-199.

Bruns, A., and Berthe-Corti, L. (1999) Int J Syst Bacteriol. **49**:441-448.

Button, D. K., Robertson, B. R., Lepp, P. W., and Schmidt, T. M. (1998) Appl Environ Microbiol. **64**:4467-4476.

Button, D. K., Robertson, B. R., McIntosh, D., and Juttner, F. (1991) Appl. Environ. Microbiol. **57**:2514-2522.

Cerniglia, C. E. (1992) Biodegradation. **3**:351-368.

Chakrabarty, A. M. (1992) Biotechnology. **24**:535-545.

Chakrabarty, A. M., Chou, G., and Gunsalus, I. C. (1973) Proc.Natl. Acad. Sci. U S A. **70**:1137-1140.

Churchill, S. A., Harper, J. P., and Churchill, P. F. (1999) Appl. Environ. Microbiol. **65**:549-552.

Connors, M. A., and Barnsley, E. A. (1982) J. Bacteriol. **149**:1096-1101.

Dagher, F., Deziel, E., Lirette, P., Paquette, G., Bisaillon, J. G., and Villemur, R. (1997) Can. J. Microbiol. **43**:368-377.

Daly, K., Dixon, A. C., Swannell, R. P., Lepo, J. E., and Head, I. M. (1997) J Appl. Microbiol. **83**:421-429.

Damaj, M., and Ahmad, D. (1996) Biochem. Biophys. Res. Commun. **218**:908-915.

Desai, J. D., and Banat, I. M. (1997) Microbiol. Mol. Biol. Rev. **61**:47-64.

Dunn, N. W., and Gunsalus, I. C. (1973) J. Bacteriol. **114**:974-979.

Dyksterhouse, S. E., Gray, J. P., Herwig, R. P., Lara, J. C., and Staley, J. T. (1995) Int J. Syst. Bacteriol. **45**:116-123.

Egland, P. G., Pelletier, D. A., Dispensa, M., Gibson, J., and Harwood, C. S. (1997) Proc. Natl. Acad. Sci. U S A. **94**:6484-6489.

Ensley, B. D., Gibson, D. T., and Laborde, A. L. (1982) J. Bacteriol. **149**:948-954.

Ensley, B. D., Gibson, D. T., and Laborde, A. L. (1990) Adv Microb Physiol. **31**:1-69.

Ensley, B. D., Gibson, D. T., and Laborde, A. L. (1996) J. Med. Chem. **39**:4065-4072.

Ensley, B. D., Gibson, D. T., and Laborde, A. L. (1998) Int. J. Syst. Bacteriol. **48**:339-348.

Favaro, R., Bernasconi, C., Passini, N., Bertoni, G., Bestetti, G., Galli, E., and Deho, G. (1996) Gene. **182**:189-193.

Field, J. A., Stams, A. J., Kato, M., and Schraa, G. (1995) Antonie van Leeuwenhoek. **67**:47-77

Fiorenza, S., and Ward, C. H. (1997) J. Ind. Microbiol. Biotechnol. **18**:140-151.

Foght, J. M., and Westlake, D. W. (1988) Can. J. Microbiol. **34**:1135-1141.

314

Franklin, F. C., Bagdasarian, M., Bagdasarian, M. M., and Timmis, K. N. (1981) Proc. Natl. Acad. Sci. U S A. **78**:7458-7462.

Frantz, B., and Chakrabarty, A. M. (1986) in: J. R. Sokatch (ed.) *The Bacteria: A Treatise on Structure and Function.* vol. **10**. pp. 295-323.

Furukawa, K., Simon, J. R., and Chakrabarty, A. M. (1983) J. Bacteriol. **154**:1356-1362.

Gauthier, M. J., Lafay, B., Christen, R., Fernandez, L., Acquaviva, M., Bonin, P., and Bertrand, J. C. (1992) Int. J. Syst. Bacteriol. **42**:568-576.

Geiselbrecht, A. D., Hedlund, B. P., Tichi, M. A., and Staley, J. T. (1998) Appl. Environ. Microbiol. **64**:4703-4710.

Geiselbrecht, A. D., Herwig, R. P., Deming, J. W., and Staley, J. T. (1996) Appl. Environ. Microbiol.. **62**:3344-3349.

Gibson, D. T. (1967) Science. **161**:1093-1097.

Gibson, S., McGuire, R., and Rees, D. C. (1991) Genet. Eng. **13**:183-203.

Gibson, S., McGuire, R., and Rees, D. C. (1996) J. Med.Chem. **39**:4065-4072.

Glaser, J. A. (1991) in: R. E. Hinchee and R. F. Olfenbuttel (eds.) *On Site Bioreclamation,* The Valdez experience, Butterworth-Bienemann, Stoneham, Massachusetts. pp. 336-384.

Govindaswami, M., Feldhake, D. J., Kinkle, B. K., Mindell, D. P., and Loper, J. C. (1995) Appl. Environ. Microbiol. **61**:3221-3226.

Grimberg, S. J., Stringfellow, W. T., and Aitken, M. D. (1996) Appl. Environ. Microbiol. **62**:2387-2392.

Hanson, K. G., Nigam, A., Kapadia, M., and Desai, A. J. (1997) Curr. Microbiol. **35**:191-193.

Harayama, S., Leppik, R. A., Rekik, M., Mermod, N., Lehrbach, P. R., Reineke, W., and Timmis, K. N. (1986) J. Bacteriol. **167**:455-461.

Hedlund, B. P., Geiselbrecht, A. D., Bair, T. J., and Staley, J. T. (1999) Appl. Environ. Microbiol. **65**:251-259.

Heitkamp, M. A., Freeman, J. P., and Cerniglia, C. E. (1987) Appl. Environ. Microbiol. **53**:129-136.

Herman, D. C., Fedorak, P. M., MacKinnon, M. D., and Costerton, J. W. (1994) Can. J. Microbiol. **40**:467-477.

Huu, N. B., Denner, E. B., Ha, D. T., Wanner, G., and Stan-Lotter, H. (1999) Int. J. Syst. Bacteriol. **49**:367-375.

Isken, S., and de Bont, J. A. (1998) Extremophiles. 2:229-238.

Johnson, G. R., and Olsen, R. H. (1997) Appl. Environ. Microbiol. **63**:4047-4052.

Khan, A. A., Wang, R. F., Cao, W. W., Franklin, W., and Cerniglia, C. E. (1996) Int. J. Syst. Bacteriol. **46**:466-469.

Kobatake, E., Niimi, T., Haruyama, T., Ikariyama, Y., and Aizawa, M. (1995) Biosens. Bioelectron. **10**:601-605.

Korda, A., Santas, P., Tenente, A., and Santas, R. (1997) Appl. Microbiol. Biotechnol. **48**:677-686.

Kuritz, T., and Wolk, C. P. (1995) Appl. Environ. Microbiol. **61**:234-238.

Laurie, A. D., and Lloyd-Jones, G. (1999) J. Bacteriol. **181**:531-540.

Leahy, J. G., and Colwell, R. R. (1990) Microbiol. Rev. **54**:305-315.

Lindstrom, J. E., Prince, R. C., Clark, J. C., Grossman, M. J., Yeager, T. R., Braddock, J. F., and Brown, E. J. (1991) Appl. Environ. Microbiol. **57**:2514-2522.

Meulenberg, R., Rijnaarts, H. H., Doddema, H. J., and Field, J. A. (1997) FEMS Microbiol. Lett. **152**:45-49.

Miller, E. C., and Miller, J. A. (1981) Cancer. **47**:2327-2345.

Morgan, P., and Watkinson, R. J. (1989) Crit. Rev. Biotechnol. **8**:305-333.

Navon-Venezia, S., Zosim, Z., Gottlieb, A., Legmann, R., Carmeli, S., Ron, E. Z., and Rosenberg, E. (1995) Appl. Environ. Microbiol. **61**:3240-3244.

Nieboer, M., Kingma, J., and Witholt, B. (1993) Mol. Microbiol. **8**:1039-1051.

Osborne, D. J., Pickup, R. W., and Williams, P. A. (1988) J. Gen Microbiol. **134**:2965-2975.

Prince, R. C. (1993) Crit. Rev. Microbiol. **19**:217-242.

Rheinwald, J. G., Chakrabarty, A. M., and Gunsalus, I. C. (1973) Proc. Natl. Acad. Sci. U S A. **70**:885-889.

Rosenberg, E. (1991) in: A. Ballows (ed.) *The Prokaryotes.* Springer-Verlag, Berlin. pp. 441-459.

Rosenberg, E., Barkay, T., Navon-Venezia, S., and Ron, E. Z. (1998) in: *Novel Approaches for Bioremediation of Organic Pollution.* R.Fass (ed.), Kluwer Academic/ Plenum Publishers. pp. 171-180

Rosenberg, E., Legman, R., Kushmaro, A., Adler, E., Abir, H., and Ron, E. Z. (1997) J. Biotechnol. **51**:273-278.

Rosenberg, E., Navon-Venezia, S., Zilber-Rosenberg, I., and Ron, E. Z. (1998) Rate-limiting steps in the microbial degradation of petroleum hydrocarbons. in: H. Rubin, N. Narkis and J. Carberry (eds.). *Soil and Aquifer Pollution*. Springer-Verlag, Berlin. pp.159-172.

Rosenberg, E., and Ron, E. Z. (1996) in: R. L. D. L. Crawford (ed.), *Bioremediation: Principles and Applications*. Cambridge University Press. pp. 100-125.

Rosenberg, E., and Ron, E. Z. (1997) Curr. Opin. Biotechnol. **8**:313-316.

Rosenberg, E., and Ron, E. Z. (1998) in: D. Kaplan (ed.) *Biopolymers from Renewable Sources*. Springer-Verlag, Berlin. pp. 281-291.

Rosenberg, M., and Rosenberg, E. (1981) J Bacteriol. **148**:51-57.

Sanseverino, J., Applegate, B. M., King, J. M., and Sayler, G. S. (1993) Appl. Environ. Microbiol. **59**:1931-1937.

Sepic, E., Trier, C., and Leskovsek, H. (1996) Analyst. **121**:1451-1456.

Shields, M. S., Reagin, M. J., Gerger, R. R., Campbell, R., and Somerville, C. (1995) Appl. Environ. Microbiol. **61**:1352-1356.

Simon, M. J., Osslund, T. D., Saunders, R., Ensley, B. D., Suggs, S., Harcourt, A., Suen, W. C., Cruden, D. L., Gibson, D. T., and Zylstra, G. J. (1993) Gene. **127**:31-37.

Sinclair, M. I., Maxwell, P. C., Lyon, B. R., and Holloway, B. W. (1986) J. Bacteriol. **168**:1302-1308.

Singer, M. E., and Finnerty, W. R. (1984) in: R. M. Atlas (ed.) *Petroleum Microbiology*. Macmillan Publishing Co., New York. pp. 299-354.

Sticher, P., Jaspers, M. C., Stemmler, K., Harms, H., Zehnder, A. J., and van der Meer, J. R. (1997) Appl. Environ. Microbiol.. **63**:4053-4060.

Suzuki, M., Hayakawa, T., Shaw, J. P., Rekik, M., and Harayama, S. (1991) J Bacteriol. **173**:1690-1695.

Swannell, R. P., Lee, K., and McDonagh, M. (1996) Microbiol. Rev. **60**:342-365.

Tan, H. M. (1999) Appl. Microbiol. Biotechnol. **51**:1-12.

Tsuda, M., Minegishi, K., and Iino, T. (1989) J Bacteriol. **171**:1386-1393.

van Beilen, J. B., Wubbolts, M. G., and Witholt, B. (1994) Biodegradation 5:161-174.

Wang, R. F., Cao, W. W., and Cerniglia, C. E. (1995) FEMS Microbiol. Lett. **130**:75-80.

Wang, Y., Lau, P. C., and Button, D. K. (1996) Appl. Environ. Microbiol. **62**:2169-2173.

Warhurst, A. M., and Fewson, C. A. (1994) Crit. Rev. Biotechnol. **14**:29-73.

Weber, F. J., Isken, S., and de Bont, J. A. (1994) Microbiol. **140**:2013-2017.

Whyte, L. G., Bourbonniere, L., and Greer, C. W. (1997) Appl. Environ. Microbiol. **63**:3719-3723.

Whyte, L. G., Hawari, J., Zhou, E., Bourbonniere, L., Inniss, W. E., and Greer, C. W. (1998) Appl. Environ. Microbiol. **64**:2578-2584.

Williams, P. A., and Sayers, J. R. (1994) Biodegradation. **5**:195-217.

Willumsen, P. A., Karlson, U., and Pritchard, P. H. (1998) Appl. Microbiol. Biotech. **50**:475-483.

Yakimov, M. M., Golyshin, P. N., Lang, S., Moore, E. R., Abraham, W. R., Lunsdorf, H., and Timmis, K. N. (1998) Int. J. Syst. Bacteriol. **48**:339-348.

Yrjala, K., Paulin, L., and Romantschuk, M. (1997) FEMS Microbiol. Lett. **154**:403-408.

Zylstra, G. J., and Gibson, D. T. (1991) Genet. Eng. **13**:183-203.

Biodata of Anna A. Gorbushina contributor (with co-author Wolfgang Krumbein) of the chapter *"Rock Dwelling Fungal Communities: Diversity of Life Styles and Colony Structure."*

Dr. Anna Gorbushina is an Assistant Professor at the Carl von Ossietzky University of Oldenburg, Germany. She is a mycologist and geomicrobiologist. She earned her Ph.D. in 1997 from the Komarov Institute of Botany of the Russian Academy of Sciences. Her fields of interest and research focus on poikilotrophic fungi and their morphogenesis under extreme changes of environment. She presently studies the influence of fungal communities on biodeterioration and biomineralization of desert rocks and monuments. Dr. Gorbushina and Professor Wolfgang Krumbein are the authors of the chapter "*The Poikilotrophic Microorganism and its Environments: Microbial Strategies of Establishment, Growth and Survival*", published in *Enigmatic Microorganisms and Life in Extreme Environments"* edited by J. Seckbach (1999 Kluwer Academic Publishers). E-mail: anna@africa.geomic.uni-oldenburg.de

Biodata of Wolfgang E. Krumbein co-author (with Anna Gorbushina) of the chapter *"Rock Dwelling Fungal Communities: Diversity of Life Styles and Colony Structure."*

Dr. Wolfgang E. Krumbein has been Professor of Geomicrobiology at the University of Oldenburg since 1974. He received his Ph.D. from the University of Würzburg (Germany) in 1966. He created the interdisciplinary Institute for Chemistry and Biology of the Marine Environment (ICBM) with a major focus on geomicrobiology and geophysiology. Dr. Krumbein published 375 papers in the fields of Biomineralization and biodeterioration of minerals and rocks. He also edited 11 books in this field. Recently his major interest focuses on the physical and mechanical effects of microorganisms on rock surfaces. Dr. Gorbushina and Dr. Krumbein are the authors of the chapter "*The Poikilotrophic Microorganism and its Environments: Microbial Strategies of Establishment, Growth and Survival*", published in *Enigmatic Microorganisms and Life in Extreme Environments"* edited by J. Seckbach (1999 Kluwer Academic Publishers). E-mail: wek@africa.geomic.uni-oldenburg.de.

J. Seckbach (ed.), Journey to Diverse Microbial Worlds, 317-334.
© 2000 *Kluwer Academic Publishers. Printed in the Netherlands.*

ROCK DWELLING FUNGAL COMMUNITIES:
Diversity of Life Styles and Colony Structure

ANNA A. GORBUSHINA AND WOLFGANG E. KRUMBEIN
Geomicrobiology, ICBM,
Carl von Ossietzky University Oldenburg
P.O. Box 2503, D-26111 Oldenburg, Germany

1. Introduction

In microbiology, that branch of biology dealing with microscopically small representatives of all three domains (Archaea, Bacteria and Eukarya), the need for biodiversity research and analysis is even greater than in any other biological science (Amann, 1999). In all groups of microorganisms, so far only a very small part of the whole species diversity is recognized, described, and characterized. Micromycetes cannot be named an exception, because their small size and the complex isolation techniques pose severe difficulties for assessment and monitoring of microscopic fungal communities (Cannon, 1998). The importance of micromycetes in many environments has been documented, but many features of this interesting group are still escaping our attention.

As it was shown by Hirsch (1986), even ordinary places like soils are often extremely rich in strange microbial forms or are presenting an unknown or non-detected diversity spectrum. Techniques appropriately modified for specific environments or even completely new ones are needed to detect (Torsvik et al., 1996) and recognize the presence of hitherto undescribed microorganisms. Even more so as with bacteria, fungal diversity in many habitats is not yet studied and is still presenting a big challenge for investigators. Particularly scarce is the existing information on the biodiversity in hostile and changing environments, where the detection and analysis of micromycete communities is even more difficult. Unknown nutritional and environmental demands of these organisms are presenting actual methodological drawbacks in biodiversity studies of fungal communities from many poorly investigated environments.

As an example of a very special and in terms of Earth history very ancient and very widespread environment with quite numerous fungal inhabitants one could name the surfaces of bare rocks. This habitat has drawn attention relatively recently because of the search for extraterrestrial life and studies on the durability of man-made monuments. For these reasons the description and investigation of the rock environment is already quite advanced. The subaerially exposed rock surface-as any interface environment- has a large reactive surface with the atmosphere and is a good support for biofilm

319

development. Energy sources and nutrients coming from the atmosphere with soil and dust particles aggregate at the interface with the solid rock substrate, giving immediate advantages to a wide spectrum of microorganisms fixed on its surface.

Although the rock inhabiting fungal community is a relatively recent object of investigation in comparison to epilithic lichens, cyanobacteria, and algae, there are quite some data allowing for an analysis of this open system experiencing sudden and unpredictable changes of environmental parameters. These changes among which the most drastic are expressed by exposing a freshly sculptured statue to a park atmosphere or excavating an ancient Kouros and exposing it to the modern polluted atmosphere of Athens or Caesarea are bringing about certain shifts in the fungal populations, which will be considered and discussed. In this chapter it is also intended to show the functional diversity of fungi on the rock surface and to discuss the diversity of their adaptive response to non-optimal growth conditions. Changes in the fungal flora and in its development under the influence of atmospheric exposure for exactly definable periods of time will be also demonstrated.

2. Rock Surface Environment And The Potential Fungal Participation In A Rock Inhabiting Microbial Community

The rock itself is not presenting a source of organic or inorganic nutrients, but just a dwelling place, where microorganisms are gaining shelter and living space, but not major "food" sources. The remaining potential supplies of energy sources for the chemoorganotrophic micromycetes thus are (i) the establishment and decay of phototrophic microorganisms with their EPS and (ii) may be more important organic matter deposited in molecular or particulate form on the bare rock surface. Sporadic nutrient support originating from atmospheric organic pollution is giving a sufficient basis to the development of an unusual fungal community diverse in species composition.

Growth and activities of micromycetes on rock surfaces and within the upper layers of rocks have been a discussed topic over several years now because atmospheric organic pollution is drastically increasing the amount and extent of chemoorganotroph fungal biofilms. The process of the deposition and accumulation of fungal propagules on the rock surface is actually a relatively fast one: after the removal of a thick altered layer of stone, fungal CFU counts of a freshly exposed rock surface were shown to increase from 0.9×10^3 to 0.75×10^6 in a short time of 11 weeks (Krumbein, 1973). This shows us the speed of fungal spore and hyphal fragments arrival on the surface and potential biomass, which is permanently trying to establish itself on the rock.

However, the pressure of environmental functions on the habitat is quite remarkable, because in addition to the lack of nutrients it includes a broad spectrum of different unfavorable factors. Extreme changes in temperature and sun irradiation, supply of nutrients and water on the rock surface is bringing about drastic changes in community structure. In such a situation the biodiversity of the community is unavoidably fluctuating. This necessitates recording the changes and looking at the seasonally and episodically dominant different micromycetes permanently and/or temporarily presenting on the rock surface. In addition ephemeral growth of species basically unsuitable for such environments has to be separated from the true and insisting inhabitant of the rock surface.

The diversity of the endolithic and epilithic micromycetes community was studied at several occasions (e.g. Gromov, 1959; Braams, 1992; Grote, 1992; Urzì et al., 1995). These studies have brought about a first realization of the fact that in this widely distributed habitat a chemoorganoheterotrophic fungal community is well-developed and plays a significant role in interactions with the substrate in question. It was repeatedly demonstrated, that micromycetes growth on the rock surface belongs to the process of a normal development of rock microflora and that fungi are important factors in rock weathering (Krumbein, 1966; Braams, 1992) and in the soil formation process (Gromov, 1963) in the temperate climate zone. The first records were connected with moderate climates and building surfaces, but later the interest has shifted to more stressed environments: for instance, Krumbein (1969) and Krumbein and Jens (1981) reported of fungi involved in biopitting, bio-exfoliation and biological erosion fronts in desert rocks of the Negev desert. Later micromycetes were found in different and extreme climates from hot deserts and the Antarctic (Friedmann and Ocampo-Friedmann, 1984) expanding to mineral surfaces like building walls in middle Europe.

The most severe conditions of exposure (e.g. deserts) are yielding a fungal flora demonstrating very unusual melanized growth forms, which were referred to as microcolonial fungi or MCF (Staley et al., 1982). A similar group of fungi was also discovered on rock surfaces of monuments in the Mediterranean region (Gorbushina et al. 1993; Wollenzien et al., 1995; Diakumaku et al., 1995; Sterflinger, 1995).

These fungal species from the rock environment have such low growth velocity in vitro combined with a special growth pattern that they have not been extensively characterized. The fungi occurring on and in the rock environment as microcolonies cannot be regarded as the most ordinary group of micromycetes because their slow way of growth in short cells with thick melanized walls is supplying them with very special morphology and, unfortunately, resistance to cultivation attempts. It is even unclear to which reasons the problems in isolation and cultivation are related. (1) Some of the organisms may be dormant, (2) some may have life-cycles with long periods without growth or cell division, and (3) they may be just growing so slowly that it needs 10 years to get a colony of pin-head size. Morphological investigations of these slow growing fungi do not yield satisfactory results because of very few morphological characters available for analysis. It was even suggested (Wollenzien et al., 1995) that these stress-induced growth patterns are making very different fungi morphologically similar on the rock surface and that it could be that different kinds of stone may each have a characteristic resident flora. However, later studies have shown that some of the strains isolated from Delian and Crimean monuments are demonstrating a remarkable similarity in ARDRA patterns allowing to cluster geographically distant strains into one group of at least generic level (Sterflinger and Gorbushina, 1997). Taxonomically some isolates of these slow growing fungi were described by Sterflinger et al. (1997) and their affiliation to two new species of one genus *Coniosporium* has been shown. Thus it was already proved that the work on isolation and characterization of fungal biodiversity of such habitats has brought a full description of two previously unknown species in a relatively short time.

3. Methods Used for Assessment of Micromycete Diversity in Rock Dwelling Communities

As one of the first steps in fungal rock community description one could name microscopical observation. Microscopy of the samples in question includes light and scanning electron microscopy of the material surface and petrographic thin sections, as well as of fungal colonies or fragments of them. Different structures belonging to different microorganisms are established in the surface layers of the rock material. The analysis of their distribution and morphology is comparable to the direct count of microbial cells in the sample: we cannot draw any conclusion of their viability, but can say which microorganisms are dominating the substrate. A typical example is shown in Figs. 1 and 2, where remarkable differences in the fungal growth are expressed in the structure of fungal colonies growing on the rock surface. In Fig. 1 fungal hyphae are developed to their full extent and are demonstrating a normal structure of any fungal colony growing on any substrate. Fast growing hyphae are spreading over the surface forming a biofilm in the crevices and on the surface itself. In a stressed environment of air-exposed rock surface, however, different structures are developed by fungi. Fungal cells in Fig. 2 are presenting a typical growth pattern for the rock surface - round cells aggregated into small colonies of various shape defined more by the material structure than by fungus itself. Instead of normal cylindrical hyphae (compare Fig. 1) spherical to subspherical often polyhedral cells as if mimicking the shape of the minerals are developed. It is this type of growth that causes sometimes the difficulties in distinction of some rock inhabiting fungal colonies (*Lichenothelia* sp. in this case) from cyanobacterial colonies (Danin, 1992). This growth type is not what one might imagine being normal for fungi, and still it is the predominant form of fungal growth on the rock surface. This type of fungal growth is referred to as "yeast-like" because their hyphae or separate cells are branching by means of holoblastic generation of a new cell, which is more typical for budding yeasts. Because of their melanized cell wall these dematiaceous (darkly pigmented) fungi are named "black yeast" (von Arx et al., 1977) and are characterized by thick melanized cell walls, and a combination of normal hyphal growth and yeast-like phases or structures at some stages of their life cycle. This group of fungi with its yeast-like or microcolonial growth patterns was registered on the surface of different rocks in different climates (Staley et al., 1982; Henssen, 1987; Gorbushina et al., 1993; Wollenzien et al., 1995; Sterflinger and Gorbushina, 1997).

Most amazing is the fact that fungal colonies of this special growth pattern are penetrating the rock substrate to a considerable depth (Fig. 3). The proportions of visible and hidden parts do not allow a comparison with an iceberg, but such a penetrative "rooting" of fungal colonies within the rock extending down from the surface is really phenomenal. This helps to realize that these darkly pigmented fungal yeast-like structures composed of short spherical to subspherical cells are representing perhaps the main form of fungal existence on the rock substrate. This fact is showing that dematiaceous fungi with yeast-like growth patterns are persistent inhabitants of rocks and especially the rock surface because their form of growth is allowing them to stand the pressure of unfavorable conditions and develop physico-mechanical means of following down with bio-erosion fronts. This may be a very ancient evolutionary pattern as already said in the introduction.

Fig. 1 and 2. Remarkable differences in the structure of fungal colonies growing on the rock surface. Fig. 1 is representing a surface of freshly excavated marble in Chersonesus museum in Crimea (1 to 3 years after excavation) with fast growing hyphae spreading over the surface (Bar 10 μm). Fig. 2. Fungal colonies on a monument with longer duration of air-exposure (approximately 30 years) are representing clump-like structures adhering to the surface and finding shelter between the crystals (Bar 12 μm).

A remarkable difficulty in the assessment of microscopic fungal diversity derives from the fact that the detection of many fungi in the environment is traditionally based on isolation and is made difficult by the normal drawbacks of an indirect isolation procedure. The latter can be improved by changing the isolation strategy: e.g. by using different media, special incubation conditions or new direct isolation methods, when only colonies or microcolonies really present in the sample are chosen for the next transfer. The main aim of this approach is to create in vitro conditions, which are selecting for the most adapted species and are not favoring growth of allochthonous microorganisms. This level of biodiversity recording means that even some quite well-investigated environments might harbour a selection of unusual microorganisms, which are resisting normal isolation-based detection procedures.

Lists of species composition are one outcoming record of a careful selection of isolation methods. The conditions of the assay will have a marked effect on which of the organisms reproduce and are therefore registered as present in this habitat. Two isolation techniques namely (1) dilution plating and (2) direct transfer of single fungal cells or colonies are usually used for the rock substrate. Both methods have their specific bias and therefore the generated data should be interpreted with recognition of the methodological selectivity.

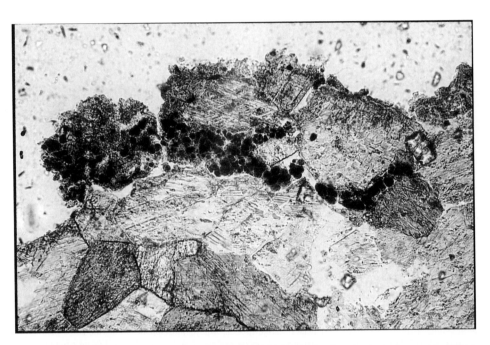

Fig. 3. Petrographic thin section of marble sample with deeply penetrating fungal colony. Aggregates of short spherical to subspherical pigmented fungal cells with different degree of disintegration are representing perhaps the main form of fungal existence on the rock substrate.

The dilution plating method (originating from soil microbiology) is stimulating the development of colonies from spores and yeast cells, but also gives a possibility to isolate those taxa that are present as mycelial filaments in the deeper layers of stone. Other methods of selective isolation (Wollenzien et al., 1995) were developed quite recently based on the knowledge of fungal growth forms. Direct transfer of visible colonies from the rock surface to special selective media aim at selecting well-established colonies which are permanently present on the surface. On agar plates these microcolonial fungi are easily out-competed by faster growing ones. This technique, however, cannot be applied to the fungal community dwelling in the depth of the substrate. These species so far are only approached by unselective isolation methods (e.g. suspension-dilution plating) or by the tedious procedure of opening the rock mechanically and screening it microscopically using micromanipulators). This transforms the microbiologist into a miniature fossil collector or speleologist.

On the other hand, new molecular retrieval approaches allow for the study of diversity of microbial communities without the need of cultivating them (Torsvik et al., 1990, 1996; Amann et al., 1995). These recent investigations help us to realize that our level of information on microbial diversity is in need of future development. Such approaches for fungi are now in the process of development (Kowalchuk et al., 1997; van Hannen et al., 1998) and need to be applied also to the fungal components of ecosystems.

4. Diversity of the Fungal Communities on the Rock Surface

Fungal diversity on different monuments from upper surface layers is recorded by species lists from rocks in the Mediterranean regions. Already these first studies have shown the marked diversity of micromycetes isolated from the rock surface (Table 1) and certain differences in cultural behavior and physiological properties of the isolates. For example, colonies of some of the isolates exhibited very low growth velocity, characteristic thick melanized cell walls, and abundant storage products inside the cells. Slow growth pattern is connected with the way of cell division, which is resembling holoblastic budding more than normal hyphal tip growth. This way the growth zones are situated not only on the growing edge of the colony, but are more diffusely scattered all over the colony mass (Gorbushina, 1997).

In addition to these relatively short lists of fungal inhabitants from Mediterranean climates, rocks in the northern Europe are showing much more fungal diversity of a special kind – many varied species of soil fungi are found there. According to Braams (1992) and Gorbushina and co-authors (in press), the most frequent isolates were represented by *Exophiala jeanselmei*, *Aureobasidium pullulans* and *Phialophora* sp. (which belong to the group of black often yeast-like fungi) in addition to classical soil genera like *Penicillium*, *Cladosporium*, *Phoma* and *Paecilomyces,* to name a few.

These diversity lists are, however, remarkably influenced by the isolation methods used. The studies of fungal rock flora carried out with the help of classical plate dilution techniques very rarely include the special group of microcolonial or yeast-like fungi because these fungi need special isolation approaches, which were used for the rock substrate only recently. In investigations done only by means of selective isolation (Wollenzien et al., 1995; Sterflinger, 1995), only species of slow growing melanized poikilotroph fungi were isolated. On the contrary, isolation only by dilution plating procedure (Anagnostidis et al., 1992) is giving the impression of predominance of fast growing soil and plant fungi also on the rock substrate. The only way to understand the fungal inhabitants of the rock environment is the combination of both methods with addition of statistical analyses of dominance and other parameters. The species list is presenting the full range of biodiversity, and the dominance analysis in combination with direct microscopy brings us an understanding which species are able to persistently grow on the rock surface over longer periods of time.

326

TABLE 1. Fungal diversity on stone monument surfaces. Bold letters are used for fungi demonstrating yeast-like growth pattern

Messina, marble, (Wolenzien et al., 1995)	Acropolis, marble, (Anagnostidis et al., 1992)	Delos, marble, granite (Sterflinger, 1995)	Acropolis, marble, (Gorbushina et al., 1997)	Korfu, calcarenite, (Urzi, unpubl.)	Chersonesus, marble, (Gorbushina et al., 1997)
Hormonema dematioides	Alternaria sp.	**Capnobotryella**-like	Alternaria alternata	Diplodia mutila	**Acrodictys sp.**
Sarcinomyces petricola	A. alternata	Cladosporium sphaerospermum	A. tenuissima	Diplodia sp.	Alternaria alternata
Cladosporium sp.	Aspergillus spp.	Cladosporium sp.	Aspergillus fumigatus	Microsphaeropsis sp.	Aspergillus terreus
Phoma sp.	A. niger	**Exophiala dermatitidis**	Ascochytulina deflectens	Ochroconis humicola	**Aureobasidium pullulans**
Ulocladium sp.	**Aureobasidium sp.**	**Exophiala salmonis**	**Aureobasidium pullulans**	O. anelli	Botrytis cinerea
	Chalaropsis sp.	**Lichenothelia sp.**	Cladosporium cladosporioides	Phoma putaminum	Cladosporium cladosporioides
	Chrysosporium sp.	Coniosporium-like	Cl. sphaerospermum	Phoma sp.	C. herbarum
	Coniothyrium sp.	**Coniosporium perforans**	C. tenuissimum	Trichoderma sp.	C. sphaerospermum
	Fusarium sp.	**Coniosporium appolinii**	Epicoccum purpurascens	**Sarcinomyces petricola**	**Coniosporium sp.**
	Gliocladium roseum	**Phaeosclera**-like	**Exophiala jeanselmei**		**Coniosporium**-like
	Hormonema dematioides	**Sarcinomyces crustaceous**	**E. moniliae**		Epicoccum purpurascens
	Penicillium sp.	Stemphylium sp.	Fusarium sp.		**Exophiala jeanselmei**
	Peyronella sp.	**Taeniolltella**-like	Geotrichum candidum		**E. moniliae**
	Phoma glomerata		**Coniosporium-like**		Penicillium nigricans
	Ph. putaminum		Penicillium verrucosum		Paecilomyces variotii
	Ph. leveilei		Phoma eupyrena		**Phaeococcus sp.**
	Phoma sp.		Ph. exigua		Phoma glomerata
	Pithomyces sp.		Ph. glomerata		Rhizopus stolonifer
	Stachybotrys sp.		Ph. lingam		Stemphylium botryosum
	Trichoderma sp.		Ph. leveilei		Trichoderma viride
	Torula sp.		Ph. putaminum		**Trimmatostroma abietis**
	Ulocladium chartarum		**Scytalidium sp.**		Ulocladium chartarum
	U. botrytis		Ulocladium chartarum		

5. Changes in Species Diversity with Increased Duration of Atmosphere Exposure.

Events following the changes from soil-covered to air-exposed rock surface are worth careful observation. Certain alterations in the appearance of the rock surface immediately after its excavation from the soil and 100 year after are indicating changes in community and in its influence on the substrate. In the case of the classical monuments we have a good basis of comparative studies. Once a monument as e.g. the sanctuary of Delos or the ancient city of Caesarea is excavated, its surface is at first cleaned by archaeologists and conservators and then exposed to all kinds of atmospheric influences. Thus the best experimental site for studies of fungal successions on the rock surface is provided. These differences are presented on Figs. 1 and 2, where the same marble type of the same building in a Greek colony in Chersonesus was excavated at an interval of more than one decade between the events. This way the only contrasting factor is the time of air exposure, but the differences in the growth patterns of fungal colonies are quite remarkable.

On freshly excavated rock surfaces the fungal species composition isolated by plate dilution and by direct microscopy are very close to the normal soil flora. This fact is easily explained if one looks at the surface of a freshly excavated and not yet very well cleaned monument surface (Fig. 1). Ample nutrients are still present on the rock as soil particles and allow the growth of quite numerous fungal hyphae. For a long time the nutrient input in soil was considered too low to maintain the development of active fungal communities, but this legend was rejected quite some years ago and the picture in Fig.1 is another confirmation of this known fact. In comparison to soil the bare surface of the natural rock is a really extremely nutrient poor environment. Nutrients, which are present in organic and inorganic soil and air-borne particles on the rock surface, are slowly cleaned, washed, or blown away by rain and wind respectively. Hereby a kind of a virgin environment is created which could be compared to the plains and rocks of Mars or ancient Earth.

In this unpredictable and hostile environment fungi with different growth patterns (Fig. 2) are settling and establishing themselves as persistent poikilotroph inhabitants of the rock surface. The flora of freshly excavated surfaces is replaced by the flora of a rock surface exposed to the atmosphere for a longer time-period with the slowly established critical conditions of ephemeral supply of nutrients and water (Table 2). At this stage of the air exposed rock community development considerable differences have also been noted between species isolated from surfaces by plate dilution techniques and direct needle isolation. As already mentioned above, the plate dilution method is favoring the development of colonies from spores, which could be incapable of a true establishment on the rock surface. Direct transfer is giving only well established colonies, and in a mature community with the presence of well-adapted species, there should be developed differences between established species and air-transported propagules. This fact confirms the assumption that the fungal community on air exposed stone surfaces is well developed with a biodiversity of a special character.

TABLE 2. Changes in mycobiota of the marble surface with the duration of open air-exposure. Bold letters are used for fungi demonstrating yeast-like growth pattern

Sample	Air exposure	Fungal species
Ch 30	15 years	*Alternaria alternata,* ***Aureobasidium pullulans,*** *Cladosporium cladosporioides,* ***C. herbarum,*** ***C. sphaerospermum,*** *Mucor hiemalis, Penicillium chrysogenum, Phoma eupyrena, Ph. exigua, Ph. glomerata, Ulocladium chartarum,* ***Scytalidium sp.,*** *Stemphilium botryosum, Trichoderma viride*
Ch 5	60 years	*A. alternata, A. tenuissima,* ***Aureobasidium pullulans,*** ***C. cladosporioides, C. sphaerospermum, Coniosporium sp., Exophiala jeanselmei, Coniosporium*** - like, ***Trimmatostroma abietis, Rhinocladiella atrovirens,*** *Phoma glomerata, Ulocladium chartarum*
Ch 15	100 years	***Acrodictys sp., Coniosporium sp., Coniosporium*** -like, ***Phaeococcus sp., Phaeosclera sp.***

The process of the establishment of this special flora, however, is complex and needs a long period of selective changes and pressure. With time the organic content of the bare rock surface is changed and the soil-derived nutrients are exhausted. At this stage poikilotrophic micromycetes adapted to irregularly occurring extreme changes of environmental conditions are slowly emerging as the major group inhabiting the rock (Gorbushina and Krumbein, 1999).

Changes in the nutrient conditions and a drastic change in the environmental conditions with the time of air exposure are responsible for the changes in fungal diversity. Increased environmental pressure is bringing about a directional change in the composition (Table 2) and spatial patterns of species (Fig. 1 and 2) comprising the community, i.e. succession. During this process the rock is open to the action of wind and rain taking down all particles and slowly cleaning the surface. A hostile surface emerges that is not offering nutrients or shelter for chemoorganotrophic organisms. Under these conditions only well-adapted organisms can grow and develop visible colonies. De novo and persistent growth of vegetative mycelium of black yeasts that is recognizable on the rock is a sign of a new and successful adaptation strategy.

Hostility of the conditions is increasing, but simultaneously a new chance is given to those propagules that fail to establish themselves in a highly competitive and overpopulated soil environment. They are able to stand a much broader spectrum of environmental pressure with all its fluctuations and unpredictability, but they cannot cope with the prolonged presence of combative soil species adapted to richer and less extreme environmental conditions.

Fungal succession in the rock environment is operating on the most detailed level of ecological organization - on the level of individual cells. More protected, but slow growing cells are getting their chance of establishment when the fast colonizers - competitive and combative species - loose their chance of development and survival under the given conditions. Differential rates of colony growth combined with the single cell survival ability represent an inherent character that defines the outcome of intraspecific competition in this case.

Differences in the isolated fungal flora might be also interpreted as a function of existing environmental conditions at the moment of isolation. Many differences were noted in fungal floras isolated in different seasons of the year. In humid periods there were more typical soil species coming up (Wollenzien, unpublished; de Leo et al., 1996). This indicates that these species are able to grow on the rock substrate under milder environmental conditions, but they are not belonging to the respective system as a permanent component. As their population size varies with the change of environmental conditions and available nutrients, these fungi could be referred to as allochthonous (Lengeler et al., 1999).

Similarities in the spectrum of fungal isolates and even species identity to air-borne environmental fungal propagules also indicate for the connection between the fungal flora on the rock surface and the fungal flora of the air and surrounding substrates (Simmons, 1981). De Leo and co-authors (1996) have proved also that in addition to these similarities there are always some species of fungi that are not found in the samples of surrounding soil and plant material. These results were once more supported by a very detailed statistical investigations done on the rock surface of marble monuments in St. Petersburg and Crimea (Bogomolova et al., 1997).

6. Diversity Of Growth Patterns As Reflected In Colony Structure

Considering the potential of colonization of air-exposed substrates, fungi possess the most suitable vegetative body structure among all existing organisms. This modular expanding structure allows them to change reversibly and fast between different functional growth modes (Gregory, 1984). Their ability of forming vegetative modifications and spores is especially fitted for surviving unfavorable conditions. Thus the fungal body represents a flexible structure capable of exploring the substrate, spreading and anchoring on its surface, changing its environment or at least decreasing the impact of environmental conditions on its body by hardening the cell walls of or embedding pigments into the outer cells of a colony.

As already mentioned above, subaerial exposure is bringing about the permanent arrival of new fungal spores or propagules on the rock surface. Not all of them are capable of establishing themselves on the surface, but any fungal spore is capable of growth (sometimes only for a short period) without any external support. Succession and establishment are two directly interconnected factors (Frankland, 1998). Those incapable of establishing themselves under difficult and frequently shifting environmental conditions will fail to form long-existing structures. Other propagules suited for this environment will give rise to vegetative cells, capable of further survival on the rock surface.

The life styles of a rock dwelling fungal community are imposed (1) by "normal" stresses or (2) by unpredictable and unexpected changes typical for this particular environment. The normal soil flora, which is frequently observed on rock surfaces at the initial steps of exposure (Fig. 1 and Table 2) or at ample water and nutrient supply could be termed combative or ruderal and is basically characterized by copiotrophic behavior (i.e. growing and metabolising fast at high energy source, nutrient and water levels but being forced into sporulation or survival structures formation when the conditions change). At conditions characterized by "normal" stresses as a lack of nutrients, energy sources and extreme dryness, however, which are only sporadically interrupted by short periods of wetness and supply of organics from the atmosphere the newly described poikilotroph strategy will be more and more established on and in the rocks (Gorbushina and Krumbein, 1999).

The group of black yeasts isolated form the rock in different climates in vitro have salinity, desiccation, temperature (Palmer et al., 1987, Sterflinger and Krumbein, 1995; Sterflinger and Kleen, 1998) and nutrient content (Gorbushina, 1997) tolerances that are much broader than their actual distribution as demonstrated by isolation experiments. This shows their broad survival potential and ecophysiological tolerances, but also presents a challenge to search for this group of fungi in some more ordinary environments, where they do not develop spectacular colonies, but where they are patiently waiting for their chance. Poikilotrophy is a special type of microbial metabolism, which is typical for changing and unpredictable air-exposed bare rock surfaces, but could be used for survival in other more normal habitats with more stable conditions but strong competitive biotic pressure.

Life styles of fungi are expressed mostly on the level of colony formation. This could be well explained by the fact that a modular organism is developing separate modules in the course of its interaction with the environment. The colony formed under unfavorable conditions is a result of such modular development and is gaining its appearance through the formation of protected stress-tolerant modules. The increase in such stress-tolerant modules is indicating the fitness of the fungus to survive the life in the discussed conditions. This way the species with the most versatile genetic equipment for module formation and stabilization will out-compete other competitors in the rock environment.

In order to achieve better results than just registration of isolated phenomenological observations, it becomes necessary to create a frame for discussion. A theory of life strategies provides the essential functional support for studies in the field of fungal ecology and ecophysiology (Cooke and Whips, 1993).

It is in the rock inhabiting fungal community where the strategies of C-selection (combative), R-selection (ruderal), and S-selection (stress-tolerant) are developed in their full extent. The tendencies of committing themselves to reproduction at the moment when adverse conditions occur (combative or ruderal strategy) or of spending their life in surviving and preparing one single reproductive act (stress-tolerant) are getting even more expressed on the rock surface. Black yeast with their slow growth pattern and non-specific vegetative stress-tolerance are still the most extraordinary between these different dispersal strategies. This survival strategy of melanized microscopic fungi with extremely low radial extension rate, high mycelial density, yeast-like growth pattern and low probability of spore formation is referred to as poikilotroph.

Hostile environments leave no chances for species, which are unable to adapt to them, but there could be different ways of living there. For ruderal and combative fungi, which are normal representatives of the soil flora, there is the way of fast development and fast fading away after producing spores if there is enough time and conditions for it. Sustainable development, however, demands for the ability to withstand the presence of stress conditions or other organisms, which militate against their survival (Rayner and Webber, 1984).

The poikilotroph principle (Gorbushina and Krumbein, 1999) in contrast describes an organism type and strategy which aim at as little response as possible to any kind of more or less drastic change (or stress). The poikilotroph organism under laboratory conditions shows stress tolerance to all kinds of negative environmental influences.

TABLE 3. Rock dwelling fungi - diversity of structural colonial forms and exhibited survival strategies

Type of isolates	Growth characteristics at non-optimal temperature	Growth characteristics under nutrient stress	Strategy
Fast growing, Spreading	Slow radial extension rate, mycelium more dense, higher spore density	Fast radial extension rate, scarce mycelium development, low spore density	Exploitation of the nutrient or energy resource, then sporulation, dormancy or spreading
Relatively slow Growing, pleomorphic	Low radial extension rate, high mycelial density,	Medium radial extension rate, dense mycelium, high density of unicellular units (spores or chlamydospores)	Exploitation of the resource combined with stress tolerance
Microcolonial (poikilotroph type)	Extremely low radial extension rate, high mycelial density, low probability of spore formation	Extremely low radial extension rate, high mycelial density, low probability of spore formation	Vegetative structures adapted to slow continuous growth under extreme changes of environmental conditions (non-specific stress tolerance)

But in the natural environment it does not really respond to stress, it does not compete, it does not really develop and exhibit escape or reproduction strategies. It just continues to live and waits. On the contrary: it will invest energy only in protection of each individual vegetative (!) cell not in any kind of unnecessary metabolic and morphogenetic response. Examples of this poikilotroph strategy are represented most elegantly in black yeasts or microcolonial fungi, which are creating a number of individuals large enough for survival (microcolony) and continue dwelling in one of the most changing environments.

All changes in the community development (including seasonal or disturbance-following shifts) are more or less sudden and they are to be reflected in the diversity studies. In such a complex system a status of stable equilibrium is subsisting, since the pressure of unfavorable external conditions is bringing the rock surface to the same starting point once and once again. This ensures that the established poikilotroph fungal population on the surface will withstand also a competition with fast growing ephemeral ruderal fungal flora. For this relatively brief "ruderal vegetation" period, poikilotroph fungi are also able to wait until combative and ruderal species have finished their life cycle in spore formation, leaving a certain amount of organic substances behind. Succession arisen by the sporadic changes in nutrient content or environmental conditions in the habitat in question are always coming to the once established diversity of rock inhabiting fungi, which are continuously dwelling on the rock surface for longer time periods.

Modern fungal ecology has to deal preferably with these autochthonous indigenous fungi in a given environment. Poikilotroph fungi are permanently present in the environment, they shape themselves to the given conditions, they grow slowly, but they die rarely and survive (and metabolize) under all circumstances and conditions given in the extremely variable and extremely changing epilithic and endolithic environment.

7. Diversity of Life Styles as a Diversity of Impact on the Substrate.

The detrimental effects of these slow growing and stress-tolerant fungi on their own environment, the rock material, have been convincingly demonstrated. Apart from their melanization which is then "transmitted" to the material as a more or less dark biopatina, these fungi have very remarkable penetration capacities (Diakumaku, 1995; Sterflinger, 1995; Dornieden et al., 1997). Both of these factors are contributing to the creation of very special interactions between the substrate and these organisms (Dornieden et al., 1997).

Temporarily present species as ruderal or combative ones arrive and quickly spread over the substrate with thin hyphae, followed by immediate sporulation. After this brief period they supply the biomass to the community. They are part of the fungal biodiversity of the rock substrate, because their coming and going belongs to the development and establishment of the system.

Fast growing ruderal or combative species are also contributing to the changes of the rock substrate. The hyphae of these species are spreading faster over bigger areas of the substrate and are also more rapidly and actively metabolising. Ruderal and combative species are frequently synthesizing more extracellular substances and are metabolically more active than slow growing fungi. Only episodically are these fungi coming to full metabolic activity, but in these short periods they could produce a considerable change on the substrate surface. Chemical substances secreted by these fungi include organic acids, carotenoids and melanoid pigments and many other products of a stressed metabolism.

8. Conclusions

The rock inhabiting fungal community is a permanently open system experiencing all sudden and unpredictable changes from a normal to a fully „unsheltered" environment. This is unavoidably causing respective changes in the fungal population. Air-borne propagules under suitable conditions are starting their development for an ephemeral life-span. These propagules sporadically getting chances for growth have nevertheless considerable influence on the substrate and should not be ignored as members of the rock dwelling community.

The permanently rock dwelling fungal community is, however, dominated by a specific group of poikilotrophic micromycetes adapted to irregularly occurring extreme changes of environmental conditions. These fungi are persistently maintaining peculiar vegetative growth patterns on the rock surface - where they penetrate deeply into the material in question- and in culture.

The theoretical distribution of poikilotrophic melanized yeast-like micromycetes should include also temperate zones and well buffered environments. However, in these milder climatic or more favorable nutrient conditions the frequency of isolation of faster growing fungi is much higher then from the rock surface. Isolation of fungi and simultaneous molecular analysis could be an instrumental help for recognizing the unculturable species. If one considers that an appropriate method for black yeast isolation from rock surfaces was developed only quite recently (Wollenzien et al., 1995), it is evident that their isolation from more densely populated and more favorable environments represents an even more difficult task.

There is a natural connection existing between the fungal flora on the rock surface and the fungal flora of the air and surrounding substrates (soil and plants). The species in question are present in the environment surrounding the rock substrate (Simmons, 1981). This fact allows us to state that the realized niches of slow growing rock inhabiting microcolonial poikilotroph fungi are determined by competition with other fungal species employing different growth and life strategies. On the rock surface this group is successfully escaping the competition with fast growing soil species.

On the rock surface fungal succession is operating on the most detailed level of ecological organization - on the level of the inherent characters of strains that define the outcome of interspecific competition by adaptations and survival strategies of the single cell or clone instead of spreading and propagation by spores as in many other kinds of environment and fungal life modes.

9. References

Amann, R.I. (1999) *Biologen heute* VDBiol **441**, 5-6.

Amann, R.I., Ludwig, W. and Schleifer, K.-H. (1995) *Microbiol. Rev.* **59**, 143-169.

Anagnostidis, K., Gehrmann, C.K., Groß, M., Krumbein, W.E., Lisi, S., Pantazidou, A., Urzi, C. and Zagari, M. (1992) in: D. Decrouez, J. Chamay and F. Zezza (eds.) II. Int. Symp. for the Conservation of Monuments in the Mediteranean Basin, Musée d'Histoire Naturelle, Geneve, pp. 305-325.

Bogomolova, E.V., Vlasov, D.Y., Sagulenko, E.S. (1997) *Mikologia i Fitopatologia* **31**, 9-15.

Braams, J. (1992) Ph.D. thesis, Oldenburg, 104 p.

Cannon, P.F. (1998) Proc. *6th International mycology congress*, Jerusalem, pp.7.

Cooke, R.C. and Whips, J.M. (1993) Ecophysiology of fungi, Blackwell Scientific Publications. Oxford. UK. P. 337.

Danin, A. (1992) *Proc. 5th Intern. Conf. On Environm. Quality and ecosystem stability*, Jerusalem, pp. 675-681.

De Leo, F., Criseo, G., Urzi, C. (1996) *Proc. 8th Int. Congress on Deterioration and Conservation of Stone*, pp. 625-630.

Diakumaku, E. (1995) Ph.D. Thesis, Oldenburg, 137 p.

Diakumaku, E.; Gorbushina, A.A.; Krumbein, W.E., Panina, L. and S. Soukharjevski (1995) *Sci. Total Environm.* **167**, 295-304.

Dornieden, Th., Gorbushina, A. A., Krumbein, W.E. (1997) *Int. J. for Restoration of Buildings and Monuments* **3**, 441-456.

Frankland, J. (1998) *Mycol. Res.* **102**, 1-15.

Friedmann, E.I. and Ocampo-Friedmann, R. (1984) in: M.J. Klug and C.A. Reddy (eds.) *Current Perspectives in Microbial Ecology*, ASM Washington, pp. 177-185.

Gorbushina A.A. (1997) Ph. D. Thesis, St. Petersburg, 148p.

334

Gorbushina A.A. and Krumbein W.E. (1999) in: Oren A. (ed.) *Microbiology and Biogeochemistry of Hypersaline Environments*, CRC Press, Boca Raton, FL., pp. 75-86.

Gorbushina, A. A., Krumbein, W. E., Vlasov, D. Yu. (1997) in: A. Moropoulou, F. Zezza, E. Kollias and I. Papachristodoulou (eds.). *Proc. IV. Int. Symp. on the Conservation of Monuments in the Mediterranean*, Vol. 4, Technical Chamber of Greece, Athens, pp. 262-270.

Gorbushina, A.A., Krumbein, W.E., Hamman, C.H., Panina, L., Soukharjevski, S. and Wollenzien, U. (1993) *Geomicrobiol. J.* **11**, 205-222.

Gorbushina, A.A., Lyalikova, N.N., Vlasov, D.Y., Khizhnjak, T.V. (2000) *Mykologia i Fitopatologia*. (in press)

Gorbushina, A.A., Panina, L.K., Vlasov, D.Y., Krumbein, W.E. (1996) *Mikologia i Fitopatologia* **30**, 23-28.

Gregory, P.H. (1984) in: D.H. Jennings and A.D.M. Rayner (eds.) *The Ecology and Physiology of the Fungal Mycelium* pp. 1-22.

Gromov, B.V. (1959) *Vestnik LGU* **4**, 146-155.

Gromov, B.V. (1963) *Vestnik LGU* **3**, 69-77.

Grote, G. (1992) Ph.D. Thesis Oldenburg, 335p.

Henssen, A. (1987) *Bibliotheca Lichenol.* **25**, 257-293.

Hirsch, P. (1986) in: F. Medusar and M. Gantar (eds.) Perspectives in microbial ecology, Ljubljana, pp. 138-142.

Kowalchuk, G.A., Gerards, S., Woldendorp, J.W. (1997) *Appl. Envir. Microbiol.* **63**, 3858-3865.

Krumbein, W.E. (1966) Ph. D. Thesis, Würzburg, 106 p.

Krumbein, W. E. (1969) *Geol. Rdsch.* **58**, 333-363.

Krumbein, W.E. (1973) *Deutsche Kunst- und Denkmalpflege* **31**, 54-71.

Krumbein, W. E. and Jens, K. (1981) *Oecologia* **50**, 25-38.

Lengeler, J.W., Drews G., Shlegel H.G. (1999) *Biology of Prokaryotes*, Thieme, Stuttgart, New York.

Palmer, F. E., Emery, D. R., Stemmler, J., and Staley, J. T. (1987) *New Phytol.* **107**, 155-162.

Rayner, A.D.M. and Webber, J.F. (1984) in: D.H. Jennings and A.D.M. Rayner (eds.) *The Ecology and Physiology of the Fungal Mycelium* pp. 383-417.

Simmons, E. G. (1981) *Mycotaxon* **13**, 407-411.

Staley, J. T., Palmer, F. and Adams, J. B. (1982) *Science*, **215**, 1093-1095.

Sterflinger, K. (1995) PhD thesis, Oldenburg, 138 p.

Sterflinger, K. and Gorbushina, A. A. (1997) *Syst. Appl. Microbiol.* **20**, 329-335.

Sterflinger, K. and Kleen, N. (1998) Proc. 6th International mycology congress, Jerusalem, p.170.

Sterflinger, K. and Krumbein, W. E. (1995) *Botanica Acta* **108**, 490-496.

Sterflinger, K., De Baere, R., de Hoog, G.S., De Wachter, R.D., Krumbein, W.E. and Haase, G. (1997) *Antonie van Leeuwenhoek* **72**, 349-363.

Torsvik V., Salte, K., Sørheim, R., Goksør J. (1990) *Appl. Environm. Microbiol.* **56**, 776-781.

Torsvik, V, Sørheim, R., Goksør J. (1996) *J. Industr. Microbiol.* **17**, 170-178.

Urzi, C., Wollenzien, U., Criseo, G., Krumbein, W.E. (1995) in: D. Alsopp, R.R. Colwell and D.L. Hawksworth (eds.) *Microbial Diversity and Ecosystem Function*, Egham, UK. pp. 289-302.

van Hannen, E.J., van Agterveld, M.P., Gons, H.J., Laanbroek, H.J. (1998) *J. Phycol.* **34**, 206-213.

von Arx, J.A., de Miranda, L.R., Smith, M. Th. and Yarrow, D. (1977) *Studies in Mycology* **14**, 1-37.

Wollenzien, U., de Hoog, G.S., Krumbein, W.E. and Urzi, C. (1995) *Sci.Total Environm.* **167**, 287-294.

Biodata of **Mikal Heldal** author of "*Morphology as a Parameter for Diversity in Bacteria Populations.*"

Dr. Mikal Heldal is a senior scientist at the Department of Microbiology at the University of Bergen, Norway. He obtained his Cand. Real. Degree in 1973 at the above University. His research experience during 25 years is in three main areas of interest: Ecotoxicology and toxicity testing using micro algae; ii) Studies of aquatic microbial ecosystems including development of analytical electron microscopy for studies of elemental composition of single cells: and iii) Studies of viral impacts on microbial communities in marine environments. During the last two decades he has focused on nutrient cycling in marine environments and the ecological impact of virus in marine microbial communities. Dr. Heldal has over 50 publications.
E-Mail: mikal.heldal@in.uib.no

J. Seckbach (ed.), Journey to Diverse Microbial Worlds, 335-346.
© 2000 *Kluwer Academic Publishers. Printed in the Netherlands.*

MORPHOLOGY AS A PARAMETER FOR DIVERSITY IN BACTERIAL POPULATIONS.

MIKAL HELDAL
Department of Microbiology, University of Bergen,
P.O. Box 7800, N-5020 Bergen, Norway.

1. Introduction

In microbiology there is a long tradition for describing morphological variation as part of a species affiliation or classification. Size, shape and flagellation are among the characters frequently used in species descriptions. From studies in the light microscope most bacteria reveal relatively few details, e.g. flagella are not visible without specific staining. Such studies are mainly related to pure cultures of isolated bacteria. This poses the question of how representative such pure cultures are for the community from which they were isolated. In this paper the diversity parameters of size and morphology are focused, and time - space changes in communities together with evaluation of species richness in its simplest form is presented.

For natural marine environments the discrepancy between total counts and viable (colony) counts is normally in the range of 10^2-10^3 bacteria ml^{-1} which means that 99% to 99.9% of the bacteria in sea water will not appear as colony forming units. Since such cells are inaccessible by traditional methods for identification, their taxonomic position and/or physiological ability to grow on solid media is still debated. Many of these cells could be dead (Zweifel and Hagstrøm 1995), they could belong to a few very abundant species, or they might be viable but nonculturable cells (Roszak and Colwell 1987).

Recently it has been suggested that the Bacteria domain consists of about 40 divisions, of which the majority are poorly represented by cultured organisms (Hugenholtz et al. 1998). These authors also stated that 13 of the divisions are based solely on environmental sequences, i.e., no cultivated species have been obtained which represent any of these groups. It has been suggested that the species richness of bacteria should be calculated for microhabitats related in size to that of the organisms (Pedrós-Alió 1993). The majority of native aquatic bacteria have average volumes of less than 1 μm^3 (Fagerbakke et al. 1996) and at cell densities of 10^6 ml^{-1} there is still more than 10^6 μm^3 of water available per cell. If the cell density at the species level is less than 1 ml^{-1} the species richness could be extremely high but still reasonable. On the other hand the species evenness seem to be low in marine environments. Recent studies have shown a range of only 2-20 numerically dominant species (Rehnstam et al. 1993, Høfle and Brettar 1995, Øvreås et al. 1997, Pinhassi et al. 1997, Murray et al. 1998).

Thingstad and Lignell (1997) predicted on a theoretical basis that a limited number ($\leq$ 100) of dominating bacterial species are present at any given time in aquatic environments, considering major growth rate and biomass controlling factors. Based on modeling, these authors also looked at factors controlling microbial ecosystems in terms of biomass and species composition. Viral lysis seems to control the steady state diversity of the bacterial community, even if the bacterial abundance is controlled by protist grazing.

In 1989 Berg et al. reported high numbers of free virus-like particles from surface water samples. During the last decade it has become clear that virus infecting both prokaryotes and eukaryotes may influence a range of biogeochemical and ecological processes. Among these processes the viral control of biodiversity and species distribution is of great interest. In general the total counts of virus is about 10 times (5-25) the bacterial total count in aquatic environments.

For survival of a laboratory grown bacterial culture with a lytic virus present, an equilibrium seems to be found at a host density of $\sim 10^4$ cells ml^{-1} (Lenski 1988). In natural microbial communities viruses may bind non-specifically to various particles, and thus the host/virus equilibrium level could be different. If we assume that $\sim 10^4$ cells ml^{-1} is a reasonable host density for species under lytic viral control, we might find about 100 – 1,000 dominant host species during the parts of the year when the total bacterial counts are close to 10^6 ml^{-1}.

The bacterial mortality caused by viruses has been estimated to 10-50% of the total bacterial mortality in surface water (Suttle and Chen 1992, Fuhrman and Noble 1995, Steward et al. 1996, Weinbauer and Høfle 1998). From these and similar studies, taking into account the general low species evenness reported, we may anticipate a relatively rapid turnover of the "species profile" and frequent changes of the dominating bacterial populations in surface waters. We should further find relatively few dominating viruses in these environments. In fact this has been indicated through recent studies based on pulsed field electrophoresis of dsDNA from environmental samples (Ruth Anne Sandaa, personal communication).

Legend for Figure 1. (see separate page)

Electron micrograph (Transmission Electron Microscope, TEM) from the start of a mesocosm experiment in Knebel Vig, Aarhus Bay, Denmark, May 11, 1991. Cells/particles were harvested from glutaraldehyde (2%) fixed samples directly onto 400 mesh Ni grids by centrifugation. Arrowheads show various morpho-types of bacteria. Tailed viruses (large arrows) and tail-less virus-like particles (small arrows) are also shown.

2. Experimental work

Bacterial cells harvested from water samples directly on electron microscope (EM)-grids by centrifugation will, even in unfixed and unstained samples, at magnifications of 3,000-30,000 show a considerably higher number of details related to size and morphology. This information can be extended by transmission electron microscope (TEM) - X-ray microanalysis of single cells, giving quantitative amounts of most major elements on a per cell basis. Nutrient availability may be related to elemental ratios in cells e.g., N:C, P:C, S:C; and the level of cellular activity may be related to the K:Na

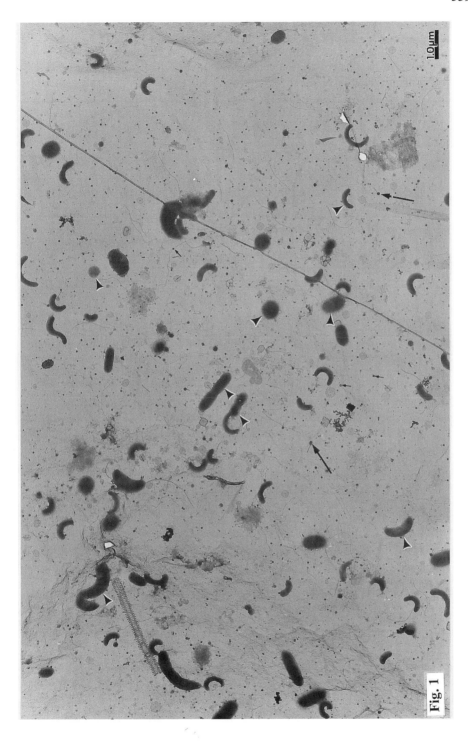

Fig. 1

ratios (Oren et al. 1997). In Table 1 some of the information obtainable from TEM-X-ray studies are summarized. Parts of the physiology related parameters are related to quantitative single cell analyses of C:N:P of native bacteria (Heldal et al. 1997). Other physiology related parameters are frequency of dividing cells (Hagstrøm et al. 1979), ionic cell content (Fagerbakke et al. 1996, Oren et al. 1997), and dry weight of cells. Fagerbakke et al. (1996) introduced the concept of "elemental" diversity based on analyses of the 8-10 major elements in cells. Results obtained indicated both a species related, and a growth condition related index for cultured and for native bacteria.

Table 1. Information obtained from whole cell preparations - TEM

Morphology	Physiology
Cellular size	Nutrient availability
Cellular shape	
Cell division	Growth rate
Spores, vesicles, inclusions, viruses, etc.	
Cell envelope (incl. sheath, capsules etc.)	Nutrient balance
Flagella	Locomotion
Prosthecae, form and chemical composition	Growth conditions

In Fig. 1-5 some examples from a mesocosm experiment in Denmark in May 1991 are given. Some details of the experiment were given in Bratbak et al. 1992. The water samples were fixed immediately in glutaraldehyde (2 % final concentration), and stored at room temperature in the dark until preparation on grids early June 1991. The cells/viruses were harvested directly onto Ni grids (400 mesh) by centrifugation (100,000g, 2.5 h) in a Beckman ultracentrifuge equipped with a rotor for swing out buckets (Børsheim et al. 1990). The grids were air dried, positive stained with 2% uranyl acetate, and viewed on a Jeol 100-CX Transmission Electron Microscope at 80 kV accelerating voltage.

Legends for Figures 2 and 3 (see separate pages).

Figure 2.
Electron micrograph from day 12 (May 22) of the mesocosm experiment. Preparation of samples and identification of bacteria and viruses are as shown in Fig. 1.

Figure 3.
Electron micrograph from day 18 (May 28) of the mesocosm experiment. Preparation of samples and identification of bacteria and viruses are as shown in Fig. 1. Note the increase of virus-like particles and algal scales in the background, and an increase in organic material covering the grid background.

3. Results

Figure 1 shows the particle fractions at the start of the experiment, from the control bag; no nutrient added. Based on morphology some 6-10 types of bacteria (arrowheads) may be identified. Curved rods (vibrio types of cells) represent dominating groups in this sample. On the day 11 of the experiment, a marked shift in the bacteria population

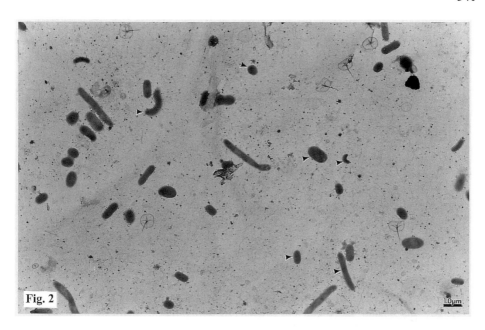

Fig. 2

1.0μm

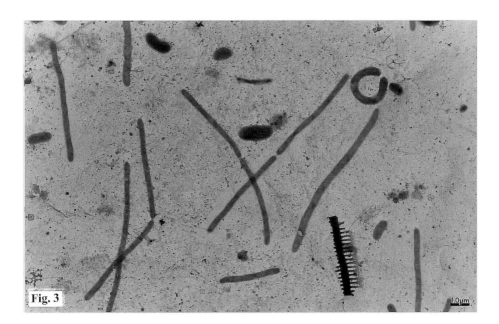

Fig. 3

1.0μm

has been recognized: most of the vibrio type cells were lost and rod shaped bacteria became dominating (Fig. 2). Still 5-8 different morphotypes have been observed. A sample from day 17 of the experiment is shown in Fig. 3. A nearly complete change in the bacterial community is seen: long filamentous bacteria are dominating. Of these filamentous bacteria the difference in cell widths may indicate at least two different populations. In total about 6 different morphotypes are present. Virus- like particles are frequently present both as tailed viruses (large arrows) and tail-less virus-like particles (small arrows, Fig. 1) both at day 1 and 11 (Fig. 2). In the sample from day 17 (Fig. 3) a relatively high number of virus-like particles is seen, but apparently few tailed viruses. In addition, the grid film is covered with organic material, giving a more electron dense background. For all these samples very few bacteria having intact flagella are observed.

In Fig. 4 the whole experiment is summarized: samples from day 3, 6, 9, 12, 15, and 21 are shown in Figs. 4a, 4b, 4c, 4d, 4e, 4f and 4g respectively. As shown in Figs. 4a and 4b relatively few cells are harvested compared to the start of the experiment (Fig.1). Similar experiments have also indicated an initial rapid decline of some bacterial populations (unpublished observations). The day 9 community (Fig. 4c) is markedly changed towards the situation shown in Fig 4d (identical with Fig. 2). Samples from day 15 and 21 (Figs. 4e and 4f) show the same trend as pointed out for day 18 (Fig. 3): a marked dominance of filamentous bacteria. Comparative samples from nutrient amended bags show a similar trend. Fig. 4g shows a day 12 sample from a bag enriched with nitrate as nitrogen source, and a more detailed image (Fig. 4h) shows that some bacteria were surrounded by virus-like particles.

In Figs. 5a and 5b, samples from day 18 of two parallel bags (N-enriched) are shown. The similarity in morphotypes of bacteria with that of the control bag (Fig. 3) is striking. In all bags filamentous bacteria were dominating, with some variability in cell density. Samples from day 18 and 21 also showed that the filamentous bacteria probably were under viral attack, with some bacteria containing large amounts of virus-like particles (Fig. 5c).

Legends for Figures 4 and 5 (presented on separated pages)

Figure 4.
Electron micrographs of samples from day 3, 6, 9, 12, 15, and 21 are shown in Figs. 4a, 4b, 4c, 4d, 4e, 4f and 4g respectively. In Fig. 4g a day 12 sample from a bag enriched with a nitrogen source is shown, and Fig. 4h shows a more detailed image from 4g where some bacteria are surrounded by virus-like particles.

Figure 5.
Electron micrograph from day 18 (May 28) of the mesocosm experiment. Samples from two different nutrient enriched bags (bag N17 and bag N18) are shown in 5a and 5b. Fig. 5c shows a filamentous bacterium with clusters of virus-like particles (arrowheads) inside the cell (from bag N17, day 18).

4. Conclusions

From these observations I conclude that the species evenness seems to be low in marine environments. By and large only 2-20 dominating species might be observed in coastal marine waters by a rough estimate based on cell size and morphology. How fast the

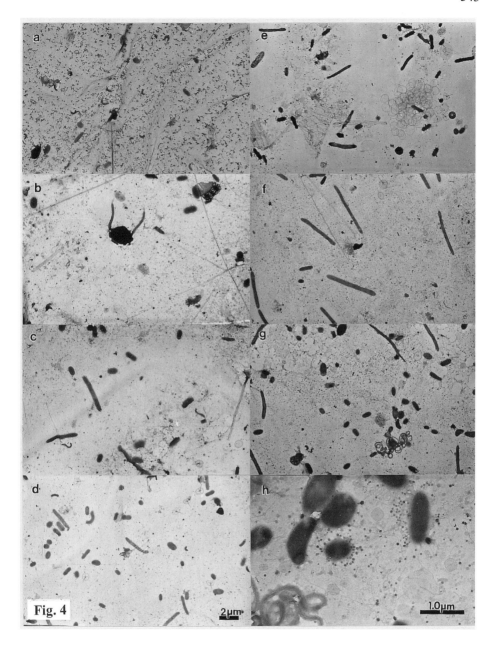

Fig. 4

344

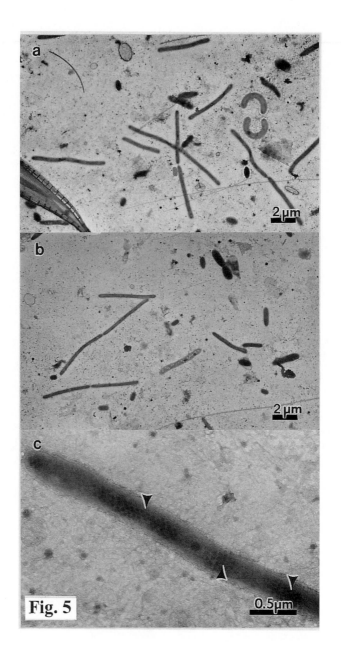

Fig. 5

changes in population profile occur in microbial communities is still an open but crucial question. The results presented here indicate that major changes may occur in a time span of less than 3-5 days. One can argue that the decline of dominating bloom-forming species may occur within a few hours, and that the rate limiting step in establishment of new dominating species is the re-growth. It is striking that the total viral counts are changing about as much within hours as during an annual cycle in coastal waters (Bratbak et al. 1996, Rodrigues et al. (submitted)). My conclusion will be that viruses are important for population changes in microbial communities, but other factors which may influence growth, such as nutrient availability and grazing are still difficult to rule out. Šimek et al. (1997) have shown grazing to favor growth of large grazing-resistant tread-like cells under certain conditions. The observed content of virus-like particles in filamentous bacteria towards the end of the experiments (Fig. 5c) indicate a marked population shift within a few days. Viral lysis and grazing pressure may thus explain population changes presented here.

In Table 1 the information obtained from TEM studies of whole cell preparations of native bacteria is summarized. The results presented above is based solely on cell size and morphology. Other characteristics like flagellation, content of gas vesicles, vacuoles and nutrient storage bodies (e.g. polyphosphate and β-hydroxyalkanoate) are not included in this study, mainly because such studies should be undertaken on non-fixed cells. By combining TEM studies and light microscopy techniques we should be able to get a more throughout understanding of the dynamic patterns governing microbial communities.

5. References

Berg, Ø., Børsheim, K.Y., Bratbak, G. and Heldal, M. (1989) Nature 340:467-468.

Bratbak, G., Heldal, M., Thingstad, T.F., Riemann, B. and Haslund, O.H. (1992) Mar. Ecol. Prog. Ser. 83:273-280.

Børsheim, K.Y., Bratbak, G. and Heldal, M. (1990) Appl. Environ. Microbiol. 56:352-356.

Bratbak, G., Heldal, M., Thingstad, T.F. and Tuomi, P. (1996) FEMS Microbiol. Ecol. 19:263-269.

Fagerbakke, K.M., Heldal, M. and Norland, S. (1996) Aquat. Microb. Ecol. 10:15-27.

Fuhrman, J.A. and Noble, R.T. (1995) Limnol. Oceanogr. 40:1236-1242.

Hagstrøm, Å., Larsson, U., Hørstedt, P. and Normark, S. (1979) Appl. Environ. Microbiol. 37:805-812.

Heldal, M., Norland, S., Fagerbakke, K.M., Thingstad, F. and Bratbak, G. (1997) Mar. Poll. Bull. 33:1-6.

Høfle, M. and Brettar, I. (1995) Limnol. Ocanogr. 40:868-874.

Hugenholtz, P., Goebel, B.M. and Pace, N.R. (1998) J. Bacteriol. 180:4765-4774.

Lenski, R.E. (1988). Adv. Microbial. Ecology 10:1-44.

Murray, A. E., Preston, C.M., Massana, R., Taylor, R.T., Blakis, A., Wu, K. and DeLong, E.F. (1998) Appl. Environ. Microbiol. 64:2585-2595.

Oren, A., Heldal, M. and Norland, S. (1997) Can. J. Microbiol. 43:588-592.

Øvreås, L., Forney, L., Daae, F.L. and Torsvik, V. (1997) Appl. Environ. Microbiol. 63:3367-3373.

Pedrós-Alió, C. (1993). Diversity of Bacterioplankton. TREE 8:86-90.

Pinhassi, J., Zweifel, U. L. and Hagstrøm, Å. (1997) Appl. Environ. Microbiol. 63:3359-3366.

Rehnstam, A.-S., Bäcman, S., Smith, D.C., Azam, F. and Hagstrøm, Å.. (1993) FEMS Microbiol. Ecol. 102:161-166.

Rodrigues, F., Fernandez, E., Head, R.N., Bratbak, G., Heldal, M. and Harris, R.P. (2000) J. Plankton Res. (Submitted)

Roszak, D.B. and Colwell, R.R. (1987) Microbiol. Rev. 51:365-379.

346

Šimek, K., Vrba, J., Perenthaler, J., Posch, T., Hartmann, P., Nedoma, J. and Psenner, R. (1997) Appl. Environ. Microbiol. 63:587-595.

Steward, G.F., Smith, D.C. and Azam, F. (1996) Mar. Ecol. Prog. Ser. 131:287-300.

Suttle, C.A. and Chen, F. (1992) Appl. Environ. Microbiol. 58:3721-3729.

Thingstad, T.F. and Lignell, R. (1997) Aquat. Microb. Ecol. 13:19-27.

Weinbauer, M.G. and Høfle, M. (1998) Appl. Environ. Microbiol. 64:431-438.

Zweifel, U.L. and Hagstrøm, Å. (1995) Appl. Environ. Microbiol. 61:2180-2185.

Biodata of **Arieh Zaritsky** and his team contributors of *Surviving Escherichia coli in Good Shape: The many Faces of Bacillary Bacteria.*

Dining in Amsterdam, August 1997. From left to right: Conrad L. Woldringh (b. 1940), Robert H. Pritchard (1930), Arieh Zaritsky (1942), Itzhak Fishov (1949)

The team consists of three generations of scientists who have spent thirty years exploring what the bacterial cell is trying to tell us about itself. Arieh Zaritsky came to the lab of his mentor Robert Pritchard at Leicester University soon after the cell-cycle model was announced by colleagues and friends Charles Helmstetter and Steve Cooper late in 1968. Immersion in the Leicester cell-cycle tumult resulted in thymine-limitation and step methodology dissociating rates of replication and growth under constant nutritional conditions. They were not aware of Conrad Woldringh's simultaneous struggle with *E. coli* cell and nucleoid shapes in Amsterdam. A fortuitous encounter in Lunteren (1974), just when our respective research lines crossed yielded a long lasting friendship associated with productive collaboration. Among other joint ventures, Conrad and Arieh organized the first two EMBO Workshops on The Bacterial Cell Cycle, in 1980 (Noordwijkerhout, with Nanne Nanninga) and 1984 (Sede-Boker, with Eliora Ron), following the 1978 example of Bob (Leicester, with Kurt Nordström, on plasmid replication). Itzhak Fishov complemented the team upon emigrating to Be'er-Sheva from Moscow in 1991. The ring was closed when Itzhak visited Conrad and met Bob at the Chorin workshop (1997) and at Leicester (visiting Vic Norris).

Many of the raw ideas stemming from the so-called "Copenhagen school" (led by Ole Maaløe) shaped our own. Bob visited Copenhagen as a plant geneticist in 1953 and Arieh (1972), as a bacterial physiologist. Together with the multi-faceted artistic talents and microscopic approach of Conrad and the biophysical skills of Itzhak, we clarified and crystallized ideas about the bacterial cell, some of which are summarized here.

For more about three of us, see our respective Web Sites:

http://www.bgu.ac.il/life/zaritsky.html (Arieh Zaritsky)
http://wwwmc.bio.uva.nl/~conrad/ (Conrad L. Woldringh)
http://www.bgu.ac.il/life/fishov.html (Itzhak Fishov)

J. Seckbach (ed.), Journey to Diverse Microbial Worlds, 347-364.
© 2000 *Kluwer Academic Publishers. Printed in the Netherlands.*

SURVIVING *ESCHERICHIA COLI* IN GOOD SHAPE
The Many Faces of Bacillary Bacteria

ARIEH ZARITSKY[1], CONRAD L. WOLDRINGH[2], ROBERT H. PRITCHARD[3] AND ITZHAK FISHOV[1]

[1] *Department of Life Sciences, Ben-Gurion University of the Negev*
P. O. Box 653, Be'er-Sheva 84105, Israel
[2] *Section of Molecular Cytology, Institute for Molecular and Cell Biology*
BioCentrum, University of Amsterdam, Kruislaan 316, 1098 SM
Amsterdam, The Netherlands
[3] *8, Knighton Grange Rd., Leicester LE2 2LE, United Kingdom*

1. Introduction

One characteristic defining a bacterial species is its cell shape (van Leeuwenhoek, 1684). Our work deals with simple symmetrical shapes that can be defined by the ratio between two dimensions, length L and diameter $2R$. There are two such possible forms, an approximation to a prolate (including a sphere, where $L/2R = 1$) and a cylinder (the length of which equals $L-2R > 0$) with hemispherical polar caps. Cells of the latter (bacilli) are often transformed into spheroids (cocci), such as upon entrance to the stationary phase (in gram negative species) or sporulation (in gram positives). The most familiar rod-shaped bacterium, which can easily become sphere-like, is *Escherichia coli*. Being the species of preference in the investigations conducted during the last 3 decades in our respective laboratories, it is the focus of this Chapter. The general laws derived are valid for other bacterial species and seem universal, albeit with slight modifications. Related subjects that are not thoroughly discussed here have frequently been reviewed in the literature (e.g., Kornberg and Baker, 1992; Koch, 1995; Messer and Weigel, 1996; Höltje, 1998; Nanninga, 1998).

We start by defining states of bacterial culture growth and describe the dependence of cell growth (mass and surface) and division and of nucleoid duplication and segregation on nutritional conditions. We specify the peculiar properties of thymine (a unique DNA precursor) and how they can be exploited to study chromosome replication and cell duplication, viability and shape. The possible induction of membrane heterogeneity by the bacterial nucleoid and the role of membrane domains in regulating cell shape and cycle events are discussed. Morphological modifications obtained by various means are compared, and a "unifying model" that attempts to couple these and other physiological parameters is advanced.

2. The Bacterial Cell Cycle

2.1. STATES OF GROWTH AND TRANSITIONS

"Steady-state growth" was defined as a situation in which "the distribution of each intensive random variable" (any attribute of individual cells) is time-invariant (Painter and Marr, 1968). The more limited term "balanced growth" is a condition in which "every extensive property of the system" (attributes of the whole population) "increases by the same factor over a time interval" (Campbell, 1957). A balanced culture is also in "exponential growth" if the factor is constant over time, and it is in "steady-state growth" if that same factor applies to cell number as well (Fishov *et al.*, 1995). The confusion over these terms in the realm of microbiology was a major reason for the poor understanding of the mechanisms governing bacterial adaptability and shape determination (Henrici, 1928; Pritchard, 1974).

The first systematic study relating cell size and macromolecular composition to the nutritional conditions (Schaechter *et al.*, 1958) opened up the field of bacterial physiology but yielded a limited amount of information. It was followed by studies of transitions between growth rates (Kjeldgaard *et al.*, 1958), which led to the discovery of the "rate maintenance" phenomenon. Delay in adjustment of the rate of protein synthesis during nutritional up-shifts was later explained by the need to raise ribosomal concentrations (Maaløe and Kjeldgaard, 1966). The longer and constant-time delay in the adjustment of cell division rate was understood ten years later, when Helmstetter *et al.* (1968) came up with their cell cycle control (C-H) model. This classical model stemmed from an extensive series of experiments with age-selected synchronous (not synchronized) cultures in a "quasi steady state" (Campbell, 1957; Maaløe, 1963; Cooper, 1991; Fishov *et al.*, 1995).

2.2. GROWTH, CHROMOSOME REPLICATION AND CELL SHAPE

2.2.1. *Dissociation between Rates of Growth and of Chromosome Replication*
The C-H model (Helmstetter *et al.*, 1968; Helmstetter, 1996) relates chromosome replication to cell growth and division. It readily explains the exponential dependence of average cell size (M_{avg}) and macromolecular composition on the culture doubling time τ as follows. Chromosome replication initiates when cell mass reaches a threshold, constant value per *oriC* (M_i), and proceeds at a constant rate to terminate C min later, irrespective of τ over a wide range of growth rates μ (inversely related to τ). The time D between replication-termination and the subsequent cell division is also relatively constant. Thus, cells growing faster (with shorter τ) are larger ($M_{avg} = M_i \ln2 \ 2^{(C+D)/\tau}$) because more mass accumulates before they divide during the constant time ($C+D$ min) following initiation at a constant M_i (Helmstetter *et al.*, 1968; Donachie, 1968; Pritchard *et al.*, 1969). During growth in rich media supporting $\tau < C$, the cell compensates by a strategy not used in eukaryotes: it initiates a new round of replication before the proceeding round has terminated. The idea that the rate of DNA synthesis is not tightly coupled to growth rate nor to rate of chain elongation (Helmstetter *et al.*, 1968) had existed earlier (Maaløe, 1961; Pritchard, 1965), and gained support from autoradiography (Cairns, 1963), genetics (Oishi *et al.*, 1964), and physiological (Pritchard and Lark, 1964) techniques. The amount (in genome equivalents) of DNA per cell, $G_{avg} = (\tau/C \ln2) [2^{(C+D)/\tau} - 2^{D/\tau}]$, thus increases with μ. On the other hand, DNA

concentration (Zaritsky and Pritchard, 1973), $G_{avg}/M_{avg} = [\tau/M_i \; C \; (\ln 2)^2] \; (1-2^{-C/\tau})$, decreases with μ because G_{avg} changes slower than M_{avg}. A dynamic, heuristic version of the model with some of its consequences and implications, including specific mutants with modified values of M_i, C and D, was developed by Norbert Vischer in Amsterdam and can be downloaded from   ftp://simon.bio.uva.nl/pub2/

2.2.2. Cell Dimensions
Under steady-state of exponential growth, a cylindrical bacterium extends during the cell cycle by elongation only (Marr et al., 1966; Trueba and Woldringh, 1980). A larger cell obtained in richer medium supporting faster growth rate, might be expected to accommodate its excess mass (or volume, because cellular density is also essentially constant (Rosenberger et al., 1978a)) in the length dimension. Unpredictably, cell diameter also changes with μ (Schaechter et al., 1958) such that its axial ratio remains relatively constant (Zaritsky, 1975a). The molecular mechanism responsible for the systematic change of cell diameter is still unknown (Cooper, 1991), but slow 'adaptation' to new dimensions following nutritional shifts (Woldringh et al., 1980; Zaritsky et al., 1993; Woldringh et al., 1995a) suggest that remodeling of the peptidoglycan sacculus is involved. It could be argued that the increase of cell diameter with growth rate is caused by drastic alterations in de-repression pattern during growth at different media (Kumar, 1976). An experimental leverage was given by the discovery of a simple means to manipulate the number of replication "positions" without changing the growth rate (Pritchard and Zaritsky, 1970).

A replication "position" n, defined (Sueoka and Yoshikawa, 1965) as a set of forks moving on a chromosome following a simultaneous initiation event at oriC, is equal to the ratio C/τ ($= C\mu$) (Zaritsky, 1975b). Values larger than 1 are allowed by the multi-forked replication strategy (Helmstetter et al., 1968). It was useful to find conditions in which n can be modified at will over a wide range without affecting cellular growth rate, by manipulating the concentration of thymine in thymineless mutants (Zaritsky and Pritchard, 1971; Ephrati-Elizur and Borenstein, 1971). Thus, instead of the 'natural' but complex change in one component of n (τ), it became possible to vary the other component (C) using exceedingly simple procedure (Pritchard and Zaritsky, 1970).

2.2.3. Models for Cell Shape Determination
Cell length has been considered for a long time to be used as a 'ruler' that can somehow trigger division when reaching two "unit cell lengths" (Donachie and Begg, 1989; Koch and Höltje, 1995). Other models have been proposed that predict the observed changes in cell shape (relative dimensions), based on the relationships referred to above. They assume that cell surface area (and also length under steady-state conditions) increases linearly (Previc, 1970) with a discrete doubling in rate at a particular point in the cell cycle (Zaritsky and Pritchard, 1973; Pritchard, 1974). Two possible explanations for a linear increase in surface area were proposed. New envelope material could be laid down at one or more discrete annular sites (like building a chimney) which double in number at a particular point in the cell cycle. Alternatively, a rate-limiting envelope component could be produced from a constitutive gene, output of it doubling when the gene replicated. If the site (or gene copies) doubled late in the cell cycle (triggered by termination, or because the presumed gene was located near the chromosome terminus), an increase in mass accumulation could only be accommodated without a change in density by an increase in cell diameter during the transition. It was postulated that the

change in diameter was a physical response to the increase in turgor that would occur during a transition to a higher growth rate, implying that diameter is passively derived (reviewed in Zaritsky et al., 1982). Other models dealt with cell elongation only, neglecting the second (width) dimension. They tied elongation rate to the number per cell of oriCs, replication forks or terCs (Donachie et al., 1973; Sargent, 1975).

Extensive statistical analysis performed could not distinguish between the various models due to similarity of their predictions for cell dimensions under steady-state of exponential growth (Grover et al., 1977; 1980; Rosenberger et al., 1978b). They do predict substantial differences for changes in dimensions during transitions between steady-states (Grover et al., 1980). The data are best fit to the model that presupposes a zonal surface synthesis at a rate proportional to the instantaneous μ which doubles when the hypothetical controlling gene is replicating d min before cell division (Pritchard, 1974; Woldringh et al., 1980; Zaritsky et al., 1982). A serious reservation stems from the unreasonably long estimated d (ca. 40 min); assembly of FtsZ ring has never been observed that early (Den Blaauwen et al., 1999).

Based on extensive observations of nucleoids under various growth conditions (Woldringh, 1976; Woldringh et al., 1977), it occurred to us that replicating chromosomes with $n > 1$ (i.e., multi-forked) need more space in the width dimension to segregate properly before the cell can divide (Zaritsky and Woldringh, 1987). We have thus proposed that the nucleoid complexity (i.e., its physical size and shape) actively determined cell diameter, predicting a relationship between cell diameter and number of genome equivalents per terminus (DNA content/nucleoid) (Woldringh et al., 1990), $G/T = G_{ter}/T_{ter} = (\tau/C \ln2)(2^{C/\tau}-1)$. (The average number of termini (completed chromosomes) per cell is $T_{ter} = 2^{D/\tau}$ (Zaritsky and Pritchard, 1973).) The available data on average diameter as a function of either τ or C are not sufficiently precise to distinguish between the different models. Nevertheless, we shall discuss below the possibility that the self-organizing properties of bacterial DNA and its direct relationship to the inner membrane may influence cell shape and division.

3. Thymine Metabolism—Cell Division, Size and Composition, Viability and Shape

3.1. THYMINE METABOLISM

The unique role of thymine in cell metabolism derives from the fact that it is a building block of one macromolecule only, DNA, and is not included as a constituent in any other cellular component, though its derivatives (e.g., thymidine-rhamnose) are used in lipopolysaccharide biosynthesis (Ohkawa, 1976). However, elucidation of thymine metabolism and estimation of pool sizes did not throw any light on the regulation of chromosome replication. On the contrary, it seems possible that it is the regime of replication that influences nucleotide concentration. Thus initiation of replication would draw on the pool reducing its size and termination might result in a temporary rise.

Since it is not a normal metabolite in E. coli and Bacillus subtilis, these species have not evolved an active system to take thymine up (Rinehart and Copeland, 1973). It is a breakdown product and can only be incorporated into DNA through salvage pathway in the presence of deoxyribonucleotides or in thyA mutants. Thymine auxotrophs lack thymidylate synthetase thus cannot produce any thymidine nucleotide endogenously; they incorporate thymine by converting it to thymidine and subsequently to dTMP using deoxyribose-1-phosphate (Pritchard, 1974; Ahmad *et al.*, 1998). A pool of the latter is generated from dUMP, which is prevented from conversion to dTMP by the mutation.

Manipulating the concentration of thymine and its derivatives used as DNA precursors in *thyA* mutants is a powerful tool to study nucleoid replication and partition, as well as the presumed coupling between these processes with cell division and shape.

3.2. THYMINELESS DEATH AND RELATED PHENOMENA

3.2.1. *Thymine Starvation and Thymineless Death*
Thymineless death (TLD) was discovered during studies of T-even phages (Watt and Cohen, 1953). Under thymine deficiency in an otherwise complete medium, cells lose the ability to form colonies on agar plates exponentially (following a brief delay) (Barner and Cohen, 1954; Cohen and Barner, 1955). This energy-dependent killing phenomenon (Freifelder and Maaløe, 1964) was ascribed to irreversible lesions induced under unbalanced growth (Cohen and Barner, 1954): inhibition of DNA synthesis while other processes continue normally. The rate of killing is retarded by concomitant inhibition of RNA or protein synthesis (Barner and Cohen, 1957, 1958). Nalidixic acid (Deitz *et al.*, 1966), methionine starvation (Breitman *et al.*, 1971), cytosine arabinoside (Atkinson and Stacey, 1968), and 5-fluorouracil (5-FU) (Cohen *et al.*, 1958) are examples of other means reducing colony-forming ability by inhibiting DNA synthesis. Low levels of cytosine arabinoside not causing TLD shorten the lag before onset of death upon thymine removal. Thymidine analogs cannot be assumed to affect DNA metabolism exclusively: osmotic damage and lysis provoked by 5-FU (Tomasz and Borek, 1962) may be related to conversion of glucose to the cell wall constituent rhamnose, via thymidine diphospho derivatives.

Ahmad *et al.* (1998) has recently reviewed the multitude of molecular and cellular reactions to thymine starvation (leading to TLD). Major examples are cell filamentation (Bazill, 1964; Donachie, 1969), mutagenesis (Smith *et al.*, 1973), DNA breakdown (Freifelder, 1969), structural changes (Nakayama *et al.*, 1994) and lack of methylation (Freifelder, 1967), induction of plasmids and prophages (Korn and Weissbach, 1962; Melechen and Skaar, 1962; Mennigmann, 1964). Integration of the information about these effects may eventually explain the immediate reasons for TLD.

3.2.2. *"Liquid Holding Recovery" and "Resurrection" of Viability*
Whatever the primary target is, TLD depends on the definition of bacterial "death". Operationally, it is loss of colony-forming ability on an agar plate. However, the definition of colony forming units is ambiguous: a bacterial culture can lose a substantial fraction of its viable cells without reduced metabolism under certain conditions, such as upon spreading *lon* mutants from a minimal growth medium on rich agar plates (e.g., Walker and Pardee, 1967; Berg *et al.*, 1976). Failure to generate a colony under one set of conditions is thus not sufficient to define a cell as dead. Furthermore, the number of particles (determined by an electronic counter) during TLD remains constant. It would seem that among the necessary conditions for a cell to be defined as "dead" is lysis or loss of its single DNA complement.

The most striking discrepancy is found when thymine is restored to a thymine-starved culture (Barner and Cohen 1956; Donachie and Hobbs, 1967). The number of colonies increases much quicker than by continued division rate of surviving bacteria. The kinetics of divisions demonstrates that a fraction of cells that lost their colony-forming ability during the starvation period regains this ability while held in liquid medium (thus defined as 'sensitive to plating'). A possible explanation for this plating sensitivity may be found in our previous proposal (2.2.3.) that during transitions between growth rates there is an imbalance between the rate of mass increase and the increase in surface area (and volume). This imbalance could become extreme in the case of thymine starvation because the rate of cell envelope synthesis would be frozen but mass continues to increase quasi-exponentially for some time. This could generate sufficient turgor stress to lead to cell fragility.

3.2.3. *Does Cyt1Aa Exert TLD on E. coli Cells?*

The TLD-like phenomenon (Douek *et al.*, 1992) caused by expressing in *E. coli* of Cyt1Aa (a component of the mosquito larvicidal activity from the entomopathogenic bacterium *B. thuringiensis* subsp. *israelensis*) has recently been associated with compaction of the nucleoid (Manasherob *et al.*, in preparation). Thymine starvation leading to TLD, on the other hand, does not result in compact nucleoids (Woldringh *et al.*, 1994, and see section 6.1.). The lethal effect of Cyt1Aa is associated with growth inhibition, apparently due to perforation of the cytoplasmic membrane, does not imitate TLD, and can thus not be used as a lever to understand the killing mechanism of thymine starvation. Taken together, both observations support the transertion model (Norris, 1995; Woldringh *et al.*, 1995b; Binenbaum *et al.*, 1999), to be dealt with below.

3.3. THYMINE LIMITATION VS. THYMINE STARVATION

As with precursors of other macromolecules, cultures of *thy* mutants continue to grow and multiply indefinitely at the same rate when the medium is supplemented with a wide range of thymine concentrations. However, manipulating the intracellular levels of thymine and its metabolites by varying the external concentration (Pritchard, 1974) is possible due to lack of active transport for this DNA precursor (Rinehart and Copeland, 1973). The resultant physiological state, defined as "thymine limitation", is completely different than that reached by thymine starvation. Thymine limitation reduces the rate of DNA chain elongation without altering the overall rate of DNA synthesis under steady state conditions since this is determined by the frequency of initiation of rounds of replication. The assumption that thymine concentrations could be changed with impunity provided there was no change in growth rate led many scientists to misinterpret their data. For example, to isolate or follow high specific radioactive DNA and save on the radioisotope, exceedingly low concentrations have been used during the labeling period (e.g., Maaløe and Rasmussen, 1963), thus leading to flawed conclusions (e.g., Lark and Lark, 1965). The discrepancies were resolved by systematic investigations relating the length of C to the external concentration supplied (Pritchard and Zaritsky, 1970; Beacham *et al.*, 1971; Zaritsky and Pritchard, 1971; 1973; Zaritsky, 1971; 1975b; Pritchard, 1974; Bremer *et al*, 1977; Molina *et al.*, 1998). This series of studies confirmed the concept, that the rate of DNA chain elongation is not coupled to growth rate, and that prokaryotes regulate DNA synthesis by the initiation rather than elongation process (Maaløe and Kjeldgaard, 1966). Helmstetter *et al* (1968) first demonstrated a constant C over a wide range of μ's in wild-type strains, while our studies manipulated C without changing μ in *thy* mutants. This technique has since been proved useful for several purposes (e.g., Bird *et al.*, 1972; Zaritsky and Pritchard, 1973;

Pritchard *et al.*, 1975; Chandler and Pritchard, 1975; Meacock *et al.*, 1977; Woldringh *et al.*, 1994; Hadas *et al.*, 1997; Zaritsky *et al.*, 1999a, b).

Slowing the rate at which replication fork traverses the chromosome to terminate by short supply of thymine metabolites (Beacham *et al.*, 1971) delays subsequent cell division. As happens at fast growth rates, cells are consequently larger with higher DNA content (though lower DNA concentration), and the nucleoid is more complex, containing a larger number of *oriC* and forks thus more DNA (Zaritsky and Pritchard, 1973; Pritchard, 1974). Filaments were anticipated as found under thymine starvation leading to TLD (Bazill, 1964; Donachie, 1969). But surprisingly, as with faster growth rates, the increase in cell size was accommodated by an increase in cell diameter (Zaritsky and Pritchard, 1973; Zaritsky and Woldringh, 1978). The connection between cell diameter and the nucleoid complexity, membrane heterogeneity and peptidoglycan synthesis will be discussed below.

4. The Transertion Model, Membrane Domains, and Cell Cycle Events

The cytoplasmic membrane has been implicated in crucial cell cycle events (Marvin, 1968; Funnell, 1993; Nanninga, 1998). For instance, rejuvenation of DnaA following replication initiation (Crooke *et al.*, 1991) requires acidic phospholipids (Castuma *et al.*, 1993), and assembly of FtsZ in a "cytokinetic ring" between daughter chromosomes anchored to the membrane preceding division (Lutkenhaus, 1993) needs phosphatidylethanolamine (Mileykovskaya *et al.*, 1998).

On a more general level, the highest activities of transcription, translation and insertion or transport ("transertion") of inner and outer membrane proteins expressed from many different genes scattered on the chromosome could be envisaged to occur around the nucleoid (Norris, 1995; Woldringh *et al.*, 1995b; Binenbaum *et al.*, 1999). This view is depicted schematically in Figure 1B. The "transertion" activity could affect enzymes involved in peptidoglycan synthesis, resulting in decreased rate at the surface surrounding the nucleoid, as has been observed by autoradiography (Mulder and Woldringh, 1991). Such a crowding effect can also result in membrane domains determined by the position(s) of nucleoid(s) and its (their) interactions with the plasma membrane, as previously suggested by Norris (1992). Sequestration of newly-replicated, hemi-methylated *oriC* is one such possibility that is widely entertained in current literature (Ogden *et al.*, 1988; Campbell and Kleckner, 1990; Bogan and Helmstetter, 1996). It may thus be the mechanism exploited to block a second round of replication-initiation (so-called "stacking") during ca. 10 min following an initiation event ("eclipse period" or inter-initiations "dead time") despite accumulated capacity for initiation (Zaritsky, 1975b).

The crowding effect may be the mechanism used to prevent premature cell division, as suggested in the nucleoid occlusion model (Mulder and Woldringh, 1989). When sister nucleoids segregate, a new less crowded membrane domain with a different phospholipid and protein composition seems to be created between them, which can signal the assembly of the FtsZ ring and the subsequent recruitment of cytoplasmic (FtsA) and membrane-bound proteins (ZipA, FtsQ, FtsL and PbpB) (Fig. 1D). These proteins have been shown by immuno-fluorescence studies (Ma *et al.*, 1996; Yu *et al.*, 1998) to be sequentially recruited to the divisome (Nanninga, 1998). An indicator for the appearance of such membrane domains at the site of separating nucleoids has recently been obtained with the fluorochrome FM4-64 (Fishov and Woldringh, 1999).

5. Peptidoglycan Synthesis, Assembly of the Division Ring and Cell Shape

5.1. DETERMINATION OR MAINTENANCE OF CELL SHAPE

Bacterial cell shape must have genetic determinants because it is characteristic for any strain and perpetuates between generations. The mechanism(s) to accomplish this is (are) completely unknown. The rigid peptidoglycan sacculus maintains cell shape, but is it the shape-determining molecule as well? One view is that the sacculus plays the role of a simple template that is duplicated (Goodell and Schwarz, 1975). It has been schematized in Höeltje's (1998) "three-for-one" growth model, proposing that the holoenzyme murein-replicase duplicates the sacculus according to a "make-before-break strategy" (Koch, 1982). The complex synthesizes three new murein strands and attaches them on both sides of a docking strand via the peptide cross-bridges. The docking strand is simultaneously degraded by transglycosylases and endopeptidases incorporated in the back of the sliding complex. This mode of shape determination is only possible, however, if the glycan strands lie in an ordered fashion and their lengths can be precisely copied by the murein synthesizing machinery (as in the Hoop theory of Cooper, 1989). This model for bacterial morphogenesis does not solve the problem of division mechanism nor of changes observed in cell diameter. A murein-based model becomes very difficult to imagine if the glycan strands are perpendicularly oriented, as recently suggested by Dmitriev *et al.* (1999). According to such a model, the necessity for order of glycan strands functioning as a template has vanished. Our preferred, alternative view, that cell shape is determined by physico-chemical interactions between cytoplasm and the nucleoid, will be described below. Obviously, the genetic background constrains these interactions through variations in composition of cellular cytoplasm and structures.

5.2. PENICILLIN-BINDING PROTEINS

The final steps in peptidoglycan synthesis and maturation are catalyzed by 12 penicillin-binding proteins (PBPs) (Spratt, 1978; 1983; Henderson *et al.*, 1997). Inactivating them either by specific β-lactam antibiotics or growth at the restrictive temperature of *ts* mutants is instructive for understanding their respective roles in bacterial physiology. For example, inactivating PBP2 yields spheroidal cells, while PBP3 is obligatory for division. However, among the series of 192 mutants with all possible combinations of deletions of eight PBP genes that has recently been constructed (Denome *et al.*, 1999), several mutants continued to grow as enlarged spheres when both genes for PBPs 2 and 3 were specifically inactivated. The only lethal combinations were those lacking PBPs 1a and 1b.

Cell division is inhibited by inactivating PBP3, by growing mutants at restrictive conditions or in the presence of specific antibiotics (e.g., furazlocillin or cephalexin). Existing constrictions (formed at the cell center) in the growing filaments are deformed into so-called blunt constrictions. The extra peptidoglycan synthesis needed for the constriction indeed takes place at these positions (Wientjes and Nanninga, 1989). In addition, FtsZ rings can form in the absence of PBP3 activity (Pogliano *et al.*, 1997). It thus seems that the enzymes responsible for the initial peptidoglycan synthesizing activity are penicillin insensitive (Nanninga, 1991).

5.3. ASSEMBLY OF THE DIVISION RING

The abundant cytoplasmic protein FtsZ has been shown to polymerize as a ring surrounding the cell center to direct the division process (Bi and Lutkenhaus, 1991). Many other Fts proteins involved in the so-called "divisome" (Nanninga, 1998) lie in the periplasm anchored to the membrane and some are cytoplasmic. The interactions between them (spatial and temporal) are currently being intensively studied (e.g., Ghigo *et al.*, 1999; and see Fig. 1D), but the consensus is that FtsZ is the first to be active and recruits the others to the division site (e.g., Taschner *et al.*, 1988; Addinall and Lutkenhaus, 1996; Liu *et al.*, 1999).

6. Nucleoid Replication and Segregation—Cell Division and Shape

6.1. COUPLING BETWEEN REPLICATION AND DIVISION

Coupling between cell division and replication of the genetic-information complement (in this case, the bacterial chromosome) is a necessary condition for survival of a species. The mechanism involved has still not been deciphered. The C-H model (Helmstetter *et al.*, 1968) raises an attractive idea, that termination of a round of chromosome replication triggers cell division by an abrupt rise of the local concentrations of dNTPs upon instantaneous cessation of their use in replication. This coupling is envisaged to be a consequence of the inter-relationships and connections between metabolism of DNA and cell wall precursors (Woldringh *et al.*, 1991).

Our approach couples the timing of cell division and location of the division ring (septum in gram positive species, constriction in gram negatives) to certain physical parameters of the cell and the nucleoid by influencing local cellular biochemistry. This view is difficult to envisage but represents a flexible and dynamic mechanism for morphogenesis. The model assumes that the self-organizing properties of DNA generate a compact, spherical nucleoid through the interactions of DNA supercoils with themselves and with the cytoplasmic proteins exerting molecular crowding forces (Woldringh and Odijk, 1999). Depending on its biosynthetic activity (concomitant transcription, translation and protein transport) the nucleoid assumes a prolate shape, which during multifork replication can change into dumbbell and multi-lobular shapes (Woldringh *et al.*, 1995b). The "transertion" activity (Norris, 1995; Woldringh *et al.*, 1995b; Binenbaum *et al.*, 1999) could crowd out or push aside the PBPs involved in peptidoglycan synthesis. This would result in the decreased rate of murein synthesis at the surface surrounding the nucleoid, as has been observed by autoradiography (Mulder and Woldringh, 1991). The above ideas have led to the model described below and in Figure 1.

6.2. THE MODEL (Figure 1, reproduced from http://wwwmc.bio.uva.nl/~conrad/)

Our model is based on the observations that nucleoid segregation in *E. coli* (Woldringh, 1976; Van Helvoort *et al.*, 1996) as well as in *B. subtilis* (Nanninga, unpublished) takes place gradually during replication. The idea that *oriC* functions as a centromer-analog was originally proposed by Ogden *et al.* (1988) and is implicit in the theoretical considerations of Dingman (1974) and in the nucleoid-formation model of Løbner-

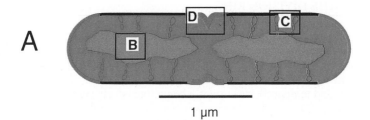

A

1 µm

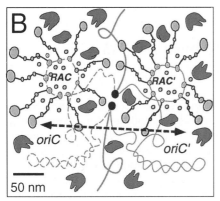

B

RAC   RAC'

oriC   oriC'

50 nm

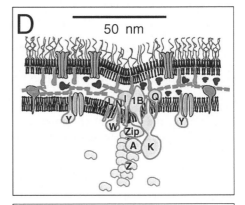

D

50 nm

N 1B Q
Y   Y
W Zip
A K
Z

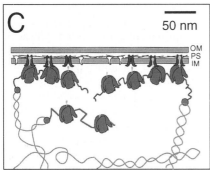

C

50 nm

OM
PS
IM

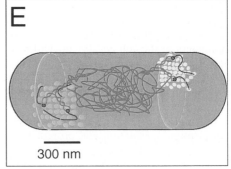

E

300 nm

Figure 1 A schematic representation of the model relating nucleoid segregation to cell division.

A. Overview of an *E. coli* cell grown in glucose minimal medium (length 3 µm, diameter 0.7 µm). Three membrane domains generated by the transertion activity of the nucleoid have been distinguished: polar, lateral and central. Boxes indicate three processes that have been depicted in more detail in B, C and D.

B. Initial displacement of the newly replicated sister origins (*oriC* and *oriC'*) within the nucleoid. The replicated daughter strands are indicated by thin, dashed and full lines. The origins (represented as loops) are assumed to be pushed apart by formation of two ribosome assembly compartments (RAC and RAC') around the *rrnC* operon. An imaginary nucleoid segregation axis (NSA, indicated by the dashed line) is assumed to develop between the RACs. The filledblack circles represent the two replication forks.

C. Formation of a proteolipid domain by transertion. Co-transcriptional and co-translational insertion of membrane proteins (transertion) indirectly links DNA to the inner membrane (IM) via RNA polymerase, mRNA, ribosomes, nascent polypeptides and (here) the Sec translocase. RNA polymerase drags the DNA at the apical loop of its supercoils, thereby forcing it to rotate (ten Heggeler-Bordier *et al.*, 1992). PS, periplasmic space; OM, outer membrane.

D. The divisome. After FtsZ ring formation between the segregating nucleoids, the following membrane proteins have been recruited: FtsA, ZipA, FtsI (PBP3), FtsQ, FtsN, FtsW, FtsL, FtsK and PBP1B. FtsY forms part of the translocase complex.

E. Formation of two proteo-lipid membrane domains (300 x 300 nm) by the transertion mechanism. The two domains, located at the front and back of the cell, consist of 50 inserted proteins (center indicated by the circles, proteins not to scale) transcribed from two different genes. Such transertion structures exist temporarily for many different genes coding for membrane proteins.

-Olesen and Kuempel (1992). The analogy with a centromer is only applicable if the replicated *oriC* is the first region to be separated. Since there is no structural analog for the eukaryotic spindle to give directionality to this initial outward movement, we have postulated that an imaginary Nucleoid Segregation Axis (NSA) is formed between the two sister *oriC*s (Woldringh and Nanninga, 1985; Woldringh *et al.*, 1994; 1995a). One possibility for such a segregation axis is formation of two Ribosomal Assembly Compartments (RACs) that push away the newly-replicated *oriC*s and may thus be regarded as a nucleolar-like structure.

6.2.1. Ribosomal Assembly Compartments and Nucleoid Segregation Axis (Fig. 1B, C)
According to this concept, diffusion of new ribosomes is limited by their functioning on nearby mRNAs. The centers of the two new RACs are considered as the ends of an imaginary line, the NSA, which extends with mass growth and pushes the nascent DNA outward. The lifetime of a RAC is equal to inter-initiation time (τ), but an old one is not just "dissolved" and replaced by new ones starting from size zero; it persists, but with a duplicated "active center". During this "reorganization", the replicated *oriC*s are pushed apart, thus used as the initial force in their partitioning (Fig. 1B). Following this initial separation, newly replicated DNA-loops are sequentially pulled into the different cell halves by the transertion activity (Fig. 1C). Instantaneous replacement of an old RAC by two new ones is necessary because a RAC is the motor that drives cell growth continuously. If duplication is inhibited (e.g., by inhibiting DNA synthesis), the RACs continue to enlarge "indefinitely" pulling the nucleoid apart into small lobules (Woldringh *et al.*, 1994).

The model further assumes that secondary axes develop during multifork replication under conditions of rapid growth. Initially these lie perpendicular to the previous axis (which is the long cell axis when $C < \tau$). In a cylindrical cell, the NSA is forced to rotate over $90°$ during its lifetime to parallel the cell's long axis. It then induces a division plane perpendicular to the longest axis (Woldringh et al., 1995a). Thus, even after reinitiation, the old axis continues to rotate as if initiation had not occurred.

The initial strand separation by formation of the nucleolar-like RACs requires the presence of rRNA genes (*rrn*) near *oriC*. This requirement is fulfilled in many genomes (Cole and Saint Girons, 1994). The RAC idea has recently gained support from a completely new experimental angle: in a systematic study inactivating sequentially from one to all seven existing rRNA (*rrn*) operons, Asai *et al.* (1999) observed "a pronounced morphological change during exponential growth; the cells became more and more elongated....not completely reversed in a Δ7 strain..., suggesting that cellular processes other than *rrn* gene dosage are still perturbed". One of those presumed "other processes" may be the initial separation of *oriC*s.

6.2.2. The Partitioning System
Another mechanism to partition sister replicons, which is found on both chromosomes and plasmids, consists of a cis-acting DNA-site and two trans-acting proteins (Par). In *B. subtilis*, several DNA sites (*parS*) have been found near *oriC*, with high affinity for SpoOJ (Lin and Grossman, 1998). It has been suggested (Glaser *et al.*, 1997; Wu and Errington, 1997) that SpoOJ forms in *B. subtilis* a nucleoprotein complex displacing the segregating origins. The above two mechanisms, i.e., formation of RACs and the Par system, may function together and complement each other. It should be noted, however, that Par is not found on the *E. coli* chromosome (although it exists on its P and F

plasmids), whereas *Caulobacter crescentus* lacks an rRNA gene close to its origin. The *par* system in the latter appears to be essential (Marczynski *et al.*, 1990; Mohl and Gober, 1997), whereas mutations in the *par*-like genes of *B. subtilis*, in which an *rrn* occurs close to *oriC*, only lead to mild effects (Webb *et al.*, 1997).

6.2.3. *Summary and corollaries*

Soon after initiation of replication, an initial displacement of the newly replicated daughter strands occurs by either one or a combination of the two mechanisms described above, the RACs (Fig. 1B) or the *par*-system. This displacement creates a direction of movement through an imaginary axis (NSA), which helps to distinguish replicated DNA regions of the two daughter strands. Subsequently, these regions are pulled to either cell half by the transertion mechanism (Fig. 1C): a promoter-containing DNA loop is translated co-transcriptionally. If it encodes plasma membrane or excreted protein, the nascent peptide is targeted by one of the known insertion or translocation mechanisms. Co-translational insertion anchors the polysome to the membrane, together with the RNA polymerase, through which the DNA tracks. The membrane-pulled DNA-loop increases the likelihood that other RNA polymerase-mRNA-ribosome complexes find an available anchoring site there. Numerous locally inserting (integral) membrane proteins with high affinity to specific phospholipids may form a distinct proteo-lipid domain (Fig. 1E).

In rapidly growing cells, the transertion-mediated pulling force acting on the replicating DNA causes its separation into lobules (as happens during replication inhibition). The perpendicular and tilted planes in which the lobules often lie confirm the existence of axes along which DNA moves. Although these structures represent characteristic shapes, recognizable in most cells of a rapidly growing population, they do not necessarily reflect the position and rotation of the hypothetical segregation axes between the RACs.

By minimal energetic considerations, the NSA initially lies perpendicular to the plane of the previous one (Zaritsky *et al.*, 1999a, b). During extension, it is forced by the rigid cell envelope to rotate until it parallels cell length. The division plane is eventually positioned perpendicularly to that of the NSA when the signal for constriction is activated, presumably around the time of replication-termination at *terC*.

6.3. TESTING THE MODEL BY MULTIPLE CONSTRICTIONS

Successive divisions are usually parallel in bacillary bacteria, whether growing under steady-state conditions or following filamentation (brought about by various means; Taschner *et al.*, 1988). In spheroids containing multi-forked replicating nucleoids, where the spatial constraint (the cylindrical rigid wall) forcing them to rotate is relieved, NSA would continue to extend in the same plane in which the duplicated *oriC*s were pushed away by the RACs following initiation. The segregating *oriC*s would be located under such conditions in the apexes of a regular tetrahedron, implying a perpendicular axis to the previous axis and to each other (as in Fig. 6c of Zaritsky *et al.*, 1999a). Our model predicts that resultant secondary constrictions are laid down in planes perpendicular to the previous and to their sister division planes, in 3 dimensions. This prediction can be confirmed by obtaining successive constrictions simultaneously on the same cell before separation to two daughters, a condition requiring enhancement (at least temporarily) of the division signals. We have recently developed a method to meet both requirements, to relieve the constraint and to enhance division signals (Zaritsky *et al.*, 1999a, b).

A brief inhibition of PBP2 activity, transforming the bacillary cell into a coccus, can be obtained either by a *pbpA ts* mutant in the restrictive conditions or by mecillinam, the latter avoids complications incurred by the temperature shift. Enhanced frequency of division signals was achieved by manipulating the nucleoid replication rate using the defined transitions ("steps") between various concentrations of thymine. Secondary constrictions in such cells were visualized by scanning electron microscopy and by confocal scanning laser microscopy of cells stained with FM 4-64 (Zaritsky *et al.*, 1999a). More recently, division rings were probed by specific anti-FtsZ monoclonal antibodies. Only partial rings (arcs) were observed, probably due to shortage of FtsZ to complete them over the wide cells, but their planes were tilted (Zaritsky *et al.*, 1999b), supporting our model.

The ultimate solution will follow FtsZ ring assembly in such multi-constricted cells by visualizing the rings with FtsZ-GFP fusion proteins in vivo. To this end, the chimeric construct *ftsZ-gfp* is cloned on the multi-copy pRRE7 for high expression (Einav *et al.*, in preparation).

7. Conclusions

The self-organizing properties of bacterial DNA and its indirect relationship to the inner membrane may determine cell shape and division.

In this view, cell diameter, and thus cell shape, may be related to the nucleoid complexity, membrane heterogeneity and peptidoglycan synthesis, determined by the vertical displacement of the segregation axes from the longitudinal axis of the cell.

There seem to be reciprocal inter-relationships between two forces as follows:
(a) the cylindrical cell wall forcing the NSA to rotate, and (b) the increased amount of DNA per nucleoid, which leads to increased membrane crowding (more genes coding for envelope proteins expressed in the vicinity of the nucleoid).

The two forces probably slow down peptidoglycan synthesis, causing decreased surface/volume ratio and thus, increased diameter.

8. Acknowledgments

A handful of devoted technicians were instrumental to achieve the progress in our labs. In particular, we wish to thank Terry Lymn (Leicester), Monica Einav (Be'er-Sheva) and Peter Huls, Evelien Pas and Rob de Raaij (Amsterdam) for great help. Developing software for cell measurements by Norbert Vischer was invaluable in enhancing research in Amsterdam. Too many students to mention here are responsible for many ideas and most work described. All four of us were aided by several EMBO Fellowships at various times during the last 30 years.

362

9. References

Addinall, S.G. and Lutkenhaus, J. (1996) *Mol. Microbiol.* **22**, 231-237.
Ahmad, S.I., Kirk, S.H. and Eisenstark, A. (1998) *Annu. Rev. Microbiol.* **52**, 591-625.
Asai, T., Condon, C, Voulgaris, J., Zaporojets, D., Shen, B., Al-Omar, M., Squires, C. and Squires, C.L. (1999) *J. Bacteriol.* **181**, 3803-3809.
Atkinson, C. and Stacey, K.A. (1968) *Biochim. Biophys. Acta* **166**, 705-707.
Barner, H.D. and Cohen, S.S. (1954) *J. Bacteriol.* **68**, 80-88.
Barner, H.D. and Cohen, S.S. (1956) *J. Bacteriol.* **72**, 115-123.
Barner, H.D. and Cohen, S.S. (1957) *J. Bacteriol.* **74**, 350-355.
Barner, H.D. and Cohen, S.S. (1958) *Biochim. Biophys. Acta* **30**, 12-20.
Bazill, C.W. (1964) *Nature* **226**, 346-349.
Beacham, I.R., Beacham, K., Zaritsky, A. and Pritchard, R. H. (1971) *J. Mol. Biol.* **60**, 75-86.
Berg, P.E., Gayda, R., Avni, H., Zehnbauer, B. and Markovitz, A. (1976) *Proc. Natl. Acad. Sci. USA* **73**, 697-701.
Bi, E. and Lutkenhaus, J. (1991) *Nature* **354**, 161-164.
Binenbaum, Z., Parola, A.H., Zaritsky, A. and Fishov, I. (1999) *Mol. Microbiol.* **32**, 1173-1182.
Bird, R.E., Louarn, J, Martuscelli J. and Caro, L. (1972) *J. Mol. Biol.* **70**, 549-566.
Bogan, J.A. and Helmstetter, C.E. (1996) *J. Bacteriol.* **178**, 3201-3206.
Breitman T.R., Finkelman, A. and Rabinovitz, M. (1971) *J. Bacteriol.* **108**, 1168-1173.
Bremer, H., Young, R. and Churchward, G. (1977) *J. Bacteriol.* **130**, 92-99.
Cairns, J. (1963) *J. Mol. Biol.* **6**, 208-213.
Campbell, A. (1957) *Bacteriol. Rev.* **21**, 263-272.
Campbell, J.L. and Kleckner, N. (1990) *Cell* **62**, 967-979.
Castuma, C.E., Crooke, E. and Kornberg, A. (1993) *J. Biol. Chem.* **268**, 24665-24668.
Chandler, M.G. and Pritchard, R.H. (1975) *Mol. Gen. Genet.* **138**, 127-141.
Cohen, S.S. and Barner, H.D. (1954) *Proc. Natl. Acad. Sci. USA* **40**, 885-893.
Cohen, S.S. and Barner, H.D. (1955) *J. Bacteriol.* **69**, 59-66.
Cohen, S.S., Flaks, J.G., Barner, H.D., Loeb, M.R. and Lichtenstein, J. (1958) *Proc. Natl. Acad. Sci. USA* **44**, 1004-1012.
Cole, S.T. and Saint Girons, I. (1994) *FEMS Microbiol. Rev.* **14**, 139-160.
Cooper, S. (1989) *J. Bacteriol.* **171**, 5239-5243.
Cooper, S. (1991) *Bacterial Growth and Dvision: Biochemistry and Regulation of Prokaryotic and Eukaryotic Division Cycles.* San Diego, CA: Academic Press, Inc.
Crooke, E., Hwang, D.S., Skarstad, K., Thony, B. and Kornberg, A. (1991) *Res. Microbiol.* **142**, 127-130.
Deitz, W.H., Cook, T.M. and Goss, W.A. (1966) *J. Bacteriol.* **91**, 768-773.
Den Blaauwen, T., Buddelmeijer, N., Aarsman, M.E.G., Hameete, C.M. and Nanninga, N. (1999) *J. Bacteriol.* **181**, 5167-5175.
Denome, S.A., Elf, P.K., Henderson, T.A., Nelson, D.E. and Young, K.D. (1999) *J. Bacteriol.* **181**, 3981-3993.
Dingman, C.W. (1974) *J. Theoret. Biol.* **43**, 187-195.
Dmitriev, B.A., Ehlers, S. and Rietschel, E.Th. (1999) *Med. Microbiol. Immunol.* **187**, 173-181.
Donachie, W.D. (1968) *Nature* **219**, 1077-1079.
Donachie, W.D. (1969) *J. Bacteriol.* **100**, 260-268.
Donachie, W.D. and Hobbs, D.G. (1967) *Biochem. Biophys. Res. Commun.* **29**, 172-177.
Donachie, W.D., Jones, N.C. and Teather, R. (1973) *Symp. Soc. Gen. Microbiol.* **23**, 9-44.
Donachie, W.D. and Begg, K.J. (1989) *J. Bacteriol.* **171**, 4633-4639.
Douek, J., Einav, M. and Zaritsky, A. (1992) *Mol. Gen. Genet.* **232**, 162-165.
Ephrati-Elizur, E. and Borenstein, S. (1971) *J. Bacteriol.* **106**, 58-64.
Fishov, I., Zaritsky, A. and Grover, N.B. (1995) *Mol. Microbiol.* **15**, 789-794.
Fishov, I. and Woldringh, C.L. (1999) *Mol. Microbiol.* **32**, 1166-1172.
Freifelder, D. (1967) *J. Bacteriol.* **93**, 1732-1733.
Freifelder, D. (1969) *J. Mol. Biol.* **45**, 1-7.
Freifelder, D. and Maaløe, O. (1964) *J. Bacteriol.* **88**, 987-990.
Funnell, B.E. (1993) *Trends Cell Biol.* **3**, 20-24.
Ghigo, J.-M., Weiss, D.S., Chen, J.C., Yarrow, J.C. and Beckwith, J. (1999) *Mol. Microbiol.* **31**, 725-737.
Glaser, P., Sharpe, M.E., Raether, B., Perego, M., Ohlsen, K. and Errington, J. (1997) *Genes Develop.* **11**, 1160-1168.
Goodell, E.W. and Schwarz, U. (1975) *J. Gen. Microbiol.* **86**, 201-209.
Grover, N.B., Woldringh, C.L., Zaritsky, A. and Rosenberger, R.F. (1977) *J. Theoret. Biol.* **67**, 181-193.

Grover, N.B., Zaritsky, A., Woldringh, C.L. and Rosenberger, R.F. (1980) *J. Theoret. Biol.* **86**, 421-439.

Hadas, H., Einav, M., Fishov, I. and Zaritsky, A. (1997) *Microbiology* **143**, 179-185.

Helmstetter, C.E., Cooper, S., Pierucci, O. and Revelas, L. (1968). *Cold Spr. Harb. Symp. Quant. Biol.* **33**, 807-822.

Helmstetter, C.E. (1996) In, F.C. Neidhardt, R. Curtiss III, J.L. Ingraham, E.C.C. Lin, K.B. Low, B. Magasanik, W.S. Reznikoff, M. Riley, M. Schaechter, and H.E. Umbarger (eds.) *Escherichia coli and Salmonella: Cellular and Molecular Biology,* 2nd ed. Washington, DC: American Society for Microbiology. pp. 1627-1639.

Henderson, T.A., Young, K.D., Denome, S.A. and Elf, P.K. (1997) *J. Bacteriol.* **179**, 6112-6121.

Henrici, A.T. (1928) *Morphogenetic Variation and the Rate of Growth of Bacteria.* Bailliere, Tindall and Cox, London.

Höltje, J.-V. (1998) *Mirobiol. Mol. Biol. Rev.* **62**, 181-203.

Kjeldgaard N.O., Maaløe O. and Schaechter, M. (1958) *J. Gen. Microbiol.* **19**, 607-616.

Koch, A.L. (1982) *J. Gen. Microbiol.* **128**, 2527-2539.

Koch, A.L. (1995) *Bacterial Growth and Form.* Chapman and Hall, New York, NY.

Koch, A.L. and Höltje, J.-V (1995) *Microbiology* **141**, 3171-3180.

Korn, D. and Weissbach, A. (1962) *Biochim. Biophys. Acta* **61**, 775-790.

Kornberg, A. and Baker, T.A. (1992) *DNA Replication* (2nd edn), W.H. Freeman and Co., San Francisco, CA.

Kumar, S. (1976) *J. Bacteriol.* **125**, 545-555.

Lark, K.G. and Lark, C. (1965) *J. Mol. Biol.* **13**, 105-126.

Lin, D.C.-H. and Grossman, A.D. (1998) *Cell* **92**, 675-685.

Liu, Z., Mukherjee, A. and Lutkenhaus, J. (1999) *Mol. Microbiol.* **31**, 1853-1861.

Løbner-Olesen, A. and Kuempel, P.L. (1992) *J. Bacteriol.* **174**, 7883-7889.

Lutkenhaus, J. (1993) *Mol. Microbiol.* **9**, 403-409.

Ma, X., Ehrhardt D.W. and Margolin, W. (1996). *Proc. Natl. Acad. Sci. USA* **93**, 12998-13003.

Maaløe, O. (1961) *Cold Spr. Harb. Symp. Quant. Biol.* **26**, 45-52.

Maaløe, O. (1963) In, *The Bacteria* **4**, 1-32. Academic Press, New York, NY.

Maaløe, O. and Rasmussen, K.V. (1963) *Colloq. Intern. Centre Nat. Rech. Sci.* **124**, 165-168.

Maaløe, O. and Kjeldgaard, N.O. (1966) *Control of Macromolecular Synthesis.* Benjamin, New York.

Marczynski, G.T., Dingwall A. and Shapiro, L. (1990) *J. Mol. Biol.* **212**, 709-722.

Marr, A.G., Harvey, R.J. and Trentini, W.C. (1966) *J. Bacteriol.* **91**, 2388-2389.

Marvin, D.A. (1968) *Nature* **219**, 485-486.

Meacock, P.A., Pritchard, R.H. and Roberts, E.M. (1977) *J. Bacteriol.* **133**, 320-328.

Melechen, N.E. and Skaar, P.D. (1962) *Virology* **16**, 21-29.

Mennigmann, H.-D. (1964) *Biochem. Biophys. Res. Commun.* **16**, 373-378.

Messer, W. and Weigel, C. (1996) in: F.C. Neidhardt, R. Curtiss III, J.L. Ingraham, E.C.C. Lin, K.B. Low, B. Magasanik, W.S. Reznikoff, M. Riley, M. Schaechter, and H.E. Umbarger (eds.) *Escherichia coli and Salmonella: Cellular and Molecular Biology,* 2nd ed. Washington, DC: American Society for Microbiology. pp. 1579-1601.

Mileykovskaya, E., Sun, Q, Margolin, W. and Dowhan, W. (1998) *J. Bacteriol.* **180**, 4252-4257.

Mohl, D.A. and Gober, J.W. (1997) *Cell* **88**, 675-684.

Molina, F., Jiménez-Sánchez, A. and Guzmán, E.C. (1998) *J. Bacteriol.* **180**, 2992-2994.

Mulder, E. and Woldringh, C.L. (1989) *J. Bacteriol.* **171**, 4303-4314.

Mulder, E. and Woldringh, C.L. (1991) *J. Bacteriol.* **173**, 4751-4756.

Nakayama, K., Kusano, K., Irino, N., and Nakayama, H. (1994) *J. Mol. Biol.* **243**, 611-620.

Nanninga, N. (1991) *Mol. Microbiol.* **5**, 791-795.

Nanninga, N. (1998) *Microbiol. Mol. Biol. Rev.* **62**, 110-129.

Norris, V. (1992) *J. Theoret. Biol.* **154**, 91-107.

Norris, V. (1995) *Mol. Microbiol.* **16**, 1051-1057.

Ogden, G.B., Pratt, M.J. and Schaechter, M. (1988) *Cell* **54**, 127-135.

Ohkawa, T. (1976) *Eur. J. Biochem.* **61**, 81-91.

Oishi, M., Yoshikawa, H. and Sueoka, N. (1964) *Nature* **204**, 1069-1073.

Painter, P.R. and Marr, A.G. (1968) *Annu. Rev. Microbiol.* **22**, 519-548.

Pogliano, J., K. Pogliano, D.S. Weiss, R. Losick and J. Beckwith. (1997) *Proc. Natl. Acad. Sci. USA* **94**, 559-564.

Previc, E.P. (1970) *J. Theoret. Biol.* **27**, 471-497.

Pritchard, R.H. (1965) *Brit. Med. Bull.* **21**, 203-205.

Pritchard, R.H. (1974) *Phil. Trans. R. Soc.* **267**, 303-336.

Pritchard, R.H. and Lark, K.G. (1964) *J. Mol. Biol.* **9**, 288-307.

Pritchard, R.H., Barth, P.T. and Collins, J. (1969) *Symp. Soc. Gen. Microbiol.* **19**, 263-297.

364

Pritchard, R.H. and Zaritsky, A. (1970) *Nature* **226**, 126-131.
Pritchard, R.H., Chandler, M.G. and Collins, J. (1975) *Mol. Gen. Genet.* **138**, 143-155.
Rinehart, K.V. and Copeland, J.C. (1973) *Biochim. Biophys. Acta* **294**, 1-7.
Rosenberger, R.F., Grover, N.B., Zaritsky, A. and Woldringh, C.L. (1978a) *Nature* **271**, 244-245.
Rosenberger, R.F., Grover, N.B., Zaritsky, A. and Woldringh, C.L. (1978b) *J. Theoret. Biol.* **73**, 711-721.
Sargent, M. G. (1975) *J. Bacteriol.* **123**, 7-19.
Schaechter, M., Maalφe, O. and Kjeldgaard, N.O. (1958) *J. Gen. Microbiol.* **19**, 592-606.
Smith, M.D., Green, R.R., Ripley, L.S. and Drake, J.W. (1973) *Genetics* **74**, 393-403.
Spratt, B.G. (1978) *Nature* **274**, 713-715.
Spratt, B.G. (1983) *J. Gen. Microbiol.* **129**, 1247-1260.
Sueoka, N. and Yoshikawa, Y. (1965) *Genetics* **52**, 747-757.
Taschner, P.E., Huls, P.G., Pas, E. and Woldringh, C.L. (1988) *J. Bacteriol.* **170**, 1533-1540.
ten Heggeler-Bordier, B., Wahli, W., Adrian, M., Stasiak, A. and Dubochet, J. (1992) *EMBO J.* **11**, 667-672.
Tomasz, A. and Borek, E. (1962) *Biochemistry* **1**, 543-552.
Trueba, F. and Woldringh, C.L., (1980) *J. Bacteriol.* **142**, 869-878.
van Helvoort, J.M.L.M, Kool, J. and Woldringh C.L. (1996) *J. Bacteriol.* **178**, 4289-4293.
van Leeuwenhoek, A. (1684) *Philos. Trans.* **14**, 568.
Walker, J.R. and Pardee, A.B. (1967) *J. Bacteriol.* **93**, 107-114.
Watt, G.R. and Cohen, S.S. (1953) *Biochem. J.* **55**, 774-908.
Webb, C.D., Teleman, A., Gordon, S., Straight, A., Belmont, A., Lin, D.C.-H., Grossman, A.D., Wright, A. and Losick, R. (1997) *Cell* **88**, 667-674.
Wientjes, F.B. and N. Nanninga. (1989) *J. Bacteriol.* **171**, 3412-3419.
Woldringh, C.L. (1976) *J. Bacteriol.* **125**, 248-257.
Woldringh, C.L., de Jong, M.A., van den Berg, W. and Koppes, L. (1977) *J. Bacteriol.* **131**, 270-279.
Woldringh, C.L., Grover, N.B., Rosenberger, R.F. and Zaritsky, A. (1980) *J. Theoret. Biol.* **86**, 441-454.
Woldringh, C.L. and Nanninga, N. (1985) In, N. Nanninga (ed.), *Molecular Cytology of Escherichia coli*. Academic Press, New York.
Woldringh, C.L., Mulder, E., Valkenburg, J.A.C., Wientjes, F.B., Zaritsky, A. and Nanninga, N. (1990) *Res. Microbiol.* **141**, 39-50.
Woldringh, C.L., Mulder, E., Huls, P.G. and Vischer, N. (1991) *Res. Microbiol.* **142**, 309-320.
Woldringh, C.L., Zaritsky, A. and Grover, N.B. (1994) *J. Bacteriol.* **176**, 6030-6038.
Woldringh, C.L., Zaritsky, A. and Grover, N.B. (1995a) *Keystone Symp. Molec. Cell. Biol.*, Santa Fe, NM, Suppl. **19A**, 121 (abstract no. A2-248).
Woldringh, C.L., Jensen, P.R. and Westerhoff, H.V. (1995b) *FEMS Microbiol. Lett.* **131**, 235-242.
Woldringh, C.L. and Odijk, T. (1999) In, R.L. Charlebois (ed.) *Organization of the Prokaryotic Genome*. Chapter 10. American Society for Microbiology, Washington, D.C., pp. 171-187.
Wu, L.J. and Errington, J. (1997) *EMBO J.* **16**, 2161-2169.
Yu, X.-C., Weihe, E.K. and Margolin, W. (1998) *J. Bacteriol.* **180**, 6424-6428
Zaritsky, A. (1971) Ph.D. Thesis, the University of Leicester, UK, pp. 97.
Zaritsky, A. (1975a) *J. Theor. Biol.* **54**, 243-248.
Zaritsky, A. (1975b) *J. Bacteriol.* **122**, 841-846.
Zaritsky, A. and Pritchard, R.H. (1971) *J. Mol. Biol.* **60**, 65-74.
Zaritsky, A. and Pritchard, R.H. (1973) *J. Bacteriol.* **114**, 824-837.
Zaritsky, A. and Woldringh, C.L. (1978) *J. Bacteriol.* **135**, 581-587.
Zaritsky, A. Grover, N.B., Naaman, J., Woldringh, C.L. and Rosenberger, R.F. (1982) *Comments Mol. Cell Biophys.* **1**, 237-260.
Zaritsky, A. and Woldringh, C.L. (1987) In, M. Vicente (ed.), *Molecular Basis of Bacterial Growth and Division* (EMBO Workshop), pp. 78-80, Segovia, Spain.
Zaritsky, A., Woldringh, C.L., Helmstetter, C.E. and Grover, N.B. (1993) *J. Gen. Microbiol.* **139**, 2711-2714.
Zaritsky, A., Woldringh, C.L., Fishov, I., Vischer, N.O.E. and Einav, M. (1999a) *Microbiology* **145**, 1015-1022.
Zaritsky, A., Van Geel, A., Fishov, I., Pas, E., Einav, M. and Woldringh, C.L. (1999b) *Biochimie* **81**, 897-900.

X

ASTROBIOLOGY AND MICROBIAL CANDIDATES FOR EXTRATERRESTRIAL LIFE

Joseph Seckbach, Frances Westall and Julian Chela-Flores

For the Biodata and portraits of these authors,
see their other contributions. "Preface" and Chapter 8 (J.S.), Chapter 2
(F.W.) and Chapter 27 (J. C-F.).

J. Seckbach (ed.), Journey to Diverse Microbial Worlds, 367-375.
© 2000 *Kluwer Academic Publishers. Printed in the Netherlands.*

INTRODUCTION TO ASTROBIOLOGY:
Origin, Evolution, Distribution and Destiny of Life in the Universe

JOSEPH SECKBACH[1], FRANCES WESTALL[2] AND JULIAN CHELA-FLORES[3]

[1]*Hebrew University of Jerusalem, 91904, and* [2]*Lunar and Planetary Institute, 3600, Bay Area Boulevard, Houston, TX 77058, and* [3]*The Abdus Salam International Centre for Theoretical Physics (ICTP), Office 276, P.O.Box 586; Strada Costiera 11; 34136 Trieste, Italy, and Instituto de Estudios Avanzados, Apartado 17606, Parque Central, Caracas 1015A, Venezuela.*

1. Are There Other Inhabited Worlds?

The only life that we know about in the universe is life on our own planet Earth. We have no idea of how representative it might be of life on other planets, although in the chapter by one of us (JCF) it is conjectured that, given suitable environmental conditions, if life started somewhere else it would be constrained to take an evolutionary pathway to eukaryogenesis and multicellularity (Chela-Flores 1998; Seckbach *et al.*, 1998).

The question whether nature in an extraterrestrial context steers a predictable course is clearly an open question, but some hints from the basic laws of terrestrial biology (natural selection and a common ancestor for all the Earth biota) can be interpreted as evidence that to a large extent evolution is predictable and not contingent. Indeed, support from independent teams suggests that natural selection overrides the randomness of genetic drift; in other words, natural selection seems to be powerful enough to shape terrestrial organisms to similar ends, independent of historical contingency (Pennisi 2000, and references therein). On the other hand, some arguments militate in favor of a human-level of intelligence being reached by the conjectured universal constrained evolutionary pathway towards eukaryogenesis and multicellularity. Certainly, in an extraterrestrial environment the evolutionary steps that led to human beings would probably never repeat themselves; but that is hardly the relevant point: *the role of contingency in evolution has little bearing on the emergence of a particular biological property* (Conway-Morris 1998).

The inevitability of the emergence of particular biological properties is a phenomenon that has been recognized by students of evolution for a long time (evolutionary convergence). There is strong selective advantage for multicellularity of eukaryotic cells that have already become neurons; such an event occurred very early during multicellular evolution on Earth (Villegas *et al.*, 2000). This argument strongly advocates in favor of the existence of other human-level of intelligence elsewhere in the cosmos. However, in spite of these persuasive arguments from the life sciences, some astronomers and paleontologists, independent of the evidence from biology, still defend the opposite point of view (Brownlee and Ward 2000).

2. Origin of Life on Earth

On the other hand, in trying to unravel the mysteries of the origin(s) of life on Earth, we have unfortunately little material upon which to work. The very fact that the Earth is a living, dynamic entity (à la Gaia of James Lovelock) means that the oldest rocks which hold the key to life's origins have been destroyed by the inexorable process of plate tectonics. The oldest rocks on Earth are found in the greenstone terrain of Isua, S.W. Greenland, which is older than 3.8 billion years (b.y.). Already these ancient rocks contain an isotopic signature indicating that bacteria inhabited the environment in which the rocks were formed (Schidlowski 1988; Mojzsis *et al.*, 1996).

If full-fledged bacteria inhabited the Earth by 3.8 b.y. before the present (b.p.), then life must have started much earlier. Theoretically, it could have initiated at any time after water condensed on the Earth's surface: comets would have brought in the prebiotic molecules (Chyba and Sagan 1992) which, in as yet not-understood ways, self organized into primitive cellular structures developing, in turn, into the last common ancestor, or cenancestor (LCA). The LCA then gave rise to bacteria; but it should be underlined that what is emerging from the extensive analysis of a large number of phylogenetic trees constructed from a variety of macromolecules of life is that three primary cellular lines of evolutionary descent are established, between which extensive horizontal transfer events have taken place (Doolittle 1999, Becerra *et al*, 2000).

This critical development took place in conditions, which we would now consider inhospitable but which were normal for earliest life. The atmosphere was mildly reducing, consisting mostly of CO_2 (Pollack *et al.*, 1987), with subsequent consequences for a lower NO level (Kasting 1990). There is much discussion concerning the amount of oxygen in early Earth's atmosphere since there is evidence for at least localized pockets of subaerial oxidation (Ohmoto *et al.*, 1999). The incident sunlight would have been about 30% less than at present because the Sun's nuclear furnace had not yet got into full swing (Sagan and Mullen 1972). It is assumed that the average temperatures were warm enough to keep water liquid on Earth owing to the greenhouse effect of the CO_2 (with perhaps some CH_4; cf., Kasting 1993). Temperatures may have been higher than at present due to the heavy meteorite/cometary bombardment, which characterized the Hadean and earliest Archaean epochs until about 3.8 b.y.b.p. (Maher and Stevenson 1988). There was no ozone layer to mitigate the deleterious effects of UV radiation. In addition, the moon was much closer to the Earth (Hartmann and Davis 1975; Cameron and Ward 1976) resulting in significant tidal influences on whatever surficial environments existed at the surface. Lastly, the aforementioned period of heavy bombardment could have sterilized the Earth a number of times (Sleep *et al.*, 1989). Despite all of this, life started, developed, it flourished and remained.

3. Evolution of Life on Earth

Thus, one of the key characteristics of life, its tenacity, developed early. The earliest bacteria may have been thermophilic organisms of the domain Archaea (Woese 1987). Already Darwin stated that life evolved in a warm little pond, and most probably these thermophiles were the first organisms on Earth (Copland 1936, Seckbach 1994, 1995, 2000). The species of thermophilic Archaea, like many of the methanogens, lie near the root in the tree of life (Valentine and Boone, in Seckbach 2000; Madigan and Marrs 1997). Recently it has been well established that all life forms that cluster around the base of evolutionary and phylogenetic trees are thermophiles (Pace 1997, Stetter 1998). There is, however, a certain rebuttal to this theory that has challenged the warm/hot origin of life, proposing that the first cells were cryophilic (Galtiers *et al.,* 1999).

Alternatively, the thermophilic signal may be an artifact of bacteria having been subjected to a thermophilic "bottle-neck" in the sense that during the period of heavy bombardment, the only bacteria to survive were those which either occupied the hydrothermal niche or which had taken refuge there (Baross and Hoffman 1985; Nisbet and Fowler 1996).

Until the rise of O_2 in Earth's atmosphere and the development of ozone, most of the early microbes may have resided in sheltered, deep subterranean niches (Onstott *et al,* 1999). Thus, life could have started below the terrestrial surface, since the exposed land was inhospitable during the early history of our planet (Davies 1999).

One of the most important events in evolution was the advent of eukaryogenesis and multicellularity (Chela-Flores 2000a). The early Earth atmospheres was anoxic with significant rises in oxygen (>15% present atmospheric levels [PAL], Holland and Beukes 1990), occurring only at about 2.1 b.y.b.p. However, precursors with eukaryotic characteristics, as well as clear biochemical evidence for the existence of oxygenic cyanobacteria, have been identified in 2.7 b.y.-old shales from the Hammersley Basin, Australia (Brocks *et al.,* 1999; Summons *et al.,* 1999), before the significant 15% PAL was reached (Holland and Beukes 1990). Between 1.5 and 1.0 b.y.b.p. photosynthetic life became abundant enough to elevate atmospheric oxygen to nearly current level.

4. Distribution of Life, Here, There and Everywhere?

4.1. DISTRIBUTION OF LIFE IN THE UNIVERSE

Since the same laws of physics, chemical thermodynamics and carbon chemistry apply everywhere, there should be a high probability that other stars and satellites may harbor life, provided liquid water is available (de Duve, 1995; cf., "cosmic imperative", as discussed in Chela-Flores' chapter in this volume). Thus, the search for life (or even prebiotic conditions) on other solid bodies within our own solar system is of vital importance; equally important is the search for life in other solar systems by means for instance of the European Space Agency "Project Darwin" and the NASA initiative with "The Terrestrial Planet Finder". Both chapters within this section address these philosophical concepts, proposing practical considerations for the survivability of life under adverse conditions (McKay, in this volume) and also for the search for life, especially on Europa (Chela-Flores, in this volume).

4.2. DISTRIBUTION OF LIFE IN OUR SOLAR SYSTEM

Titan, a satellite of Saturn, with its thick, CH_4-containing atmosphere has been described as a "natural exobiology laboratory (Jakosky, 1998, p.192). In fact, the Cassini-Huygens spacecraft is due to rendezvous with Titan in 2004 to investigate this natural chemical laboratory.

The planets that occur within the "habitable zone" of our solar system are Earth, Mars and Venus. They have had similar early histories and life may have developed on all three of these planets (McKay, 1997; Jakosky, 1998).

However, whereas Earth is now warm, wet and equable (its atmosphere is low in CO_2, and high in N_2), Mars is a cold and dry desert (with an atmosphere relatively high in CO_2 and low in N_2) the general atmospheric pressure is a fraction of that of the Earth), and Venus is an inferno, as far as life is concerned. As discussed in McKay's chapter (in this volume), the upper temperature limit for life is an uncomfortable 113° C (Blockl et al., 1997). This is maximum temperature limit of *Pyrolobus fumarii* (Stetter, 1998; see chapter by Seckbach and Oren, in this volume). The 400°C at Venus' surface is therefore simply too high for liquid water and the molecular bonds of any biogenic organics would be broken down.

Likewise, McKay addresses the lower temperature limits of life, noting that there is abundant viable life in the cold dry deserts of Antarctica and the Arctic (Gilichinsky et al., 1992; McKay et al., 1994), as well as the fact that life can apparently survive for some millions of years in the Siberian tundra (Vorobyova et al., 1997). Chela-Flores (1998) underlines the fact that even eukaryotic organisms can survive in these environments. The deep ice samples at the Vostok Station, in which microorganisms were recently detected (Priscu et al. 1999, Karl et al. 1999), may be considered as an analogue of the environment present on Europa, the frozen Jovian moon which appears to have a subsurface ocean (Carr et al., 1998). Radar mapping of Lake Vostok revealed that liquid water exists below the icy crust. This water is possibly warmed up by pressure of the ice above and by geothermal sources below. Thus, these recent terrestrial observations may hold clues for the existence of life on other worlds.

The lack of water and the effects of UV radiation preclude the possibility of life at the surface of Mars (the high concentration of carbon dioxide in the Martian atmosphere screens off the UV radiation of wavelengths below 190 nm, but harmful UV-B radiation between 200 and 300 nm reaches the surface at full intensity). However, McKay (this volume) discusses the possibility of subsurface hydrothermal "islands" in the frozen aquifer of Mars, where life could potentially survive, but he also points out that it would have to contend with lethal doses of radiation from the surrounding rocks for significant periods of time.

The recent discovery of a possible biogenic signature within 3.9 b.y.-old carbonate globules in fractures of the Martian meteorite ALH84001 (McKay et al., 1996) has raised excitement among astrobiologists (Seckbach, 1997) and spawned fertile interest in this relatively new field (*viz.* NASA's Virtual Astrobiology Institute and related projects). Although the data from the Allan Hills meteorite, and also descriptions of structures having bacteriomorph shapes from the younger Martian meteorites, Nakhla and Shergotty

(McKay *et al.,* 1999) are tantalizing, there are as yet no firm conclusions concerning Martian biotics.

The problems are two-fold and need to be resolved before samples are returned from other astral bodies (Mars, cometary or asteroidal). (1) Recent work has underlined the problem of widespread, modern contamination of extraterrestrial materials (Steele *et al.,* 2000), thus complicating interpretations. Furthermore, (2) we do not yet have a well-established database of truly biogenic structures, as opposed to biomorphic structures, although this is presently being addressed (McKay et al., 2000, Westall 1999).

One other aspect of the initial research by McKay *et al.* (1996), their description of possible "nanobacterial fossils" has led to interest in the size limits of life, with the surprising results that viable life may be very much smaller than originally believed (Kajander and Ciftcioglu 1998; Uwins *et al.,* 1998; Gillet *et al.,* 2000). The new developments in microbiology and micropalaeontology suggest that the diversity of microbial life in the Universe may turn out to be much larger than that presently encountered on Earth, thus presenting new challenges to the astrobiologist.

4.3. DESTINY OF LIFE IN THE UNIVERSE

The impact of extraterrestrial life on philosophy and theology goes beyond the scope of this book, which has been confined within the limits of science. But to complete this general overview of Astrobiology, we would like to comment on its last aspect, namely the destiny of life in the universe. This topic requires going beyond the frontiers of science. Other aspects of our culture have to be brought into a comprehensive dialogue, rather than within an intercultural debate (Chela-Flores, 2000b).

5. References

Baross, J.A. and Hoffman, S.E. (1985) *Origins of Life*, **15**, 327-345.

Becerra, A, Silva, E. Lloret, L. Velasco, A.M. and Lazcano, A. (2000) in: Chela-Flores, J., Lemarchand, G.A. and Oro (eds.) Astrobiology, Proceedings of the Iberoamerian School of Astrobiology, Caracas, 1999. Kluwer Academic Publishers: Dordrecht, The Netherlands. (To be published).

Blochl, E., Rachel, R., Burggraf, S., Hafenbradl, D., Jannasch, H.W., and Stetter, K.O. (1997) *Extremophiles* **1**, 14-21.

Brocks, J.J., Logan, G.A., Buick, R. and Summons, R.E. (1999) *Science*, **285**, 1033-1036.

Brownlee, D. and Ward, P. (2000). *Rare Earth.* Copernicus, Springer. Verlag, Berlin (in press).

Cameron, A.G.W. and Ward, W.R. (1976) *Lunar Science VII*, pp. 120-122.

Carr, M.H., Belton, M.J.S., Chapman, C.R., Davies, M.E., Geissler, P., Greenberg, R., McEwen, A.S., Tufts, B.R., Greely, R., Sullivan, R., Head, J.W., Pappalardo, R.T., Klaasen, K.P., Johnson, T.V., Kaufman, J., Senske, D., Moore, J., Neukum, G., Schubert, G., Burns, J.A., Thomas, P. and Veverka, J. (1998) *Nature* **391**, 363-365.

Chela-Flores, J. (1998) in: J. Chela-Flores and F. Raulin. (eds.) *Chemical Evolution: Exobiology: Matter, Energy, and Information in the Origin and Evolution of Life in the Universe.* Kluwer Academic Publishers, Dordrecht, The Netherlands. pp. 229-234.

Chela-Flores, J. (2000a) *Astronom. Soc. Pacific Conf. Ser.* (in press).

Chela-Flores, J. (2000b) in: J. Chela-Flores, G.A. Lemarchand and J. Oro (eds.) *Astrobiology.* Proceedings of the Iberoamerian School of Astrobiology, Caracas, 1999. Kluwer Academic Publishers: Dordrecht, The Netherlands. (To be published).

Chyba, C. and Sagan, C. (1992) *Nature* **355**, 125-131.

Conway-Morris, S. (1998) *The Curcible of Creation. The Burgess Shale and the Rise of Animals.* Oxford University Press, New York, pp. 9-14.

374

Copland J.J. (1936) Ann. N.Y. Acad. Sci. **36**, 1-226.

Davies, P. (1999) The Fifth Miracle: The Search for the Origin and Meaning of Life. Simon & Schuster, N.Y. p.304.

de Duve, C. (1995) *Vital Dust: Life as a Cosmic Imperative*, Basic Books, New York.

Doolittle, W.F. (1999) *Science* **284**, 2124-2128.

Galtier, N., Tourasse, N. and Gouy, M. (1999) Science **286**, 155-157 and 220-221.

Gilichinsky, D.A., Vorobyova, E.A., Erokhina, L.G., Fyordorov-Dayvdov, D.G., and Chaikovskaya, N.R. (1992) *Adv. Space Res.* **12 (4)**, 255-263.

Gillet, P, Barrat, J.A., Heulin, T., Achouak, W., Lesourd, M., Guyot, F., and Benzerara, K. (2000) *Earth and Planet. Sci. Lett.* **175**, 161-167.

Hartmann, W.K. and Davis, D.R. (1975) *Icarus* **24**, 504-515.

Holland, H.D. and Beukes, N.J. (1990) *Amer. J. Sci.* **290-A**, 1-34.

Jakosky, B. (1988) *The Search for Life on Other Planets*, Cambridge Univ. Press, Cambridge. UK.

Kajander, E.O. and Ciftcioglu, N. (1998) *Proc. Natl. Acad. Sci. USA* **95**, 8274-8279.

Karl, D.M., Bird, D.F., Björkman, K., Houlihan, T., Shackelford, R. and Tupas, L. (1999) Science **286**, 2144-2147.

Kasting, J.F. (1990) Origin of Life **20**, 199-231.

Kasting, J.F. (1993) *Science* **259**, 920-926.

Madigan, M.T. and Marrs, B.L. (1997) Sci. Amer. **276**, 66-71.

Maher, K.A. and Stevenson, D.J. (1988) *Nature* **331**, 612-614.

McKay, C. P. (1997) *Origins Life Evol. Biosph*ere **27**, 263-289.

McKay, C.P., Clow, G.D., Andersen, D.T., and Wharton, R.A. (1994) *J. geophys. Res.*, **99**, 20,427-20,444.

McKay, D.S., E.K. Gibson, E.K., Thomas-Keprta, K.L., Vali, H., Romanek, C.S., Clemett, S.J., Chillier, X.D.F., Maedling, S.R., and Zare, R.N. (1996) *Science*, **273**, 487-489.

McKay, D. S., Wentworth, S. W., Thomas-Keprta, K., Westall, F. and Gibson, E.K. (1999) Lunar *and Planetary Sci. Conf.*, Houston, Abst. 1816.

McKay, D.S., Bell, M.S., Prejean, L., Gibson, E.K., Allen, C.C., Wentworth, S. W., Thomas-Keprta, K., Westall, F. and Morris-Smith, P. (2000) http://www-sn.jsc.nasa.gov/astrobiology/ biomarkers/.

Mojzsis, S.J., Arrhenius, G., McKeegan, K.D., Harrison, T.M., Nutman, A.P., and Friend, C.R. (1996) *Nature* **384**, 55-59.

Nisbet, E.G. and Fowler, C.M.R. (1996) *Nature* **382**, 404-405.

Ohmoto, H. , Rasmussen, B., Buick, R., and Holland, H.D. (1999) *Geology* **27**, 1151-1152.

Onstott et al. (1999) in: J. Seckbach (ed.) *Enigmatic Microorganisms and Life in Extreme Environments.* Kluwer Acad. Publishers. Dordrecht, The Netherlands, pp. 487-500.

Pace, N.R. (1997) Science **276**, 734-740.

Pennisi, E. (2000) *Science* **287**, 207-208.

Pollack, J.B., Kasting, J.F., Richardson, S.M., and Poliakoff, K. (1987) *Icarus* **71**, 203-224.

Priscu, J.C., Adams, E.E., Lyons, W.B., Voytek, M.A., Mogk, D.W., Brown, R.L., McKay, Ch..P., Takacs, C.D., Welch, K.A., Wolf, C.F., Kirshtein, J.D. and Avci, R. (1999) Science **286**, 2141-2144.

Sagan, C.A. and Mullen, G. (1972) *Science* **177**, 52-56.

Schidlowski, M. (1988) *Nature* **333**, 313-318.

Seckbach, J. (1994) J. Biol. Phys. **20**, 335-345.

Seckbach, J. (1995) in: C. Ponamperuma and J. Chela-Flores (eds.) *Chemical Evolution: The Structure and Model of the First Cell.* Kluwer Academic Publishers, Dordrecht, The Netherlands. pp 335-345.

Seckbach, J. (1997) in: C.B. Cosmovici, S. Bowyer and D. Werthimer (eds.) *Astronomical and Biochemical Origins and the Search for Life in the Universe.* Editrice Compositore: Bologna. pp. 511-523.

Seckbach, J. (ed) (1999) *Enigmatic Microorganisms and Life in Extreme Environments,.* Kluwer Academic Publishers, Dordrecht, The Netherlands. p 687.

Seckbach, J. (2000) *Astronom. Soc. Pacific Conf. Ser.* (in press).

Seckbach, J., Jensen, T.E., Matsuno, K., Nakamura, H., Walsh, M.M. and Chela-Flores, J. (1998) in: J. Chela-Flores and F. Raulin (eds.). *Chemical Evolution: Exobiology: Matter, Energy,and Information in the Origin and Evolution of Life in the Universe.* Kluwer Academic Publishers, Dordrecht, The Netherlands. pp. 235-240.

Sleep, N.H., Zahnle, K.J., Kasting, J.F., and Morowitz, H.J. (1989) *Nature* **342**, 139-142.

Steele, A, Goddard, D.T., Stapleton, D. Toporski, J.K.W., Peters, V., Bassinger, V. Sharples, G., Wynn-Williams, D.D., and McKay, D.S. (2000) *Meteoritics*, **35,** (in press).

Stetter, K.O. (1998) in: K. Horikoshi and W.D. Grant (eds.) *Extremophiles, Microbial Life in Extreme Environments.* Willey-Liss, N.Y. pp. 1-24.

Summons, R.E., Jahnke, L.L. Hope, J.M., and Logan, G.A. (1999) *Nature* **400**, 554-557.

Uwins, P.J.R., Webb, R.I., and Taylor, A.P. (1998) *Am. Mineral.* **83**, 1541-1550.

Villegas, R, Castillo, C. and Villegas, G.M. (2000) in: J. Chela-Flores, G.A. Lemarchand and J. Oro (eds.) *Astrobiology*. Proceedings of the Iberoamerian School of Astrobiology, Caracas, 1999. Kluwer Academic Publishers: Dordrecht, The Netherlands . (To be published).

Vorobyova, E., Soina, V., Gorlenko, M., Minkovskaya, N., Zalinova, N., Mamukelashvili, A., Gilichinsky, D., Rivkina, E. and Vishnivetskaya, T. (1997) *FEMS Microbiol. Rev.* **20**, 277-290.

Westall, F. (1999) *J. Geophysical Res.* **104**, 16,437-16,451.

Woese, C.R. (1987) *Microbiol. Rev.* **51**, 221-271.

Biodata of **Christopher P. McKay** author of the chapter entitled: *Life in the Cold and Dry Limits: Earth, Mars and Beyond.*

Dr. Christopher McKay is a Research Scientist at the Space Science Division of NASA-Ames Research Center (Moffett Field, CA. USA). He obtained his Ph.D. in Aster-Geophysics in 1982 from the University of Colorado in Boulder and has been with the NASA-ARC since that time. His current research focuses on the evolution of the solar system and the origin of Life. Dr. Chris McKay is also active in planning for future Martian missions including human settlements. His studies on Mars brought him recently to some Martian-like environments such as, to Antarctic dry valleys, to Siberian and to the Canadian Arctic. Dr. McKay received several prizes such as, the Urey Prize (1989), Arthur S. Flemming Award (1991), NASA Ames Associate Fellow Award (1994), Thomas O. Paine Memorial Award (1994) and the Astronomical Assoc. N. California Professional Award (1998).
E-mail: cmckay@mail.arc.nasa.gov

J. Seckbach (ed.), Journey to Diverse Microbial Worlds, 377-386.
© 2000 *Kluwer Academic Publishers. Printed in the Netherlands.*

LIFE IN THE COLD AND DRY LIMITS:
Earth, Mars And Beyond

CHRISTOPHER P. MCKAY
Space Science Division NASA
Ames Research Center
Moffett Field CA 94035 USA

1. Introduction

Life on Earth survives in a variety of environments with microorganisms having the most diversity in lifestyles. The limit of survival in extreme environments is of interest in understanding the limits of biochemistry and potentially useful information can be obtained from the adaptive strategies of extremophiles. The study of life in extreme environments also provides the basis for considerations of possible life on other worlds in our solar system - conditions that are extreme compared to Earth.

Liquid water is the universal requirement for life and the search for liquid water on other worlds is the operational basis for the search for life on those worlds (McKay, 1991). In our solar system the most interesting target is Mars (McKay, 1997a) because of the direct evidence that liquid water flowed across its surface for geologically long periods of time (>Myr) early in its history (McKay and Davis, 1991; Carr, 1996). Europa, a moon of Jupiter, is also of interest since it may hold a deep ocean (>100 km) under a thick ice cover (10 km). Comets are of tertiary interest in that they are known to contain ice and organic material and may also have contained liquid water at some point in their history. Thus, the ability of comets to hold dormant life warrants consideration (McKay, 1997b).

2. Limits to Life on Earth

In considering life in the cold and icy worlds of the solar system it is useful to begin with a review of the limits of life on Earth. The limits of life have been understood for many years. There is a popular perception that we have discovered over the past few years entirely new categories of extreme environments in which life can survive. This is simply not the case. A comparison of our current understanding of the limits to life with a tabulation complied almost twenty years ago (Kushner, 1981) shows that the only significant change is a modest increase in the upper temperature limit for life from $100^{0}C$ to $113^{0}C$ (Blochl et al. 1997). Our understanding of the ecology of extreme environments has improved due to more complete studies of sites (such as the Antarctic dry valleys, the deep-sea hydrothermal vents, the Yellowstone geyser fields, etc.). On the other hand, the basic physiological limits to the survival and growth of individual organisms have been understood for many decades. Table 1 shows a current list of the limits of life (McKay, 1998).

Table 1. Limits to Life (McKay, 1998).

Parameter	Limit	Note
Origin of Life	Similar to Survival	Based on early Earth
Lower Temperature	-15°C	Liquid water
Upper Temperature	113°C	Thermal denaturing of proteins
Low Light	10^{-4} Full sun light (noon)	Algae under ice and deep sea
	0.2 mE $m^{-2}s^{-2}$	
pH	1 -11	
Salinity	Saturated NaCl	Depends on the kind of salt
Water Activity *	0.6	0.8 for bacteria
Underground	?	
Radiation	1-2 Mrad	Maybe higher for dry or frozen state
Dormancy	10-100 Myr	Radiation limited

* 1 ratioed to pure water = 1

3. Metabolism and Dormancy in Permafrost

Recent studies of microorganisms in Siberian permafrost have provided insights into the growth and long-term survival of microorganisms in the frozen state. The oldest continuously frozen permafrost in Siberia is about 3.5 Myr old and there is good evidence that the material has remained frozen over this time (Gilichinsky et al. 1992). Samples aseptically extracted from drill holes in this permafrost contain bacteria at levels of up to 10^8 culturable units per gram soil (Vorobyova et al. 1997).

The current temperature of the permafrost is -10°C and one key question is the state of the bacteria at this low temperature. Rivkina et al. (2000) have experimentally addressed this question by incubating permafrost samples at temperatures from +5°C to -20°C. To detect microbial activity ^{14}C-labeled acetate was mixed in with the permafrost at the start of the experiments. Subsequently, counts were determined from lipids extracted from replicates and compared to a baseline established after initial preparation. The initial mixing was accomplished with the samples held at sub-zero temperatures. It is important to realize that this mixing refreshes the environment surrounding any microorganisms in the sample. Two features of these results are of interest. First, after an initial lag period there is a rapid increase in incorporation of labeled acetate into lipids. However, this rapid uptake is not sustained and the counts level off. Rivkina et al. (2000) attribute this leveling off to the exhaustion of local resources surrounding each microorganism in the sample. This scenario explains why the final asymptotic value varies exponentially with temperature: the thin film of unfrozen water that exists in soil-ice mixtures down to -40°C decreases exponentially with decreasing temperature below zero.

These results indicate that the limitation to microbial activity over long periods of time at subfreezing temperatures is the transport of nutrients to, and removal of waste material from, the microorganisms and not the intrinsic reductions in the rate of metabolism. The bacteria found in permafrost after millions of years have spent this time in a metabolic state that is well below their potential at that temperate -- i.e. they are starving.

There must be a lower limit to the metabolic rate that the bacteria in the permafrost must maintain to remain viable indefinitely. In order to remain viable organism must overcome two factors: thermal decay and natural radiation.

At any finite temperature the second law of thermodynamics implies that entropy will increase and chemical degradation will occur in all microorganisms. Such decay would include racemization of amino acids, breaking of chemical bonds, denaturing of proteins, etc. In general, the rate of decay, Rx, in response to temperature, T, can be represented by the Arrhenius rate equation:

$$Rx = A \exp(-\Delta E/kT) \qquad (1)$$

Where ΔE is the activation energy, k is Boltzman's constant, and A is the pre-exponential term.

An important implication of equation 1 is that thermal decay is a very steep function of temperature. This can be illustrated for the case of amino acid racemizaton. Bada and McDonald (1995) show that amino acid racemization varies exponentially with temperature with a drop of a factor of 7 for every 10^0C reduction in temperature. In addition they find that liquid water accelerates the decay by a factor of over 100. Dry materials have greatly reduced racemization rates, and this presumably applies to deeply frozen material as well in which there is only ice present.

Kanavarioti and Mancinelli (1990) have followed this logic to argue that at the low temperatures expected on Mars ($< -70^0C$) biomolecules would have lifetimes against thermal decay that exceed billions of years.

To prevent cumulative degradation due to this increase in entropy an organism living in the Siberian permafrost must exert a minimal level of free energy. If the rate of entropy, S, generation with time, t, is given as dS/dt then the free energy, F, that the organism must expend to maintain steady state is strictly greater than:

$$F > T \, dS/dt \qquad (2)$$

If the rate at which a permafrost organism can utilize free energy from its environment is less than this level, that its entropy must increase with time and eventually it can no longer survive.

Currently there are no models of a bacteria cell that allow us to compute dS/dt as a function of temperature. To do so would require that we had a catalog of all possible molecules in the cell and assigned to each bond in each molecule an activation energy and decay rate as per equation 1. Using this model we would then be able to compute the rate at which an organism would experience entropy increased decay. In order to determine how this limited an organism's survival it would be necessary to understand how much thermal decay, or increase in entropy, an organism could sustain and still remain viable. An estimate of this might be obtained from the level of radiation required to render an organism unable to survive rather than just unable to reproduce.

A second factor that limits long term survival even at extremely low temperatures is natural radiation. Natural radiation is caused by gamma particles emitted by naturally occurring long-lived radioisotopes. These are thorium, uranium, and potassium. In Siberian permafrost the measured level of natural radiation deep (25 m) below the surface is 0.2 rad/yr. About half of this comes from U and Th with the rest from K.

It would be expected that crustal materials on Mars would have similar levels of U, Th, and possibly lower levels of K based on theoretical considerations of the source material that formed both planets and on direct measurements of the Martian meteorites (Laul et al. 1986).

Radiation doses of 2 Mrad are sufficient to kill most organisms in soil and 18 Mrad will kill even the most radiation resistant strains such as *Deinococcus radiodurans* (Minton, 1994; Battista, 1997). Dehydration resistance is known to enhance radiation survival in bacteria (Mattimore and Battista, 1996). Interestingly, there is experimental evidence that a considerable part of the damage done by radiation is associated with free radical formation due to liquid water (Swarts et al. 1992) and therefore in the dehydrated state damage is reduced. Recent data suggests that freezing at low temperatures (-70°C) also enhances survival by a factor of 5 presumably also because of the absence of liquid water (J. Daly, personal communication).

Taking 2 Mrad as the typical lethal dose for soil bacteria, then at a rate of 0.2 rad/yr, a lethal dose of radiation is accumulated in about 10 Myr. Thus bacterial survival in 3.5 Myr old permafrost is not a test of the ultimate survival against radiation and is not even close to the limits of *D. radiodurans*. Even if the permafrost bacteria are not able to metabolize enough to repair radiation induced damage they could still survive the 3.5 million years of accumulated damage. There might be some elimination of less resistant strains but there would not be sterilization of the soil.

4. Mars

The surface conditions on Mars today are too harsh to support any life from Earth. The essential limiting fact is the virtual absence of water in a liquid form. However, Mars was not always a desert world. There is direct evidence that liquid water flowed on the Martian surface early in the history of that planet. This makes a search for fossil evidence for life a key science objective for future Mars missions (McKay, 1997a).

The polar caps of Mars contain large expanses of ice but they are not promising targets for a search for life. The primary reason is that the ice there remains too cold, even in the summer months, for liquid water to form. The summer temperatures are held low by the cooling effect of the evaporation of the winter cap of carbon dioxide. The maximum summer temperature at the North Pole -- the warmer of the two poles – was -75°C to -65°C during the Viking mission (Kieffer et al. 1977). Deep below the polar cap surface the combination of geothermal heating and pressure might result in a liquid phase but it is not clear that there would be an energy source for life in such an environment.

The recent discoveries of extensive life deep underground on Earth may not be directly applicable to Mars. The vast majority of the deep biosphere on Earth is based on the consumption of organic material produced by photosynthetic ecosystems on the surface. Usually this consumption is coupled to oxygen also produced photosynthetically at the surface. At the time of this writing there is only one ecosystem known that represents an anaerobic chemosynthetic subsurface ecosystem (Stevens and

McKinley 1995). It has long been recognized that this sort of system, using hydrogen and carbon dioxide to form methane, is the ideal subsurface system for Mars (Boston et al. 1992). Recent work (Radu Popa, personal communication) indicates that in the presence of ferrous iron the reaction of hydrogen sulfide with carbon dioxide can be used biologically. Thus, if there is a subsurface hydrothermal region on Mars generating liquid water from the ground ice and if there is a source of hydrogen or hydrogen sulfide on Mars then it is possible that life might be present there.

While environments capable of supporting life may be present in the Martian subsurface it is not clear that life would be present. We can think of these subsurface hydrothermal sites as "islands". Biologically the essential problem is how life can spread from one subsiding island to a second emerging island separated in location and time. As discussed below it is unlikely that life could remain dormant waiting for suitable conditions over any significant period of the history of the planet.

Even if it lacks systems alive at the present time the subsurface of Mars may hold important clues to Martian life. Any microorganisms trapped in the ancient Martian permafrost would have been frozen at low temperatures, $-70^{0}C$, for billions of years. Over this long period of time these organisms would have accumulated hundreds of lethal doses of radiation due to natural levels of uranium, thorium, and potassium. However their basic biochemicals would be intact. It would be possible to recover their genetic material and amino acids -- if they had any. Most importantly it would be possible to tell if the Martian organisms were genetically related to life on Earth. This is a possibility that cannot be discounted. During the end of the late bombardment 3.8 Gyr ago there could have been significant exchange of materials between Earth and Mars. The evidence that life may have been present on the Earth by this early time certainly opens up the possibility of biological exchange of materials between the two planets. This essential biochemical information can not be obtained from fossils, biochemical material is needed.

Reaching biochemically preserved microorganisms would require access to the ancient cratered terrain near the south polar regions. Access to sediments below the depth of thermal cycling would optimize the chance that preserved material would still be present. This may require sampling below the depth of propagation associated with the thermal wave caused by the obliquity cycle.

Recent magnetic data has indicated large somewhat linear crustal magnetism in the ancient cratered terrain on Mars. Near the Hellas and Argyre impact basins these magnetic features are not present even though the terrain is heavily cratered. The presumption therefore (Connerney et al. 1999) is that terrain with the linear crustal magnetic features represents the oldest unaltered locations on Mars. The furthest south of such features located at 180 W, 70 - 80 S would be the best target for a search for preserved frozen remnants of early Martian life.

5. Europa, Comets, and Beyond

Under the ice surface of Europa there may be a layer of liquid water. The recent results from the Galileo spacecraft provide considerable evidence for an ocean of water underneath the surface ice on Europa (Carr et al. 1998). The surface of Ei crisscrossed by streaks that are slightly darker than the rest of the icy surface. If an ocean beneath a relatively thin ice layer then these streaks may represent cracks where the water has come to the surface.

Because of the thick ice crust (well over 10 km) and the evidence of hydrothermal activity, the possibility of chemosynthetic life in vents on the bottom of Europa's ocean should be considered.

Life at deep-sea vents on the Earth is often described as being independent of the surface biosphere. However, the life forms observed in abundance at these vents (crabs, worms, etc) are part of an ecosystem based on chemoautotrophic bacteria that consume H_2S outgassing from the vent and oxidize this with O_2 dissolved in the ambient seawater. This O_2 derives ultimately from photosynthesis on the surface of the Earth. On Europa, H_2S may be emanating from geothermal vents but there is not likely to be a source of O_2 in the ocean water. Without O_2 the energetics of H_2S consumption are not favorable. However H_2S can react with CO_2, in the presence of soluble ferrous iron, and this can form the basis for a chemoautotrophic ecosystem. If H_2 were present this could also form the basis for a microbial ecosystem. It is likely that carbon dioxide is present in the ocean of Europa (Crawford and Stevenson 1988).

There are many ecosystems in the Arctic and Antarctic regions that thrive and grow in water that is continuously covered by ice. In addition to the polar oceans where sea ice diatoms perform photosynthesis under the ice cover, there are perennially ice-covered lakes in the Antarctic dry valleys in which microbial mats based on photosynthesis are found in the water beneath a 4 meter ice cover. The light penetrating these thick ice covers is minimal -- about 1% of the incident light (McKay et al. 1994). Using these Earth-based systems as a guide it is possible that sunlight penetrating through the cracks (the observed streaks) in the ice of Europa could support a transient photosynthetic community (Reynolds et al. 1983).

The main problem with life on Europa is the question of its origin. Lacking a complete theory for the origin of life, and lacking any laboratory synthesis of life, we have to base our understanding of the origin of life on other planets on analogy with the Earth (Davis and McKay 1996). It has been suggested that hydrothermal vents may have been the sites for the origin of life on Earth and in this case the prospects for life in a putative ocean on Europa are improved. However, the early Earth contained many environments other than hydrothermal vents, such as surface hot springs, volcanoes, lake and ocean shores, tidal pools, and salt flats. If any of these environments were the locale for the origin of the first life on Earth then the case for an origin on Europa is weakened considerably.

Comets like asteroids are known to be rich in organic material and water -- frozen as ice and thus unsuitable for life. As a comet approaches the sun its surface is warmed considerably, but because of the low pressure this lead only to the sublimation of the ice.

There has been the suggestion that soon after their formation the interior of large comets would have been heated by short lived radioactive elements ([26]Al) to such an extent that the core would have melted (see e.g. Podolak and Prialnik, 1997). In this case there would have been a subsurface liquid water environment similar to that postulated for the present day Europa. Again the question of the origin of life in such an environment rests on the assumption that life can originate in an isolated deep dark underwater setting.

7. Conclusions

Everything we have learned about Life we have learned on Earth by studying the one example found here and there is still more to learn from life on Earth. By looking at other planets we may learn new and important lessons as well. In our solar system the most promising site for a search for life beyond the Earth is Mars. Beyond Mars, Europa, and comets provide key targets for exobiology.

8. References

Bada, J.L. and McDonald, G.D. (1995) Amino acid racemization on Mars: Implications for the preservation of biomolecules from an extinct Martian biota, Icarus 114, 139-143.

Battista, J.R. (1997) Against all odds: The survival strategies of *Deinococcus radiodurans*, Ann Rev. Microbiol. 51, 203-24.

Blochl, E., Rachel, R., Burggraf, S., Hafenbradl, D., Jannasch, H.W., and Stetter, K.O. (1997) Pyrolobus fumarii, gen. and sp. nov. represents a novel group of Archaea, extending the upper temperature limit for life at 113 C. Extremeophiles 1, 14-21.

Boston, P.J., M.V. Ivanov, and McKay, C.P. (1992) On the possibility of chemosynthetic ecosystems in subsurface habitats on Mars, Icarus 95, 300-308.

Carr, M.H. (1996) *Water on Mars*. Oxford University Press, New York.

Carr, M.H., Belton, M.J.S, Chapman, C.R., Davies, M.E., Geissler, P., Greenberg, R., McEwen, A.S., Tufts, B.R., Greeley, R., Sullivan, R., Head, J.W., Pappalardo, R.T., Klaasen, K.P., Johnson, T.W., Kaufman, J., Senske, D., Moore, J. Neukum, G., Schubert, G., Burns, J.A., Thomas, P. and Veverka, J. (1998) Evidence for a subsurface ocean on Europa, Nature 391, 363-365.

Connerney, J.E.P., Acuna, M.H., Wasilewski, P. Ness, N.F., Reme, C. Mazelle, D. Vignes, R. P. Lin, Mitchell, D. and Cloutier, P. (1999) Magnetic Lineations in the Ancient Crust of Mars, Science 284, 794-798.

Crawford, G.D. and Stevenson, D.J. (1988) Gas-driven water volcanism and resurfacing of Europa, Icarus, 73, 66-79.

Davis, W.L. and McKay, C.P. (1996) Origins of life: A comparison of theories and application to Mars, Origins Life Evol. Biosph. 26, 61-73.

Gilichinsky, D.A., Vorobyova, E.A., Erokhina, L.G., Fyordorov-Dayvdov, D.G., and Chaikovskaya, N.R. (1992). Long-term preservation of microbial ecosystems in permafrost, Adv. Space Res. 12 (4), 255-263.

Kanavarioti, A. and Mancinelli R.L. (1990) Could organic matter have been preserved on Mars for 3.5 billion years? Icarus 84, 196-202.

Kieffer, H.H., Martin, T.Z., Peterfreund, A.R., Jakosky, B.M., Miner, E.D., and Palluconi, F.D. (1977) Thermal and albedo mapping of Mars during the Viking primary mission, J. Geophys. Res. 82, 4,249-4,299.

Kushner, D., 1981. Extreme environments: Are there any limits to life? in *Comets and the Origin of Life*, C. Ponnamperuma (ed) D. Reidel, Dordrecht, pp. 241-248.

Laul, J.C., Smith, M.R., Wanke, H., Jagoutz, E., Dreibus, G., Palme, H., Spettel, B., Burchele, A., Lipschultz, M.E., and Verkouteren, R.M. (1986) Chemical systematics of the Shergotty meteorite and the composition of its parent body (Mars), Geochim. Cosmochim. Acta 50, 909-926.

Mattimore, V. and Battista, J.R. (1996) Radioresistance of Deinococcus radiodurans: Functions necessary to survive ionizing radiation are also necessary to survive prolonged desiccation, J. Bacteriol. 178, 633-637.

McKay, C.P. (1991) Urey Prize lecture: Planetary evolution and the origin of life, Icarus 91, 93-100,

McKay, C. P. (1997a) The search for life on Mars, Origins Life Evol. Biosph. 27, 263-289.

McKay, C.P. (1997b) Life in comets, in: *Comets and the Origin and Evolution of Life*, P.J. Thomas, C.F. Chyba, and C.P. McKay (eds.) Springer, New York. pp. 273-282.

McKay, C.P. (1998) Life in the Planetary Context, in: *Origins, Proceedings of the International Conference* C.E. Woodward, J.M. Shull, and H.A. Thronson, Jr. (eds.) held at Estes Park, Colorado, 19-23 May, 1997. Astronomical Society of the Pacific Conf. Series, Vol. 148, pp. 449-455.

McKay, C.P. and Davis, W.L. (1991) Duration of liquid water habitats on early Mars. Icarus 91, 214-221.

McKay, C.P., Clow, G.D., Andersen, D.T. and Wharton, R.A., Jr. (1994) Light transmission and reflection in perennially ice-covered Lake Hoare, Antarctica, J. Geophys. Res. 99, 20427-20444.

Minton, K.W. (1994) DNA repair in the extremely radioresistant bacterium Deinococcus radiodurans, Mol. Microbiol. 13, 9-15.

Podolak, M. and Prialnik, D. (1997) [26]Al and liquid water environments in comets, in: *Comets and the Origin and Evolution of Life*, P.J. Thomas, C.F. Chyba, and C.P. McKay (eds.) pp. 259-272, Springer, New York.

Reynolds, R.T., Squyres, S.W., Colburn, D.S. and McKay, C.P. (1983) On the habitability of Europa, Icarus 56, 246-254.

Rivkina E.M., Friedmann, E.I., McKay, C.P. and Gilichinsky, D.A. (2000) Bacteria in permafrost: Metabolic activity at subzero temperatures, Science, in review.

Stevens, T.O. and McKinley, J.P. (1995) Lithoautotrophic microbial ecosystems in deep basalt aquifers. Science 270, 450-454.

Swarts, S.G., Sevilla, M.D., Becker, D., Tokar, C.J., and Wheeler, K.T. (1992) Radiation-induced DNA damage as a function of hydration, Radiation Res. 129, 333-344.

Vorobyova, E., Soina, V., Gorlenko, M., Minkovskaya, N., Zalinova, N.,Mamukelashvili, A., Gilichinsky, D., Rivkina, E. and Vishnivetskaya, T. (1997) The deep cold biosphere: facts and hypothesis, FEMS Microbiol. Rev. 20, 277-290.

Biodata of **Julian Chela-Flores,** contributor of "*Terrestrial Microbes as Candidates for Survival on Mars and Europa.*" and is a co-author (with J. Seckbach and F. Westall) of "*Introduction to Astrobiology: Origin, Evolution, Distribution and Destiny of Life in the Universe.*"

Dr. Julian Chela-Flores was born in Venezuela and studied in the University of London, England, where he obtained his Ph.D. (1969) in quantum mechanics. He was a researcher at the Venezuelan Institute for Scientific Research and Professor at Simon Bolivar University (Caracas) until his retirement in 1990. He is a Fellow of: The Latin American Academy of Sciences, The Third World Academy of Sciences, the Academy of Creative Endeavors (Moscow) and a Corresponding Member of the Academia de Fisica Matematicas Y Ciencias Naturales (Caracas).

Dr. Chela-Flores current positions are: Staff Associate of the Abdus Salam International Center for Theoretical Physics, Trieste and Professor, Institute for Advanced Studies, Caracas. His particular area of expertise is Astrobiology . Dr. Chela-Flores has been the organizer of the Trieste Conferences on *Chemical Evolution and the Origin of Life*, since 1991 (and is the co-director since 1995) as well as an editor of all Proceedings from the Trieste conferences. In 1999 he co-directed the Iberoamerican School of Astrobiology in Caracas, Venezuela.

E-mail: **chelaf@ictp.trieste.it**

J. Seckbach (ed.), Journey to Diverse Microbial Worlds, 387-398.

TERRESTRIAL MICROBES AS CANDIDATES FOR SURVIVAL ON MARS AND EUROPA

JULIAN CHELA-FLORES

The Abdus Salam International Center For Theoretical Physics,
Miramare P.O.Box 586; 34136 Trieste, Italy, And Instituto De Estudios
Avanzados, Apartado 17606 Parque Central, Caracas 1015a,Venezuela.

1. Introduction

Mars and the ice-covered satellites of Jupiter are currently the most favourable sites for the search of extraterrestrial life. The motivation for the search for life in the Solar System is the evidence of liquid water in the early history of Mars and, at present, in the interior of at least two of the galilean satellites (Callisto and Europa). Hydrothermal vents on the Earth's sea floor have been found to sustain life forms that can live without direct solar energy. Similar possible geologic activity on Europa, caused by tidal heating and decay of radioactive elements, makes this Jovian moon the best target for identifying a separate evolutionary line. This search addresses the main problem remaining in astrobiology, namely, the distribution of life in the universe. We explore ideas related to Europa's likely degree of evolution, and discuss a possible experimental test. The total lack of understanding of the distribution of extraterrestrial life is particularly troublesome. Nevertheless, technical ability to search for extraterrestrial intelligence, by means of radioastronomy, has led to remarkable technological advances. In spite of this success, the theoretical bases for the distribution of life in the universe are still missing. The search for life in the Jovian satellites can provide a first step towards the still missing theoretical insight.

2. Eukaryogenesis as a cosmic imperative

We have suggested in our recent work (Chela-Flores, 1996; 1998a-c) that eukaryogenesis may be a *universal phenomenon*. We formulate the conjecture that the laws of physics and chemistry imply an 'imperative' appearance of eukaryogenesis during cosmic evolution [paraphrasing the well known sentence that "life is a cosmic imperative" (De Duve, 1995a)]:

> *Life is not only a natural consequence of the laws of physics and chemistry, but once the living process has started, then the cellular plans, or blueprints, are also of universal validity: The simplest cellular blueprint (prokaryotic) will lead to a more complex cellular blueprint (eukaryotic). Eukaryogenesis will occur inexorably because of evolutionary pressures, driven by environmental changes in planets, or satellites, where conditions may be similar to the terrestrial ones.*

In spite of the difficulty of identifying 'terrestrial-like environments', we have formulated a testable hypothesis related to the degree of extraterrestrial evolution on the pathway to intelligence. We emphasize that the conjecture has been formulated strictly at the cellular level; its relevance to the Drake Equation is discussed in Sec. 4.1.

3. A new approach to the distribution of life in the universe

Many hints suggest the conjecture of the universality of eukaryogenesis. One such hint concerns the combined action of natural selection, and the inevitable effect of symbiosis (Margulis, 1993). Besides, horizontal gene transfer (HGT) may also be a factor that drives prokaryotes into the more advanced eukaryotes (Smith *et al.*, 1987).

However, an intriguing question is whether the transition prokaryote-eukaryote will occur in a suitable planetary, or satellite environment. We dwell on eukaryogenesis, since on Earth this was the first step towards multicellularity and, subsequently, intelligence. Insights derived form biogeochemical data suggest that the prokaryotic blueprint will make its first appearance on a planet, or satellite, in a relatively short geological time (Schildowski, 1988). There are even indications that sedimentary rocks older than 3.7 giga years (Gyr) before the present (BP) from Isua contain reduced carbon of likely biogenic origin (Rosing, 1999). Additional support for an early onset of the prokaryotic blueprint was the result of analysis of banded iron formation from Isua and Akilia of some 3.8 Gyr BP (Mojzsis *et al.*, 1996). Once the period of heavy bombardment was over, some 3.9 to 4 Gyr BP, the appearance of prokaryotic life was almost instantaneous in the context of a geological time frame. We may assume that prokaryotes are bound to occur in environments where chemical and geological conditions may be similar to the terrestrial ones. This remark induces us to ask whether eukaryogenesis is a universal phenomenon, which is the bigger and deeper issue.

Our aim is to discuss experimental means for testing the conjecture of the universality of eukaryogenesis within the solar system; we also explore the implication of the conjecture in the context of the search for extraterrestrial intelligence (SETI).

4. Looking for sites where parallel evolution could occur within our Solar System

4.1. NEW EQUATIONS FOR THE DRAKE PARAMETER f_i

An early reference regarding the transition from prokaryotes to eukaryotes in relation with the Drake Equation was made by Carl Sagan. His remarks were in the context of a discussion of the SETI projects at a 1971 conference (Sagan, 1973). In order to make Sagan's general comment more specific, we consider the Drake Equation $N = f_i$, where N is the number of civilizations capable of interstellar communication, is a constant of proportionality involving several factors that we need not discuss here; f_i is the fraction of life-bearing planets or satellites where biological evolution produces an intelligent species (Drake and Sobel, 1992). We suggest that the Drake parameter f_i is itself subject to the equation:

$$f_i = \kappa_1 f_e f_m \, , \tag{1}$$

where κ_1 is a constant of proportionality, f_e, f_m denote, respectively, the fractions of planets or satellites where eukaryogenesis, or multicellularity may occur. Our conjecture motivates the search *within our own solar system* for a key factor (f_e) in the distribution of life in the universe, including intelligent life. The presence of the parameter f_m in our equation (1) can be understood, since once the eukaryotic level of evolution was reached on Earth multicellularity was bound to follow (De Duve, 1995b).

The extrapolation of the transition to multicellularity into an extraterrestrial environment is suggested by the selective advantage of organisms that go beyond the single-cell stage. Such organisms have the possibility of developing nervous systems and, eventually, brains and intelligence. At present the alternative equation which omits from the discussion the f_m parameter (by including it in the constant of proportionality) is more useful in the planning of possible experiments that may be implemented in the eventual Europa lander (Chela-Flores, 2000). If cell formation is possible in a short geological time frame, there are going to be evolutionary pressures on prokaryotes to evolve, due to symbiosis, HGT and natural selection (cf., Sec. 3. In fact, these evolutionary mechanisms are going to provide strong selective advantage to those cells that can improve gene expression by compartmentalization of their genomes. (Larger genomes would be favored, since organisms with such genetic endowment would have better capacity for survival, and hence better ability to pass their genes to their progeny.) Whether the pathway to eukaryogenesis in a Europan-like environment, or elsewhere in the cosmos, has been followed, is clearly still an open question.

4.2. CAN EXOGENOUS MICROORGANISMS EXIST?

Hydrothermal vents on the Earth's sea floor have been found to sustain life forms that can live without direct contact with solar energy. Similar possible volcanic activity on Europa, caused by its interaction with Jupiter and the other Galilean satellites, makes this Jovian moon the best target for a possible identification of a living micro-organism beyond our planet (Delaney *et al.,* 1996). An open question is whether a habitable planet, or satellite, has to be in the "habitable zone" of its star; volcanism seems to be sufficient as a source of energy for driving chemical into biological evolution. Alternatively, another possibility is the decay of radioactive elements.

The present work aims at turning the question of *distribution of life in the Universe,* from the present realm of conjectures, into a well-defined scientific discipline that could be tested in the foreseeable future within our solar system. The current question of distribution of life in the universe will inevitably be faced with an 'armada' of space missions. Such efforts should be constrained to test only clearly formulated hypotheses (cf., Sec. 1), backed by the proposal of realistic, specific, and unambiguous experiments (cf., Sec. 5). We recall the recent proposal of a space mission called the Cryobot/Hydrobot, which in principle would be capable of investigating the possible existence of life in Europa (Horvath *et al.,* 1997). Independently, in Japan there is a feasibility study of a space mission to explore the subsurface ocean of Europa and search for indicators of biological activity. Other projects will probably follow. In particular, the Japanese proposal includes instrumentation for *in situ* observation by means of a submersible of the type of the hydrobot (Raulin and Kobayashi, 1998). If successful, these missions offer excellent possibilities for testing the eukaryogenesis conjecture (cf., Sec. 1). We have maintained that an appropriate experiment to test eukaryogenesis is feasible, in spite of the evident severe payload limitations that the nature of the mission is bound to impose on us.

4.3. IS THE EARTH ANALOGOUS TO EUROPA AND EARLY MARS?

Once we give up the chauvinistic point of view that has been forced upon us by the multicellular nature of *Homo sapiens,* the similarity between the environments of Europa, early Mars and the Earth becomes evident. We are beginning to realize that if life does exist on Europa it will be mainly deep, aquatic, cryophilic and most likely unicellular. Early Mars may have been analogous to the Earth as well.

On the other hand, it is worth underlining that deep, aquatic, low temperature environments for unicellular organisms are also predominant on Earth (Prieur *et al.,* 1995). We need only recall that 70% of the Earth is covered by sea water, of which two-thirds have a temperature of around 2° C. Cryogenic conditions are widespread in our planet. More than 80% of the Earth biosphere, including the polar regions, is permanently cold (from the point of view of the mean annual temperature). The deep sea (> 1000 m) represents 88% of the Earth area covered by sea water and 75% of the total volume of the oceans; in other words, the deep sea represents 62% of the biosphere. Micro-organisms can be subject to extreme temperature fluctuations. In polar and tundra soil these fluctuations can have a lower bound of some -15° C. Finally, in taking advantage of the analogy between the Earth and Europa, we may make use of the wide experience with viable micro-organisms in permafrost, which may serve as the background against which to test conjectures. The possibility of detecting biomolecules in Europa, on the ice surface itself, rather than in the possible ocean underneath, was made recently (McKay, 1998). A possible mechanism for bringing biomolecules to Europa's surface was subsequently discussed (Chela-Flores, 1998d). Unfortunately, the Galileo Europa Mission is restricted to infrared and ultraviolet spectroscopy. Surface biogenic tests on Europa may have to wait for the further orbital mission to search for traces of putative Europan biochemistry, or signs of extant life.

4.4. CAN LIFE BEGIN IN THE TOTAL ABSENCE OF SUNLIGHT?

The answer to this question is important for the possible existence of life on Europa, or any of the other iced satellites of the Outer Solar System. Earth-bound eukaryotes depend on an oxygenic atmosphere, which was in turn produced by prokaryotic photosynthesis over billions of years. A possible scenario favoring the existence of Europan micro-organisms decouples hydrothermal-vent systems from surface photosynthesis. Indeed, experiments have already shown that chemical evolution leading to biological evolution is possible in conditions similar to those of hydrothermal vents (Huber and Wachtershauser, 1998). Further, the delivery of amino acids at hydrothermal vents is possible, either by cometary or by meteoritic delivery (Chyba, 1998). Rather than prebiotic evolution, the genesis of *a primitive cell* in the deep ocean independent of photosynthesis, is still a wider issue to be settled experimentally. We may recall some related evidence against hydrogen-based microbial eco-systems in basalt aquifers, namely ecosystems in rock formations containing water in recoverable quantiites

(Anderson *et al.,* 1998). The new experiment raises doubts on the specific mechanism proposed for life existing deep underground (Stevens and McKinley, 1995). However, the new evidence of Anderson and co-workers refers more to the specific Stevens-McKinley means of supporting microbial metabolism in the subsurface, rather than being an argument against the possible precedence of chemosynthesis before photosynthesis, which is really the wider and deeper issue to be settled. What remains to be shown in microbiology is that some barophilic and thermophilic micro-organism has a metabolism that can proceed in completely anoxic conditions, deprived from carbon and organic-nitrogen derived from surface photosynthesis. For example, such experiments probing the ability of a given micro-organism to survive in well-defined environments have already been performed; it was shown that *Cyanidium,* a primitive alga was able to thrive in a pure carbon-dioxide atmosphere (Seckbach *et al.,* 1970). Thus, the case for life's origins, either through chemosynthesis first, or through a secondary reliance on photosynthesis at hydrothermal vents (by using oxygen dissolved in the sea-water), or deep underground, are still open questions. While this situation remains unsettled, plans for experiments have to be made by the space agencies, as the technological capability is consolidated for eventual landers on Europa, Callisto, Enceladus and Triton.

5. On the ubiquity of eukaryotes in Antarctica

The relevance of the information to be retrieved from Lake Vostok can be made more evident by means of the following question:

What might be learned regarding eukaryogenesis from the possible study of viable diatoms from permafrost and deep ice?

Since the presence of biochemical traces on the Europan surface has been suggested, it becomes imperative to pursue analogous research in Lake Vostok, particularly concerning eukaryotes. Diatoms, discovered in 1702 by the microscopy pioneer Anton van Leeuwenhoek, are some of the most interesting micro-organisms to consider, given their ubiquity on Earth: In just one litter of sea water one may find as many as ten million diatoms, which may be considered the primary foodstuff of the sea. Marine species often form a brown coating on Arctic ice floes.

The ubiquity of diatoms may also be exemplified by what is known in the other Earth analogues of the Europan environment: Permanently frozen lakes in a series of dry valleys were discovered in 1905 by the British explorer Sir Robert Scott. From the point of view of geology and microbiology some of the best studied frozen lakes are in the Taylor Valley, namely Lake Fryxell and Lake Hoare; further north, Lake Vanda, in the Wright Valley, is also remarkable. Some species of diatoms (Pennales) are known to dwell under the permanently ice-covered lakes of the Antarctic dry valleys. Some further details of the Antarctic lakes are given elsewhere (Chela-Flores, 1997). Once the planetary protection protocols are duly taken into account, forthcoming knowledge of the micro flora that populates the substantial water volume of Lake Vostok will be of great value for anticipating, and testing the instrumentation requirements that might be needed. Amongst the micro-organisms that are permanently living in the Antarctic lakes there are examples of eukaryotes, a few of which are illustrated in our previous work

(Chela-Flores, 1997). We have emphasized the presence of diatoms before mentioning other examples of algae, because diatoms comprise the largest number of algae in the benthic mats of these singular biotopes; our main motivation, however, is to underline the significance of eukaryogenesis in astrobiological research. In the Antarctic biotope, eukaryotes have demonstrated to thrive in Europa-like conditions. By the mechanism explained by Wharton and co-workers (Parker *et al.,* 1982; Wharton *et al.,* 1983; Doran *et al.,* 1994), vertical transport of diatoms and other micro-organisms is possible in the permanently frozen lakes. Hence, it is hardly surprising that diatoms have been found recently in the permafrost and deep ice of the Lake Vostok region. It should be kept in mind, however, that the mechanism of vertical transport that applies in the dry valley lakes, where the ice covering the lakes measures a few meters, may not apply in the case of Lake Vostok, where the depth of the ice covering the lake is measured in kilometers. The Cryobot/Hydrobot (CH) mission (Horvath *et al.,* 1997) would benefit form the experience that has been gained in the dry valley lakes of Antarctica.

6. On the search for extraterrestrial eukaryotes

6.1. POSSIBLE EXPERIMENTS ON OR BELOW ICED SATELLITE SURFACES

To settle the question whether the iced satellites are potential sources of parallel evolution for micro-organisms, in this section and the next one we shall discuss possible experiments that may be carried out on and below the ice surface, which could be implemented by means of a landing craft, of the type of the cryobot; indeed, it seems feasible to search for extant life by means of the either of the two subsurface-oriented missions (cf., Sec. 4.2).

We have argued above that a factor in the lack of uniformity in surface brightness and color of the Europan surface may be the presence of micro-organisms, or their biomolecules. In other words, the search for extraterrestrial biochemistry, or biology on the surface of Europa ought to be a possible straightforward and evident aspect of the Europa campaign. The biogenic hypothesis can be tested, for instance, by spectroscopic search of the Europan surface ice. On the other hand, it seems reasonable to test directly for surface biochemistry, or organisms.

6.2. ON POSSIBLE BIOLOGY EXPERIMENTS

Missions to the Galilean satellites would benefit form the experience that has been gained in the dry valley lakes of Antarctica. In the biology experiment presumably the maximum size of the Cryobot would be some 10 to 15 cm diameter and 1.5 - 2 m long (its equivalent in an alternative mission would presumably be subject to similar constraints). Within the restricted space available there would be an *"in-situ* chemistry laboratory". The submersible instrumentation would aim to determine whether the ocean exhibited one set of requirements for "life we would recognize" (Horvath *et al.,* 1997). We believe that the detection of life and *its evolutionary stage* should inevitably be one of the primary goals of any exploration program aiming at the exploration of the iced satellites in the outer solar system. We wish to define in some detail the minimum equipment that is needed in the biology experiment proposed earlier (Chela-Flores, 1998b). The optical system that would be proposed for the submersible depends on the chemical composition of the putative micro-organisms. The technique relies on the material being able to induce luminescence by the application of various dyes. Fluorescent dyes are detected with a fluorescent microscope. If microscopic fluorescence is used to probe for life, some advantages and some challenges are immediately evident.

First, in contrast to the Martian search for life, the typical resolution needed for fluorescence micrographs of chromosomes is 10 m (Lodish *et al.,* 1995). Such resolution is well within the scope of a light microscope. Martian research is linked to the electron microscope, since the nodules in the Allan Hills meteorite, which are currently under discussion, have been suggested to be nanobacteria, 50-200 nm in diameter (Westall, 1999). The difficulties of going beyond a light microscope in the Hydrobot are evident. Hence, it is feasible to think in terms of a *robotic biologist'.* It would be contained within the Hydrobot and consist of simple optics and sampling arms. At later stages in the cell cycle this structure serves as the basis for further folding, ending up at the highest degree of folding observed at the metaphase chromosome. This is an extremely fortunate feature from the experimental point of view. We only need to recall that the ultimate aim of the biology experiment is to develop tests that are compatible with the reduced dimensions available. Indeed, chromosomes stain easily, in a well-defined manner. The biochemical basis for the difference between heterochromatin (the more compact structure of chromatin) and euchromatin (its less compact form), remains unknown. Heterochromatin is not only a clear hallmark of eukaryogenesis (Chela-Flores, 1998a), but it is also a unique indicator of eukaryoticity, which is amenable to the tasks that the 'mechanical biologist' will have to perform. We confine our attention to the clearest hallmark for eukaryogenesis: heterochromatic genomes that respond in an unambiguous manner to well-defined dyes, the result of which could be recorded with video equipment for later analysis, after relaying the results to an Earth-bound laboratory.

6.3. WHEN CHEMICAL DETAIL OF THE GENOME IS UNCERTAIN

Quinacrine fluorescent dye inserts itself between base pairs in the DNA helix producing the so-called Q-bands, which for the planned mission would probably suffice. We suggested earlier (Chela-Flores, 1998b) the more involved use of Giemsa stain to produce the more permanent R-bands (Alberts *et al.,* 1994). This is probably an unnecessary complication. Adjacent areas stain differently. The bands give a clear indication of slightly different modes of DNA packaging. It is the tightness of the genomic material that would be an indicator of a higher degree of evolution. The question is not so much what is the chemical detail of the genome, but what is the degree to which it has been packaged. It may be argued that gene activity is correlated with light-staining bands. (For instance, genes that are transcriptionally active are light-staining (Watson *et al.,* 1987). This aspect of the experiment is its relevant, since *it does not force upon us the requirement of previous detailed knowledge of the putative Europan biochemistry.* The main scope of the experiment is to expose eukaryoticity at the level of gene expression, whose most characteristic indicator is heterochromaticity .

7. Discussion

7.1. EXCLUSION OF REFUGES AGAINST EVOLUTION

There seems to be a strong case for the exclusion of refuges against evolution (Little *et al.*, 1997). Cambrian fauna, such as lamp shells (inarticulate brachiopods) and primitive molluscs (Monoplacophora), were maintained during Silurian times by micro-organisms that lived in hydrothermal vents. In the current Cenozoic Era these hot environments demonstrate that such fauna no longer inhabits these environments and hence has been unable to escape evolutionary pressures. Hence, this remark rules out the possibility that these deep-sea environments are refuges against evolutionary pressures. In other words, the evidence so far does not support the idea that there might be environments, where ecosystems might escape biological evolution, not even at the very bottom of deep oceans. It is then reasonable to assume that any micro-organism, in whatever environment on Earth, or elsewhere, would be inexorably subject to evolutionary pressures. On this planet eukaryotes seem to have been the consequence of over 2 Gyr of evolutionary pressures acting on the prokaryotic blueprint. The first appearance in the fossil record of eukaryotes occurred during the Proterozoic Eon, after prokaryotic communities (stromatolites) were well established in the previous Archean Eon, some 3.5 Gyr BP (Schopf, 1993). It still remains to be confirmed, or rejected, whether the Europan environment may have had liquid water in a geological time frame. In such a favorable environment a primordial Archaea community would have had sufficient time for evolutionary pressures have modelled a primordial Archaea community. If these conditions occurred on Europa, then, according to our conjecture (cf., Sec. 2), eukaryogenesis would have been inevitable. Recent observational evidence does suggest the presence of an ocean (Carr *et al.*, 1998; Khurana *et al.*, 1998).

7.2. THE POSSIBILITY OF EUKARYOGENESIS ON EUROPA AND MARS

There are several reasons why Mars may have experienced a more rapid environmental evolution towards an atmosphere rich in oxygen (McKay, 1996; 1998). This may have favored a restricted period a rapid pace of evolution of the background population of prokaryotic cells towards eukaryogenesis. Such an oxic environment is favorable to the first appearance of the eukaryotic blueprint. On Earth eukaryogenesis occurred as far back as 2 Gyr BP, according to the micropalaentological data. Clearly such a Martian "Eden" may not have lasted for long on a geologic time scale, although we should recall that we are continually improving our understanding of the geologic history of Mars, as the evidence of crater statistics for recent volcanism demonstrates (Hartmann *et al.*, 1999). Some work of "exo-palaentology" clearly remains to be done (Farmer, 1997), which could be facilitated by new technical support currently being discussed, such as the JPL aerobot, or a Martian airplane.

On the other hand, the eukaryotic transition may be a general consequence of geological changes on an Earth-like planet, or satellite coupled to the effect of natural selection. We have argued that evolution may have occurred in Europa and that the experimental test of this conjecture is feasible through a space mission. We have seen that difficult instrumentation issues are involved. The preparation of a package to search for life either on Mars or Europa is a formidable task. The method elaborated for the exploration of Mars (Kobayashi et al., 1998), which is based on fluorescence microscopy, would still require further miniaturization before it could be adapted to the case of Europa. In relation with the possibility of eukaryogenesis having occurred on Mars, it has been pointed out (McKay, 1996; 1998) that since there are several factors that may have accelerated oxygenation on Mars (less volcanic activity, smaller oceans during the first billion years), then before the environment deteriorated, Mars may have experienced eukaryogenesis earlier than when it occurred on Earth.

7.3. OUTLOOK

Unlike the situation concerning the distribution of life in the universe, two aspects of astrobiology have already sound scientific approaches. Firstly, the study of *the origin of life,* the first aspect of astrobiology, is based on the theory of chemical evolution, which is a time-honored scientific discipline. The second aspect of astrobiology, *the evolution of life,* has scientific bases provided by the two well-established insights of Darwin: "the theory of common descent", and "natural selection" as a mechanism for biological evolution. In the present work we have defended the thesis that if the conjecture in Sec. 1 were to be tested successfully within our own solar system by the biological experiment discussed in Sec. 5, it would not only show the non vanishing of the all-important parameter f_i [(cf., Eqn. (1)], one of the most controversial parameters of the Drake Equation, but at the same time it would bridge a remaining gap in astrobiology, namely the distribution of life in the universe. In other words, a preliminary test of the conjecture in Sec. 1 on the iced satellites would serve as a firm scientific basis on which to develop eventually the science of *the distribution of life in the universe.*

8. References

Alberts, B. Bray, D., Lewis, J., Raff, M., Roberts, K. and Watson, J.D. (1994) *Molecular Biology of the Cell,* 3rd ed. Garland Publishing: New York, p. 355.

Anderson, R.T., Chapelle, F.H. and Lovly, D.R. (1998) *Science 281,* 976-977.

Carr, M.H., Belton, M.J.S., Chapman, C.R., Davies, M.E., Geissler, P., Greenberg, R., McEwen, A.S., Tufts, B.R., Greely, R., Sullivan, R., Head, J.W., Pappalardo, R.T., Klaasen, K.P., Johnson, T.V., Kaufman, J., Senske, D., Moore, J., Neukum, G., Schubert, G., Burns, J.A., Thomas, P. and Veverka, J. (1998) *Nature* **391,** 363-365.

Chela-Flores, J. (1996) *Europa Ocean Conference,* (Abstracts), 5th. Capistrano Conf. San Juan Capistrano Research Institute, California, USA, p. 21.

Chela-Flores, J. (1997) in: R.B. Hoover (ed.) *Instruments, Methods and Missions for Investigation of Extraterrestrial Microorganisms,* Proc. SPIE, **3111,** pp. 262-271.

Chela-Flores, J. (1998a) *Origin Life Evol. Biosphere* **28,** 215-225.

Chela-Flores, J. (1998b) *Origins Life Evol. Biosphere* **28,** 583-596.

Chela-Flores, J. (1998c) in: R.J. Russell, W.R. Stoeger, and F.J. Ayala (eds.) *Evolutionary and Molecular Biology: Scientific Perspectives on Divine Action.* Vatican City State/Berkeley, California, Vatican Observatory and the Center for Theology and the Natural Sciences. pp. 79-99.

Chela-Flores, J. (1998d) in: J. Chela-Flores and F. Raulin (eds.) *Exobiology: Matter, Energy, and Information in the Origin and Evolution of Life in the Universe,* Kluwer Academic Publishers, Dordrecht, The Netherlands, pp. 229-234.

Chela-Flores, J. (2000) The Solar System. Astron. Soc. Pacif. Conf. Ser. (in press).

Chyba, C. (1998) *Nature* **395,** 329-330.

De Duve, C. (1995a) *Vital dust: Life as a cosmic imperative.* Basic Books: New York.

De Duve, C. (1995b) *Vital dust: Life as a cosmic imperative.* Basic Books: New York. pp. 171-175.

398

Delaney, J., Baross, J., Lilley, M. and Kelly, D. (1996) *Europa Ocean Conference*, San Juan Capistrano Research Institute, California, USA, p. 26.

Doran, P.T., Wharton, Jr., R.A. and Berry Lyons, W. (1994) *J. Paleolimnology* **10**, pp. 85-114.

Drake, F. and Sobel, D.(authors) (1992) *Is there anyone out there? The scientific search for Extraterrestrial Intelligence.* Delacorte Press: New York. pp. 45-64.

Farmer, J. D. (1997) *in*: R.B. Hoover (ed.) Instruments, Methods and Missions for Investigation of Extraterrestrial Microorganisms, The International Society for Optical Engineering, Washington USA. Proc. SPIE, 3111, pp. 200-212..

Hartmann, W.K., Malin, M., McEwen, Carr, M., Soderblom, L., Thomas, P., Danielson, E., James, P. and Veverka, J. (1999) Nature **397**, 586-589.

Horvath, J., Carsey, F., Cutts, J., Jones, J., Johnson, E., Landry, B., Lane, L., Lynch, G., Chela-Flores. J., Jeng, T.-W. and Bradley, A. (1997) in: R.B. Hoover (ed.) *Instruments, Methods and Missions for Investigation of Extraterrestrial Microorganisms*, Proc. SPIE, **3111**. pp. 490-500.

Huber, C. and Wachtershauser, G. (1988) *Science* **281**, 670-672.

Kobayashi, K., Kaneko, T., Kawasaki, Y. and Saito, T. (1998) in J. Chela-Flores and F. Raulin (eds.) *Exobiology: Matter, Energy, and Information in the Origin and Evolution of Life in the Universe*, Kluwer Academic Publishers: Dordrecht, The Netherlands, pp. 251-254.

Khurana, K.K., Kivelson, M.G., Stevenson, D.J., Schubert, G., Russell, C.T., Walker, R.J. and Polanskey, C. (1998) *Nature* **395**, 777-780.

Little, C.T.S., Herrington, R.J., Maslennikov, V.V., Morris, N.J. and Zaykov, V.V. (1997) *Nature* **385**. 146-148.

Lodish, H., Baltimore, D., Berk, A., Zipursky, S.L., Matsudaira, P. and Darnell, J. (1995) *Molecular Cell Biology*, 3rd ed., Scientific American Books, W.H. Freeman: New York, p. 354.

Margulis, L. (1993) *Symbiosis in Cell Evolution Microbial Communities in the Archean and Proterozoic Eons.* 2nd. ed. W.H. Freeman and Company: New York.

McKay, C.P. (1996) in: J. Chela-Flores and F. Raulin, F. (eds.) *Chemical Evolution: Physics of the Origin and Evolution of Life.* Kluwer Academic Publishers, Dordrecht, The Netherlands. pp. 177-184.

McKay, C.P. (1998) in: J. Chela-Flores and F. Raulin (eds.) *Exobiology: Matter, Energy, and Information in the Origin and Evolution of Life in the Universe*, Kluwer Academic Publishers, Dordrecht, The Netherlands, pp. 219-228.

Mojzsis, S.J., Arrhenius, G., McKeegan, K.D., Harrison, T.M., Nutman, A.P. and Friend, C.R. (1996) Nature 384, 55-59.

Parker, B.C., Simmons, Jr., G.M., Wharton, Jr., R.A., Seaburg, K.G. and Gordon Love, F. (1982) *J. Phycol.* **18**, 72-78.

Prieur, D., Erauso, G. and Jeanthon, C. (1995) *Planet Space Sci.* **43**. 115-122.

Raulin, F. and Kobayashi, K. (1998) *Exobiologically-oriented space methodologies.* Summary of Symposium F3.3 in: ISSOL Newsletter **25(2)** p. 9, 15.

Rosing, M.T. (1999) Science 283, 674-676

Sagan, C. (1973) in: C. Sagan (ed.) *Communication with Extraterrestrial Intelligence (CETI).* The MIT Press, Cambridge, Massachusetts . p. 113-146.

Schidlowski, M. (1988) *Nature* **333**, 313-318.

Schopf, J.W. (1993) *Science* **260**, 640-646.

Seckbach, J., Baker, F.A. and Shugarman, P. M. (1970) *Nature* **227**, 744-745.

Smith, M.W., Feng, D-F., and Doolittle, R.F. (1987) *Trends Biochem. Sci.* **17**, 489-493.

Stevens, T. O. and McKinley, J. P. (1995) *Science* **270**, 450-454.

Watson, J.D., Hopkins, N.H., Roberts, J. W., Steitz, J.A. and Weiner, A.M. (1987) *Molecular Biology of the Gene*, 4th. ed., The Benjamin Cummings Publishing Co. Menlo Park, Calif. pp. 685- 686.

Westall, F. (1999) J. Geophys. Research. **104**, 16,437-16,451.

Wharton, Jr., R.A., Parker, B.C. and Simmons, Jr., G.M. (1983) *Phycologia* **22**, 355-365.

Author Index

Subject Index

404

410